THE ONTOLOGY OF TIME

PHAENOMENOLOGICA

SERIES FOUNDED BY H.L. VAN BREDA AND PUBLISHED UNDER THE AUSPICES OF THE HUSSERL-ARCHIVES

163

ALEXEI CHERNYAKOV

THE ONTOLOGY OF TIME

Being and Time in the Philosophies of Aristotle, Husserl and Heidegger

ALEXEI CHERNYAKOV
St. Petersburg School of Religion and Philosophy
St. Petersburg, Russia

THE ONTOLOGY OF TIME

Being and Time in the Philosophies of Aristotle, Husserl and Heidegger

SPRINGER-SCIENCE+BUSINESS MEDIA, B.V.

A C.I.P. Catalogue record for this book is available from the Library of Congress.

ISBN 978-90-481-6049-5 ISBN 978-94-017-3407-3 (eBook)
DOI 10.1007/978-94-017-3407-3

Printed on acid-free paper

TABLE OF CONTENTS

CHAPTER FIVE
GOD WITHOUT BEING AND THOUGHT WITHOUT THINKER

CHAPTER SIX
SEARCHING FOR THE LOST SUBJECT

CHAPTER SEVEN
PRIMORDIAL TEMPORALITY AND ONTOLOGICAL DIFFERENCE

KEY TO SOME REFERENCES CITED IN THE TEXT

Throughout this book references to frequently cited texts are identified by the following abbreviations and code letters.

E. HUSSERL'S WORKS

Hua = "Husserliana" – Edmund Husserl, *Gesammelte Werke*, Bd. 1ff. (Den Haag: M. Nijhoff, 1950 ff.)

EP = *Erste Philosophie*. 2. Teil: *Theorie der phänomenologischen Reduktion*, hrsg. v. R. Boehm, Hua VIII (Den Haag, 1959).

EU = *Erfahrung und Urteil*, redigiert und hrsg. v. L. Landgrebe. (Hamburg: F. Meiner, 1985).

Ideen I = *Ideen zu einer reinen Phänomenologie und phänomenologischen Philosophie*, 5. Aufl. (Tübingen: Max Niemeyer Verlag, 1993).

Ideen II = *Ideen zu einer reinen Phänomenologie und phänomenologischen Philosophie*. Zweites Buch: *Phänomenologische Untersuchungen zur Konstitution*, hrsg. v. M. Biemel, Hua IV (Den Haag, 1952).

Ideen III = *Ideen zu einer reinen Phänomenologie und phänomenologischen Philosophie*. Drittes Buch: *Die Phänomenologie und die Fundamente der Wissenschaften*, hrsg. v. M. Biemel, Hua V (Den Haag, 1952).

Krisis = *Die Krisis der europäischen Wissenschaften und die transzendentale Phänomenologie. Eine Einleitung in die phänomenologische Philosophie*, hrsg. v. W. Biemel, Hua VI (Den Haag, 1954, 1962).

LU = *Logische Untersuchungen*. Zweiter Band, 2. Teil., hrsg. v. U. Panzer, Hua XIX/2 (Den Haag, Boston, Lancaster, 1984).

ZB = *Zur Phänomenologie des inneren Zeitbewußtsein* (1893–1917), hrsg. v. R. Boehm, Hua X (Den Haag, 1959).

M. HEIDEGGER'S WORKS

GA = Martin Heidegger, *Gesamtausgabe*, Bd. 1 ff. (Frankfurt a. M.: Vittorio Klostermann Verlag).

EM = *Einführung in die Metaphysik*, GA 40 (Frankfurt a. M.: V. Klostermann, 1983).
English translation by R. Manheim: *Introduction to Metaphysics* (New Haven and London: Yale University Press, 1987).*

GP = *Die Grundprobleme der Phänomenologie*, GA 24 (V. Klostermann, Frankfurt a. M., 1989).
English translation by A. Hofstadter: *The Basic Problems of Phenomenology* (Bloomington and Indianapolis: Indiana University Press, 1988).

KPM = *Kant und das Problem der Metaphysik* (Frankfurt a. M.: V. Klostermann, 1973).
English translation by R. Taft: *Kant and the Problem of Metaphysics* (Bloomington and Indianapolis: Indiana University Press, 5th ed., 1997).

MAL = *Metaphysische Anfangsgründe der Logik im Ausgang von Leibniz*, GA 26 (Frankfurt a. M.: V. Klostermann, 1978).
English translation by M. Heim: *The Metaphysical Foundations of Logic* (Bloomington and Indianapolis: Indiana University Press, 1984).

PS = *Platon: Sophistes* (Marburger Vorlesung Wintersemester 1924/25), GA 19 (Frankfurt a. M.: V. Klostermann, 1992).
English translation by R. Rojcewicz and A. Schuwer: *Plato's Sophist* (Bloomington and Indianapolis: Indiana University Press, 1997).

SZ = *Sein und Zeit*, 16. Aufl. (Tübingen: Max Niemeyer Verlag, 1986).
English translation by J. Macquarrie and E. Robinson: *Being and Time* (Oxford: Basil Blackwell, 7th ed., 1987).

OTHER ABBREVIATIONS

DN = Dionysius the Areopagite, *The Divine Names and Mystical Theology*, ed. G. M. Prochorov (St. Petersburg: Glagol, 1994). The text reproduces the critical edition: *Corpus Dionysiacum* I. Pseudo-Dyonysius Areopagita: *De divinis nominibus*, ed. B. R. Suchla. *Patristische Texte und Studien* 33 (Berlin-New York, 1990).

PG = *Patrologiae cursus completus* (J.-P. Migne). *Ser. graeca.*

PhG = G. W. F. Hegel, *Phänomenologie des Geistes*, hrsg. v. J. Hoffmeister (Berlin: Akademie Verlag, 1971).

PL = *Patrologiae cursus completus* (J.-P. Migne). *Ser. latina.*

---

* For the reasons discussed below the English translations are often modified. Page numbers are indicated according to the German editions.

INTRODUCTION

1.1. *Why the Ontology of Time?*

The intention that directs this research consists in an attempt to provide a *hermeneutic analysis* of the drastic changes, which have occurred in 20th century philosophy, in identifying the new role ascribed to the subject of time and temporality within the scope of ontology. After the fundamental works of E. Husserl, M. Heidegger, P. Ricœur, and E. Levinas, it has been understood that the traditional issue (which could be traced back to Parmenides) between being and time, between the eternal and the transient (or historical), must once again be re-examined. Time itself is recognized now as the deepest ground of ontological inquiry, which sets in motion the entire system of fundamental philosophical concepts.

This does not mean, of course, that our understanding of time did not change in the course of these fundamental transformations. In order to comprehend the new role of time within "first philosophy," the *concept of time* itself is to be subjected to a careful investigation and interpretation. It is necessary to come back to Aristotle's questions in *Physics* IV: In what sense can we ascribe being to time itself, and what is the "nature" of time as (a) being? In other words, to understand the role of time within the scope of ontology means to develop simultaneously the *ontology of time*. This is what the title of this work intends to designate. Moreover, my aim is to demonstrate that in a definite sense the postmodern *onto-logy is chrono-logy*.

To be sure, *historical* attempts to understand the "nature of time" represent a tremendous variety of possible approaches and viewpoints, and we cannot hope to look at the question from all possible angles. That is why our investigation is confined to a particular ontological tradition embracing several more or less coherent (independently of their rather broad chronological limits) ways of thought. The central figure for us is *Martin Heidegger* and his "new start" in ontology, which has generated immense transformations in the philosophical thought of our century.

1.2. *The Method*

This "new start" absorbs, nevertheless, the metaphysical tradition from its very foundation by the Greeks, and we can understand Heidegger's ontological turn only against the background of this great tradition as a peculiar transformation-in-continuation, which Heidegger himself called "destruction of the history of ontology." This phrase should be understood in a quite positive sense: de(con)struction is a necessary operation in the archaeology of thought, guided by the intention to rediscover the forgotten ways of thinking that somehow preserve their "effaced traces" within the historico-philosophical landscape. This sort of analysis can be carried out only on the condition that one takes into account not only the *synchronic* stratum of current conceptuality in its internal logical relationships, but also the *diachronic axis* of conceptual genesis as well. Philosophical concepts, topics, and motives always contain, concealed in them, traces of their development; they are interrelated not only on the synchronic plane of formal logical operations, but also (though in a non-manifest way) along all the depth of their chronology and genealogy.

Our general strategy consists in bringing together different (sometimes distant) but interrelated philosophical topics in order to clarify problems, to reveal hidden intentions behind them, and to proceed further along their pre-delineated paths. This hermeneutic analysis enables us sometimes to disclose deep and stable structural correspondences ("*homologies*") between seemingly dissimilar lines of thought.

1.3. *The Aim*

The whole research is built around the attempt, first formulated as a task by Heidegger, *to discriminate between entities (or "beings") and their being.* According to Heidegger's intention, elaboration of this distinction ("*der Unterschied von Seiendem und Sein*") makes it for the first time possible to thematize being (*das Sein*) as such, which means to build an "authentic" ontology. Heidegger coins a special *terminus technicus* — "ontological difference" — to designate the distinction between an entity or a being (*Seiendes*) and its being (*Sein*).[1] Our investigation culminates in the last chapter in a thorough analysis and "deciphering" of Heidegger's enigmatic formula: "*Der Unterschied von Sein und Seiendem ist in der Zeitigung der Zeitlichkeit gezeitigt.*"[2] The "nature of time" and the "nature" of ontological difference prove to be, to say the least, closely interrelated within the framework of Heidegger's ontology. In a sense, time "is" nothing else but the ontological difference. Yet strictly speaking time "is" not, because it is not a being among beings; it "*temporalizes itself*" *as the ontological difference.*[3] The main goal of my research then consists in the clarification and demonstration of this last thesis.

---

[1] The more integral and distinct our idea of Heidegger's philosophical task becomes, the better we understand that the ontological difference is the rock-bottom of Heidegger's philosophy. That is how E. Levinas evaluated this concept. See also Th. de Boer, *The Rationality of Transcendence* (Amsterdam: J. C. Gieben, 1997), pp. 115ff.

[2] M. Heidegger, *Die Grundprobleme der Phänomenologie*, Gesamtausgabe, Bd. 24, 2. Aufl. (Frankfurt a. M.: V. Klostermann, 1989), p. 454. Hereafter will be cited as GP with the appropriate page number. English translation by A. Hofstadter: *The Basic Problems of Phenomenology* (Bloomington & Indianapolis: Indiana University Press, 1988).

[3] The German verb "*zeitigen*" means as a phenomenological *terminus technicus* "to temporalize" = "to produce time," but retains simultaneously its common meaning of "to ripen," "to let things grow ripe for something" and thus refers indirectly to the meaning of the verb "to temporize."

As was already stated, in order to proceed along the ontological lines mentioned above, I pursue a certain hermeneutic strategy, trying to inscribe Heidegger's constructions into a broader historico-philosophical context. In doing this I partly follow Heidegger himself and choose the same *nomina actoris*. The *diachronic* dimension of our investigation (the dimension of "genealogy") leads from the "postmodern" concept of the ontological difference back to the Medieval controversies concerning the *distinctio et compositio essentiae et existentiae in ente creato*, and further on to some subtle distinctions within Aristotelian physics and metaphysics. This tracing-back, as it has been already said, is not simply a tribute to historical curiosity, but rather an attempt to *clarify the concept of ontological difference* (as primordial temporality) *via* the subsequent disclosure of its diachronic strata.

My deep conviction, which I share with many contemporary authors, is that it is the *Corpus Aristotelicum* that contains the most important clues for Heidegger's solutions. P. Ricœur writes that "a certain reappropriation of Aristotle under the guidance of Heideggerian concepts can lead back in turn to a better apprehension of the leading concepts of *Being and Time*."[4] However, this is, I believe, too modest a description of the state of affairs concerning the relationship between Heidegger and Aristotle.[5] Heidegger himself in his books and lecture courses interprets a large body of Aristotle's texts. It is not our goal in this research to evaluate, whatever the criteria of such an evaluation might be, whether Heidegger's interpretation of Aristotle is "authentic" or not. Much more important is the task of observing and studying how Heidegger's ideas, and even terminology, "grows out" of this interpretation. For me, to follow this diachronic dialog is a way to inscribe the "ontology of time" into multidimensional hermeneutic space. Within this space fundamental ontological concepts are clarified not only according to their internal ("logical") structure, or through analysis of their *synchronic* relationships with other concepts (dialectics), but simultaneously by means of disclosing the *diachronic* strata, the genetic structures of their meaning.

A *locus classicus* (or, perhaps, **the** *locus classicus*) for the whole philosophy of time is the Fourth Book of Aristotle's *Physics*. Time, according to Aristotle, is inseparably connected with *movement*; time is a definite formal moment of movement, its "number." Moreover, Aristotle's *Physics* itself is in a sense a kind of *ontological justification* of movement and time after the sentence of death passed on them by Parmenides.

[4] P. Ricœur, *Oneself as Another* (Chicago and London: University of Chicago Press, 1994), p. 311.

[5] Today we know that in the decade preceding the publication of *Being and Time* Heidegger worked at great length on Aristotle, to the point that Rémi Brague, in his excellent book *Aristote et la question du monde* (Paris: PUF, 1988, p. 55), states that "de la sorte, l'œuvre majeur de Heidegger est le substitut d'un livre sur Aristote qui ne vit pas le jour. Qui plus est, je pense que *Sein und Zeit* n'est pas seulement à la place d'un livre sur Aristote au sens où il l'aurait simplement remplacé, en traitant d'un autre sujet. Il me semble au contraire que *Sein und Zeit* se donne pour but de dégager les présupposés de quelque chose que son auteur n'appelle pas encore 'la métaphysique', mais dont il trouve la première et paradigmatique formulation dans l'œuvre d'Aristote."

1.4. *Non-Being and Time*

The opening section of the *Ontology of Time* is dedicated to the pre-history of the concept of time found in Parmenides' Poem. The verse of crucial importance for the whole history of ontology (fr. 8 DK, 5f.) states that being (τὸ ἐόν) "has not ever been and will not be, since it is now, all together, one, indivisible. For what parentage of it will you look for?" Yet, what is said here is not a total condemnation or expulsion of *Chronos*. Rather the verse is a proclamation of the predominance of the "now." Since being is perfect in its immutability and persistence, since it abides in its totality without interruption, there is no room for another time, except the permanent "now," alongside of being. Παρουσία, the presence in the present, the presence as the present — this is the only ontologically legitimate meaning of time, all the rest has no sense and is a mere invention of mortals. But then Parmenides' "now" is indistinguishable from Parmenides' "being." "Time is not nor will be another thing alongside being." (8, 36f.) *Chronos* is a redundant, misplaced, void and dangerous name. Still the "now" is tacitly recognized as a notion almost equivalent to *being* itself.

Heidegger asks in his *Kantbuch*:

What project lies at the basis of this comprehension of being? [...] What is the significance of the fact that a being in the proper sense of the term is understood as οὐσία, παρουσία, i.e., basically as "presence," an immediate and always present possession?

The answer is:

The project relative to time (*der Entwurf auf die Zeit*), for even eternity, taken as the *nunc stans*, for example, is *thoroughly* conceivable as "now" and "persistent" only on the basis of time. [...] This project reveals that *being* is synonymous with *permanence in presence*.[6]

1.5. *Time as Number and Calculating Soul*

The next step in this "self-evident projection of being onto time" (*der Entwurf auf die Zeit*) which I analyze in chapter 2 is Aristotelian time theory. Aristotle defines time as the "number of movement in relation to the before and after" (*Phys.* IV 11, 219b1f.). If we want to really understand what Aristotle means, not to read our own contemporary notions into his text, we need to reconsider step by step his concept of κίνησις and ἐνέργεια (section 1), of number (section 2), of the *before-and-after* relation, of the "now" as permanent, and the "now" as fluent (sections 4, 5). This hermeneutical work, based upon both conceptual and philological analysis, forms a considerable part of the chapter.

Still the main question that guides our discussion here is *whether we can discover a prototype of the ontological difference already within Aristotelian time theory.* Of course, it seems legitimate in searching Aristotelian sources of ontological difference to reproduce quite literally Heidegger's definition — *der Unterschied von Seiendem und Sein* — in Greek, and to ask about the distinction between τὸ ὄν and

[6] M. Heidegger, *Kant und das Problem der Metaphysik*, 4. Aufl. (Frankfurt a. M.: V. Klostermann, 1973), § 44, p. 233. Hereafter cited as KPM.

τὸ εἶναι. But, as I try to show, it is another Aristotelian dichotomy that proves to be more important, namely the distinction articulated by means of an enigmatic formula (which has been for many years a matter of peculiar interest to Aristotle scholars): ὅ ποτε ὄν ἐστι τὸ X / τὸ εἶναι τῷ X. This way to posit the distinction is most frequently used by Aristotle precisely in the Fourth Book of his *Physics* in order to clarify the basic concepts belonging to the sphere of movement and temporality; in particular, this is Aristotle's way to discriminate between the two faces of the "now": the "now" as fluent and changing, and the "now" as permanent and self-identical. I attempt to establish (by means of "reappropriation of Aristotle under the guidance of Heideggerian concepts" and also using the *Commentarium in physicorum Aristotelis libros* of Thomas Aquinas) a fundamental correlation between this distinction on the one hand, and the ontological difference on the other. The principles of this demonstration will be briefly outlined.

One can always distinguish three main dimensions in Aristotle's method. I shall call the first one *metaphysical*: it consists in tracing back a problem under consideration to the ἀρχαί, the basic principles and concept of *first philosophy*. The second one (which is actually a side of the metaphysical consideration) can be termed the *logical dimension*. First philosophy asks about entity as entity, it asks *what* entity (a being) *is* insofar as it possesses being (*Metaph.* IV 1, 1003a21). Yet the question "*what is...?*" refers to the *meaning*. Clarification of the meaning of an entity as entity requires a systematization of the ways of speaking of "entity," of the ways of pronouncing the word "is." Aristotle's creation, which we call *logic* today, was for its creator himself the science dealing with the way entity exhibits itself (its being, its beingness-essence), the way it "bears witness" (in Greek — κατηγορεῖ) to itself in the logos. Logic is always onto-logic, an aspect of ontology, and this *onto-logical parallelism* should always be taken into account when interpreting Aristotle. Finally, the *psychological* (or rather *phenomenological*) dimension is of principal importance for Aristotle: if a philosopher asks about entity as entity he must, being faithful to the Parmenidian tradition, take into consideration how the soul (in particular, the intellect of the soul, νοῦς) encounters beings. This context is extremely important especially for Aristotelian *chrono-logy*, since here Aristotle searches for the answer to the questions concerning the *nature of time*, analyzing how movement and time *are given to the soul*, how we *perceive, distinguish and recognize temporal determinations.*

The fundamental premise of Aristotelian ontology (which I discuss at length in section 3 according to the scheme of onto-logical parallelism mentioned above) can be stated as follows: *A being* (τὸ ὄν), in the sense which is fundamental for this ontology, is to be understood as a participle derived from the *copula* "is" (*S is P*). That is, *being* (τὸ ὄν, *ens*), generally speaking, is not a self-sufficient term, it needs to be completed and (thereby) refined; it is "open" towards the subject *and* towards the predicate. To acquire its full meaning, the participle "*being*" must be put between an implied subject and an implied predicate, i.e., must be inscribed into the following construction:

S — being — (*what* or *as what*) P.

A being (ens) is said about S, and in this case implies a certain P, or about P, and implies a certain S.

I suggest that a tacit prototype of the ontological difference in Aristotle's ontology can be detected on this *logical level* as the difference between the function of the copula in the closed identity formula "*X is X*," on the one hand, and the openness of the "X is —" for varied predicates. Of course, the "is" as identity (*X is X*) also indicates being in its internal duality discovered by Aristotle. This duality means, in general situations, that the *that-about-which* (the subject, ὑποκείμενον) of an assertion is shown in its being through and by means of something else (the predicate, κατηγορούμενον). Nevertheless, if we had only one unique way of manifestation, only one unique *mode* of being, we could never even pose the question about the distinction of entity (*ens*) and (its) being (*esse*). Indeed, the entity is, by definition, exactly *that-what-is*, and the "is" in its "proper philosophical sense" would always signify one and the same, for example, Parmenidian monolithic identity. It was Aristotle who first discovered the manifold meanings of being by discovering (among other thing) different figures of predication, different functions of the "is" as copula.

I demonstrate in sections 4 and 5 of the second chapter how this *logical* distinction is connected with the *phenomenological* distinction of the *noetic* and *dianoetic* ἐνέργειαι of the intellect. Though the latter terms are not to be found in Aristotle's writings, their introduction and the analysis of the corresponding concepts have solid ground within the texts of *De anima* and *Metaphysics*: the *noetic* ἐνέργεια is the presence of a *noetic* form in the intellect, which is at the same time a "perfect action" of the intellect itself; the *dianoetic* one is the "perfect action" of connecting together the subject and the predicate in a judgment.

I would like to indicate briefly how these distinctions are connected with Aristotelian time theory. One can express the being of X in two ways:

1. *X is now (and always, and essentially) X.* Being, expressed by this "is," refers to the *essence* of X, to its internal form, to that which means (and always meant) to be this X (τὸ τί ἦν [τῷ X] εἶναι). The Scholastics used to speak in this connection of *esse essentiae*.
2. *X is now* — .... X *is* now as openness to the various ways of logical manifestation: *X is Y, Z, T*.... This openness posits X in its being and this being is articulated and designated by the copula. If X is the name of a thing composed of matter and form (a *compositum*), the "is" here implies also openness to the accidental ("matter-of-fact," "here-and-now") determinations of the thing, and thus points to its *existence*, to *esse existentiae*.

The "now" must be understood differently here. In the first case, it is the permanent presence of the persistent substantial form, which is always self-identical. In the second case, it is open to the factual event-content. These "now(s)" are the two temporal forms of the *noetic* and the *dianoetic* ἐνέργειαι respectively. "Phenomenologically" speaking (i.e., from within the context of the active intellect), the distinction between these two ἐνέργειαι is the distinction of the two faces of the "now." As I attempt to demonstrate, in Aristotle's ontology this distinction serves as the chrono-logical condition for *differentiating* between a being (ὄν) and its being (τὸ εἶναι αὐτοῦ), never accomplished explicitly in the *Corpus Aristotelicum.*

The analysis of chapter 2 is fundamentally important because it "radiates" its content over the whole book. For instance, the Aristotelian onto-logical parallelism emerges once more in connection with the ontologically important medieval semantic distinction of *ens sumptum nominaliter* and *ens sumptum participaliter.* The Aristotelian notion of ἐνέργεια lies also behind our consideration of *actus essendi* within the framework of medieval metaphysics.

According to my vision, it is the concept of ἐνέργεια and its internal form that provides the clue for the whole time theory in *Phys.* IV. And not only this: I believe that the role of this Aristotelian notion, in spite of voluminous discussions, still has not been fully understood and properly estimated. I attempt to provide a thorough phenomenological interpretation of this concept and to use the reinterpreted notion of ἐνέργεια as an important conceptual tool throughout the book. In particular, it plays an essential role in discussing Husserl's theory of "*internal time consciousness*" and Heidegger's "*ecstases of temporality*" and their "*horizontal schemata.*"

In my opinion, Heidegger does not give Aristotle his due when he considers Aristotle's time theory in *Physics* as a perfect expression of the "common-sense" (*vulgär*) understanding of time by way of a sequence of "now(s)." For according to Aristotle the "now" has a double nature: it is no longer (as in Parmenides) a synonym of the permanent presence-in-the-present; it rather *articulates* time, i.e., generates the before-and-after relation; it is *open* to the factual time determinations, and that means it is open to the future. Time itself according to the *Commentary* of Thomas Aquinas could be compared to the *movement* of the "now." Nevertheless, the predominance of the "now" as *nunc stans* governs in concealment the whole of Aristotelian chrono-logy: the changing "now," the *nunc fluens*, is, like movement of the moving body, an "accidental determination" of the *nunc stans*, the "substance." The ontological priority of substance in Aristotelian ontology finds its counterpart in the chrono-logical priority of the self-identical "now" and the phenomenological priority of the *noetic* ἐνέργεια of the intellect.

1.6. *Ontology of Human Action*

The ontological concept of ἐνέργεια embraces not only the ἐνέργειαι of the intellect but also the whole "spectrum" of the ἐνέργειαι of the soul (in particular those relevant to moral action). It includes different ways of ἀληθεύειν, to "arrive at truth," to abide in "unconcealment" (as Heidegger interprets the Greek term). *The soul possesses or discloses truth* in different ways (*Eth. Nic.* VI 3, 1139b15ff.), and not only by (or in) "theoretical wisdom" (σοφία), and its *noetic* ἐνέργεια, but also by (or in) "practical wisdom" or "prudence" (φρόνησις), and the correspondent *phronetic* ἐνέργειαι. As a "place" where the truth is disclosed the soul has its own topography, depicted by Aristotle in the *Nicomachean Ethics.* Heidegger's interpretation of *Eth. Nic.* VI that prefaces his lectures (1924/25) on Plato's *The Sophist* constitutes an even more striking example than those mentioned above, demonstrating the continuity between Heidegger and Aristotle. Here we can observe an amazing kind of "transfiguration" of Aristotelian discourse into Heidegger's conceptual system, of Aristotle's Greek into Heidegger's "German." I attempt to demonstrate in chapter 4 how close the interrelation is between Heidegger's project of *Fundamental ontology* on the basis of the existential analytic of *Dasein*, on the one hand, and Aristotelian "topography" of human soul and its ways to "arrive at truth" (ἀληθεύειν) in *Eth. Nic.* VI, on the other. For Aristotle, to be sure, the disclosedness of being in *phronetic* ἐνέργειαι possesses only ontologically posterior meaning, but for Heidegger the existential counterpart of these ἐνέργειαι becomes the foundation of *fundamental ontology.* Chapter 4 is a *sui generis* point of transition from diachronic to synchronic analysis (and this is the reason why it is placed not immediately after chapter 2). The main ideas outlined here acquire their full development in chapters 6 and 7 within the framework of discussion of the *existentiale* of care and its temporal meaning.

1.7. *Distinctio et Compositio Essentiae et Existentiae*

Chapter 3 is devoted to the medieval controversies concerning the *distinctio et compositio essentiae et existentiae in ente creato.* Heidegger himself considers this distinction as being a "side" of ontological difference and comments upon it at length in his lectures on *The Basic Problems of Phenomenology.* It comes as no surprise that the preceding consideration of Aristotle's *first philosophy* paves the way for our interpretation of this Scholastic prototype of Heidegger's "ontological difference." My analysis is based first of all on the *Disputationes metaphysicae* by Francisco Suarez, a treatise that fundamentally influenced Heidegger. A careful rereading of some fragments of this text under the guidance of Heidegger's ontology reveals, quite unexpectedly, that we can discover in Suarez not only one more version of the medieval metaphysical *distinctio* mentioned above but also the striking idea of *finiteness* (and thus de-finiteness, the internal form) inherent in *existence* as such. This idea (in a *fundamental-ontological* guise) is of crucial importance for the *existential analytics of Dasein.* Heidegger says in his *Kantbuch*: "More primordial than the man himself is the finiteness of *Dasein* in him."

1.8. The Transition to the "Synchronic" Analysis

This archaeology of thought occupies the first half of the book and, I hope, sheds light on the deep backgrounds of Heidegger's intentions.[7]

Another strategy that I pursue in my research is an attempt (which is not new, to be sure) to move along the lines of Heidegger's departure from Husserl's phenomenology. What seems to be new and, I hope, productive, is a certain reappropriation of the late Husserl's phenomenology of time under the guidance of Heidegger's cri-

---

[7] Chapter 5 plays a special role, and I shall come back to it later on.

tique and revision. If "with regard to its subject-matter, phenomenology is the science of the being of entities — ontology,"[8] is it possible, looking at Husserl's phenomenology as it were from within Heidegger's context, to consider it as an advanced and self consistent *ontology of the present-at-hand*? What are the consequences of this perspective for Heidegger's *fundamental ontology*? At least, this approach leads back to a better apprehension of the paradoxes of phenomenological *egology* and the "riddles of the living present" (to use K. Held's expression) in the late Husserl.

The term "ontological difference" was mentioned for the first time (publicly) by Heidegger and discussed in the lecture course of 1927 entitled *Die Grundprobleme der Phänomenologie*. One of the main tasks of this lecture series was to outline the basic ideas of *Being and Time* and to pave the way for the promised (but never completed) second part of that treatise. The title *The Basic Problems of Phenomenology* is not accidental, because Heidegger intends in this text to re-examine the most fundamental phenomenological principles, in particular — to undermine the status of "*intentionality*" as the ultimate ground for phenomenological inquiry and the gene-concept out of which the whole *Corpus phaenomenologicum* is to be constructed. Is it possible to ask about the ground of this ultimate ground?

This program marks, of course, one particular line of Heidegger's departure from Husserl's basic principles. And it is important to bear in mind that all these ways of departure are guided by one and the same chain of questions. If in Husserl's "regional ontologies" the role of the *sphere* or *region* of "*absolute being*" is ascribed to consciousness (*Bewußtsein*), what is the concept (the "*project*") of being that underlies this conclusion? Has not the meaning of "being" already been taken for granted within the scope of Husserl's ontological clarifications, so that the analysis of the basic ontological questions failed to meet the fundamental phenomenological requirement of *Voraussetzungslosigkeit*? Does not the question about the meaning of being (*die Frage nach dem Sinn von Sein*) come first and play the most fundamental role? To what extent can the phenomenological claim to provide a firm foundation for ontology be justified?

As early as his Marburg lectures of 1925, *Prolegomena zur Geschichte des Zeitbegriffs,* Heidegger formulates his diagnosis: in phenomenology, "being" signifies "being-an-object" adequately and "originally" (*originär*) given to consciousness.[9] The being of consciousness, in its turn, is absolute (in the sense of *nulla re indiget ad existendum*), because the latter is the source of all objectifying acts and at the same time, as Husserl believed, can itself be objectified in a phenomenological reflection.[10] To be-an-object means to be an intentional correlate of objectifying consciousness. We find ourselves incapable of going beyond the meaning of being which has been taken for granted in Husserl's phenomenology unless we go beyond the intentional relation as such, unless we raise the question concerning its deeper

[8] M. Heidegger, *Sein und Zeit*, 16. Aufl. (Tübingen: Max Niemeyer Verlag, 1986), p. 37. Cited as SZ.

[9] Cf. Husserl's famous definition of "truly existent" object in the first volume of *Ideas*, § 142.

[10] In chapter 5, sect. 2 we discuss the difficulties that originate from the attempts to *objectify* the deepest source of constitution, the difficulties which the later Husserl himself clearly saw.

source and its ontological meaning. However, for Husserl (who follows Brentano in this respect) intentionality is a characteristic feature, a *proprium*, or even the *essence* of consciousness as such, and this means that the classical concept of *Bewußtsein* and its ontological role become a target of Heidegger's critique.

The most essential feature of this critique consists in Heidegger's intention, declared first in *Being and Time* (SZ 363, ftn.), to go *beyond the intentional relation* and to show that "intentionality is grounded in the ecstatic unity of *Dasein*." "Intentionality," Heidegger says, is not the title of a solution but the name of a problem. The intentional relation is characterized by Heidegger as a derivative modus, an *existential-ontological modification*, of more fundamental existential structures inherent in *Dasein*. The analysis of the concept of *existential-ontological modification* in different contexts plays an important role in the development of our research.

1.9. *Searching for the Lost Subject*

Intentionality, according to Husserl, is a relation (attitude, or "accomplishment") of the *transcendental subject*, and the main characteristic feature of the sphere of pure subjectivity. Heidegger's attempt to go beyond the intentional relation leads unavoidably to de(con)struction of the Cartesian concept of the transcendental subject as *substantia cogitans*. In chapter 6, I quote J.-L. Marion who asks: "Has phenomenology ever had a more urgent challenge to confront than the determination of what, or possibly who, succeeds to the subject?" Yet, we must not lose sight of the fact that the "modern ontology of subjectivity" has been undermined from the inside of Husserl's phenomenology itself. I mean the so-called "paradoxes of reflection" (discussed in detail in chapter 5, section 2) and the riddles of the "living present" — themes which we find in Husserl's late manuscripts. The subsequent development of these ideas by the disciples of Husserl destroyed or at least threatened the classical Cartesian notion of the *transcendental ego*. If we ask now, in view of Heidegger's project of fundamental ontology: "what, or possibly who, succeeds to the subject?", the immediate answer would be perhaps: "*Dasein* itself." Nevertheless, this answer is too vague and leads nowhere.

I attempt to answer the question concerning the successor of the transcendental subject by means of a detailed investigation of the relationship between the concept of transcendental subject (according to Descartes, Kant and Husserl), on the one hand, and the central *existentiale* of care. One of the main theses of this work reads as follows: *it is the ontological concept of care that comes to replace within the scope of Heidegger's fundamental ontology the modern concept of transcendental subject.*[11] The expression "comes to replace" acquires its rigorous meaning in the last chapter of our research. I shall proceed to this topic later in this Introduction. However, already in chapter 6 is established a peculiar form of relationship between the *existentiale* of care and the transcendental subject, which we call "*negative analogy*" (section 2). The negative analogy is a side of Heidegger's "destruction" method mentioned above. In the text of *Being and Time* there are many passages — and they are

[11] The thesis is discussed at length and substantiated in chapter 6.

of crucial importance — where the meaning of the basic concepts of the existential analytic of *Dasein* is formed by means of *opposing* them to the fundamental concepts of classical ontology (*the ontology of the present-at-hand*, as Heidegger terms it). These oppositions are not just illustrations; they perform an important function of meaning-formation. The oppositions point to a special kind of connection between the concepts opposed, and it is this form of opposing-connecting that I term "negative analogy." In particular, the relation of negative analogy ties together the *categories* in the sense of Kant's transcendental logic and the *existentialia* of *Dasein* (SZ § 9). The power of this analogy seems at first sight to be rather limited because, formally speaking, categories are the most general predicates of an object-at-hand and the *existentialia* are characteristics of the *being* of *Dasein*. But this point of view is, to put it in Heidegger's words, a result of "the oblivion of being." As it is demonstrated in chapter 2 of my book, already for Aristotle, categories are not so much the most general predicates, in a modern sense of the term, as ways of disclosing being in the element of logos, the possible ways of manifesting being in the propositional form: *X is* —. In Kant's transcendental logic, the pure concepts of understanding — the categories — are the most general laws of synthesis, i.e., the forms of pure acts of the transcendental subject. In this sense, the categories in the *Critique of Pure Reason* are to be considered ultimately as determinations of the *transcendental subject*. The latter is the principle of unity, the "center," for the system of categories.

In a sense, Heidegger in *Being and Time* remains faithful to the method of transcendental philosophy, for he approaches being *via* the *modes of Dasein's encounter* with beings, though these modes are no longer derivatives of pure consciousness and determinations of its transcendental "possessor" = *the I*. Still the *existentialia* are those characters of the being of *Dasein*, which make this encounter possible. The first and last word for this ultimate possibility are *being-in-the-world* which itself is a side of *Dasein*'s existence structured by particular *existentialia*. This integral structure is united and centered, according to Heidegger, in the *existentiale* of care. Thus, we deal here with a fundamental analogy:

transcendental subjectivity : categories (in Kantian sense) =
"care" (ontologically understood) : *existentialia*.

This "proportion" is explored in detail and its limits are established. I consider this investigation to be of considerable importance from the ontological standpoint in general, and for the goals of my book, in particular. Indeed, Heidegger's fundamental ontology is the ontology of "what-is-made" (τὸ ποιητόν) and of "what-is-done" (τὸ πρακτόν). First of all it is the ontology of πράξει ὄν,[12] the ontology of human action, and finally the ontology of the agent. "*Curo ergo exsisto*" determines a new foundation of post-modern ontology; the *existentiale* of care is the central concept of Heidegger's existential analytic of *Dasein*, just as the transcendental subject is the ultimate ontological *fundamentum* in modern philosophy since Descartes.

Nevertheless, our investigation would be superficial if from the very start we accept uncritically the modern identification of subjectivity with "I-ness" (*Ichheit*),

[12]I forge this expression in order to emphasize the ontological context of the discussion, imitating Aristotle's φύσει ὄν, τέχνῃ ὄν = τὸ ποιητόν.

to use Fichte's term. Heidegger writes in his work entitled *The European Nihilism*: "Thanks to Descartes and after Descartes the 'subject' of metaphysics becomes mainly man, the human ego. How does man assume the role of the proper subject and of the only subject? Why is this human subject superimposed on the *ego*, so that subjectivity is here equal to the sphere of the *ego*? Is subjectivity defined by the *ego* or, vice versa, the *ego* by subjectivity?" I attempt to answer this question having in view our ultimate target — *primordial temporality* as a source of the intentional relation.

The foundation of our analysis was laid in the previous chapters, especially by interpretation of the Aristotelian concept of substance as the ultimate ὑποκείμενον (chapter 2). *Subjectum* translates the Greek ὑποκείμενον as "that which lies under." Thus, the subject is that which is "thrown under" as a prior support or the deepest (sub)stratum upon which other determinations, attributes and modi, may be based. These latter determinations of a thing can be given to (or represented by) the intellect as its objects, can become present in representation. In the Scholastics, this mode of being, the being-for-intellect, has been termed *objective*. (The contemporary usage of this expression has exactly the opposite meaning.) Thus the *subjectum*, according to the formal structure of the concept, being the deepest substratum of all possible determinations, is the basic condition for objective being, for the *ob-jectum* as such. We need to keep in mind that according to this ontology the subject is the "*farthest* link" in the chain *subject — object — intellect* in relation to the faculty of cognition. After Descartes, the situation has been drastically changed, and this process finds its completion in Kant's transcendental philosophy. For Kant, the *ego* in its first and basic meaning is nothing else but the *I* of the representation "I think," which makes possible all representations as such. The entire manifold of objective determinations (as components of a representation) is now to be considered as a modification of the "I think," and ultimately as the determination of the transcendental *I*. The latter becomes thus the primordial ὑποκείμενον, the ultimate *subjectum*. "I think" means for Kant, first of all, "I combine (*Ich verbinde*)"; therefore the *ego* as the transcendental unity of self-consciousness, as the unifying principle, is the ultimate condition of the combining activity, and thus, of the integrity of the "objective" manifold given in intuition. The transcendental *ego* becomes the necessary condition of the presence of what is present, of the ob-ject. Thus, the *ego* in the sense of *personalitas transcendentalis* is the *subject as the possibility of the object.*

An important result of this chain of arguments can be formulated as follows: man "assumes the role of the proper subject and, strictly speaking, of the only subject" when (and only when) entity (*ens*) assumes the role of the proper object and being (*esse*) assumes the role of *objectivity* or *positivity*. From this the "fundamental axiom of modern ontology" is readily apparent: *to be means to be able to become an object.*[13] Or in the transcendental interpretation: to be means to be posited by the mind as an object; being means positivity. Nevertheless this axiom leads to a difficult *aporia*: the being of the transcendental *ego* itself becomes a problem, because, as the primordial

[13] This thesis is nothing but a paraphrase of "a basic presupposition" of "traditional ontology" (in the broad sense of the word), the presupposition which, according to Heidegger, governs secretly both everyday language and the history of philosophy.

source of positing, it itself can never be objectively posited; in its transcendental life it resists objectification and "avoids positivity." I discuss this paradox in connection with Husserl (and Plotinus) in chapter 5 and in connection with Kant in chapter 6.

1.10. *Care as Primordial Temporality*

In the first section of chapter 7, I proceed towards Heidegger's notion of primordial temporality, trying not to lose sight of Husserl's theory of "internal time consciousness" and "absolute temporal flow." Different "images of time" are rooted, according to Heidegger, in the ecstatic-horizontal temporality of *Dasein*, and we need to penetrate into this most fundamental level of the "time" problem. Here too Aristotle is of crucial importance. According to the definition of *Being and Time*, primordial temporality is the *ontological meaning of care*. In the *Nicomachean Ethics* we encounter a "human counterpart" of physical time (χρόνος) called καιρός:[14] καιρός is the *moment of human action*, of human decision and choice (προαίρεσις), which is inscribed in the structure of *Dasein*'s care. Heidegger's fundamental thesis concerning the role of "*Augenblick*" can be formulated as follows: not only the "present moment" of "objective" (sc. "objectified") time, but also the "now" (*das Momentanjetzt*) within the structure of objectifying acts (as they present themselves in the realm disclosed through the phenomenological reduction) is a derivative modification of the "Augenblick."

1.11. *Differentia Differens*

The last section of the book accumulates everything that has been achieved earlier. First of all, the relation of negative analogy here acquires its proper ontological meaning. Let me describe it as follows: concepts are connected by the relation of a negative analogy if the correspondent "things themselves" are connected by the relation of an *existential-ontological modification*. The term designates in Heidegger's fundamental ontology a certain change in the mode of *Dasein*'s encounter with beings, i.e., a turn in the understanding-interpreting of being. The basic example of such a modification in *Being and Time* is the thematization of an item of equipment (*Zeug*) as something present at hand (*Ding*), positing it as an object of representation, i.e., transition from uninterrupted dealing-with to theoretical looking-at. However, the content of this notion is far from being exhausted by this sort of example, which is only one element in the row of "existential-ontological modifications" having different characters. Heidegger's attempt to disclose the deepest grounds of being and to relate them to the being of a particular entity called *Dasein*, to its own existence, leads him to the concept of "derivation" (*Ableitung*) embracing all sides of the objectifying attitude: a thing, an object must be understood as a derivative of an item of equipment, the being-at-hand — as grounded in being-ready-to-hand, the apophantic "as" (*das apophantische Als*) of predication is the result of a derivation having the hermeneutic "as" (*das hermeneutische Als)* of dealing-with for its

[14] Heidegger translates — "der Augenblick," having in mind the correspondent term in Kierkegaard.

starting point. But the most fundamental modification that lies in the foundation of all others is the derivation of the temporal form belonging to the objectifying acts from the internal time of human action, and ultimately from the ecstatic-horizontal temporality of *Dasein*. In order to analyze this basic modification, we need to overview Husserl's theory of *internal time consciousness*, to interpret in this context Heidegger's claim stating the secondary, derivative role of the Husserlian temporal form in comparison to the ecstatic-horizontal temporality of *Dasein*, and to consider to what extent this claim can be justified. My thesis, already anticipated in the preceding overview and substantiated in chapter 7, is as follows: *the transcendental ego is to be considered as the* ***existential-ontological modification*** *of the primordial self-projection which is the meaning of care.*

The ecstatic character of primordial temporality, "ecstaticity" as such, is the deepest ontological (= chrono-logical) meaning of care. To be sure, the temporal grounding of pure transcendental subjectivity is proclaimed by Kant and Husserl as well. According to Husserl, the *ego cogitans* "temporalizes" itself in the absolute temporal flow. This temporalization, as I show, is nothing else but the "ever-failed" attempt of self-objectification in reflection, the endless longing (as in Plotinus) for the positivity of the Self. That is why another important thesis is closely interrelated with the preceding: quite in keeping with Heidegger's terminology *one could call the "absolute temporal flow" in the sense of Husserl's phenomenology the "existential ontological modification" of the primordial ecstatic-horizontal temporality.*

I use all the results of the preceding analysis to clarify (and hopefully to enrich) Heidegger's own discussion of the distinction between "being-what" and "being-how" (a form of ontological difference introduced in his lectures *The Basic Problems of Phenomenology*). *This difference depends, in its turn, on the difference of the temporal ecstases of primordial temporality and of the horizons belonging to them.* The primordial difference of the temporal ecstases, temporalizing itself, is a distinction before that which is distinguished. The temporal form, the possibility of distinguishing between *before* and *after* is born of this proto-difference, of the distinguishing distinction.

1.12. *God without Being and Thought without Thinker*

After Descartes the question *Sed quid igitur sum?* becomes the most important question of *first philosophy*. Husserl defined his phenomenology as transcendental *egology*. Heidegger in *Being and Time* posits the "existential question" about the *Who* of Dasein. Here temporality proves to be the hidden ground of Self-positing and Self-revealing: in Husserl as the *absolute temporal flow*, in Heidegger as the *ontological meaning of care*. Throughout this work, I refer to different historico-philosophical contexts and apply different hermeneutical strategies in order to emphasize that the concept of the human Self, the concept of being, and the concept of time are inseparably linked with one another. To this triad of philosophical concepts I add in chapter 5 one more of a theological nature, *viz.* the relationship between God and man as it has been developed, on the one hand, in Orthodox *apophatic* theology and, on the other hand, during the Scholastic controversies concerning the problem of *visio Dei*.

The first section of the chapter is devoted to the analysis of the famous passage from the treatise *On the Divine Names* by Ps.-Dionysius the Areopagite, which states that God as the source of being "has not been, will not be and never was, He did not come into being, does not and will not; more than that — He is not" (V 5). As we see, this statement not only reproduces within a new context the Parmenidian "[being] has not ever been and will not be," but also strengthens the apophatic tune of the verse by rejecting its positive part ("since it is now, all together, one, indivisible...") in the striking "He is not." In order to interpret this text I use the classical Scholia of two great Byzantine theologians, John of Scythopolis and Maximus the Confessor. The Scholium, which allows me to inscribe the fragment under consideration into the scope of phenomenological analysis, interprets the *being beyond being* (which is simultaneously the giver of being and thus the *good beyond good*) as the intellect (νοῦς) and constructs a *noetic* counterpart to the ontological text of Dionysius. It says among other things, "The autonomous good beyond good, being an intelligence and an ἐνέργεια directed (lit.— *turning round*) towards itself, abides in the state of actuality, not in the state of potentiality. Primarily it is ***thoughtlessness*** (ἀφροσύνη), and then becomes an intelligence in actuality." This commentary would remain no less enigmatic than the text of Dionysius itself unless we bridge it with Plotinus' consideration of the One. (This move is perfectly justified from the historical point of view.) The One is *beyond being* because it is, as Plotinus says, "without *whatness.*" Plotinus' method here is the method of reflection: the One is deprived of its "what" because it cannot become an object of the intellect; still the intellect pursues the One as its own ultimate *source,* and while pursuing it the intellect loses it. Only in a kind of "thoughtlessness," which means, in technical terms, the identity of form (εἶδος) and ἐνέργεια, is the intellect in a sense able to possess the One.

In the second section of the chapter, I attempt to demonstrate that the paradoxes of reflection in the writings of the late Husserl are structurally of the same nature. Let me articulate this "homology relation" by means of two important analogies or, in Latin, *proportions*:

> the divine Being beyond being : the totality of beings = the intellect beyond thinking (as pure noetic ἐνέργεια directed towards itself) : the totality of its objects = the transcendental subject : the flow of consciousness (which is at the bottom the absolute temporal flow).

The first proportion is the onto-theological analogy that we find in the Scholia; the second one is the phenomenological interpretation of the reflective procedures of Plotinus.

The relevance of such a phenomenological interpretation is supported by another text, *viz.* the commentaries of Albert the Great upon the same passages in Dionysius' treatise. As we have seen, the *noetic* parallel of the *ontic* statement "He is not" consists in the impossibility to reach by reflection the source or foundation of the intellect's activity. According to Aristotelian metaphysics as accepted by the Scholastic thinkers, this foundation (the ultimate ὑποκείμενον) cannot be anything else but an οὐσία (a "substance"). This ontological qualification is unavoidably implied by the form itself in which the problem is posed and the general metaphysical context of the epoch. In the commentary of Albert the Great the question is, whether and how the divine *substance* can be reached by the intellect, whether and how it is possible for the intellect *videre Deum per substantiam suam immediate*. This question, under the heading of *visio Dei*, has become one of the most controversial problems of medieval theology. I attempt to demonstrate that it is possible to find in Albert's commentary brilliant pieces of phenomenological analysis, which confirm the structural analogy or homology stated above.

To be sure, onto-theology considers the problems we touched upon in the reverse perspective in comparison to phenomenology: it is God, not the transcendental *ego*, who is here the center of perspective. Still the difficulties and paradoxes, which both ways of philosophizing encounter here, have the same basis, *viz*. the "main axiom of traditional ontology" — *to be means to be able to become an object*. Heidegger finds a way out of this impasse by reconsidering the foundations of ontology and reviving the ancient concept of ἀλήθεια. In particular, as it has been already said, he reproduces and develops within the ontological horizon a most important result of the *Nicomachean Ethics* — the discovery of different ways of arriving at truth.

It is my hope that the different paths outlined above come together in a certain "topos" of thought, the borders of which can be clearly seen. Here the efforts to elucidate the fundamental concepts and problems of post-modern ontology are concentrated. The result of these efforts is not (and cannot be) a "solution," but rather a clarification. Here ontology-as-chronology begins its work and poses its proper questions.

1.13. *Acknowledgments*

This book is a revised version of my doctoral thesis defended at the Philosophical Department of the Free University in Amsterdam. I owe a special tribute of gratitude to my supervisor Professor Theodore de Boer whose kind, critical and sustained interest over many years supported and directed the development of this research.

I want to thank all the members of the reading committee Professor A. P. Bos, Dr. Karin de Boer, Professor J. van der Hoeven, and Professor R. te Velde for their patient and careful study of the manuscript of my dissertation. Their critical remarks added immeasurably to its quality.

I wish to thank cordially the Rector Magnificus of the Free University in Amsterdam Professor T. Sminia, the Dean of the Faculty of Philosophy Professor Willem de Jong, the authorities of the University and the staff of the Faculty who helped me to solve in a timely manner many organizational and financial problems, including the printing of promotion copies of my thesis, preparation of the defense ceremony, and many other difficulties which looked unsolvable of a distance from St. Petersburg.

I owe a debt of gratitude to Mr. Yegor Nachinkin whose endless patience and accuracy have helped me immeasurably in my struggle for clear English.

I want to express my deep gratitude to Professor Patrick A. Heelan whose ideas stimulated my interest towards phenomenology and who helped this work to take its final shape.

CHAPTER ONE

NON-BEING AND TIME

(The prehistory of the concept of time)

1. THE CIRCLE AND THE SPHERE

Φησὶ δὲ [Ξενοφάνης] οὐσίαν θεοῦ σφαιροειδῆ [εἶναι]

Diogenes Laertius, *Vitae philosophorum* IX 19.

Whatever one may say about Heidegger's philosophy and whatever one thinks about it, there can be absolutely no doubt that today's philosophy is, one way or another, a "philosophy after Heidegger." Today's philosophical thinking cannot ignore a strange and, at first sight, pretentious "philosophical myth," stating that it is being[1] itself, that in a given epoch and historical time reveals itself in a certain way — thus and not otherwise — and simultaneously conceals itself, illuminates itself and hides itself. According to this mythologeme the occurrence of self-giving does not obey definite *a priori* laws (disclosed and established laws are only posterior glimpses of the primordial self-illumination of being). Perhaps, it is the arbitrariness of the will or incomprehensibility of fate (*Geschick*), rather than *a priori* laws, that should be considered as the analogy of this self-giving (*Sich-schicken*) of being in history (*Geschichte*). Here lies the essential character of history as the history of being (*Seinsgeschichte*), and here the word "epoch" (ἐποχή) lays bare its original meaning of "a stop," "a pause," "a suspension," "a retention." Actually, the deep sense of the "historical" and of the "epochal" is determined by the way being reveals itself during one of its many successive stops, residences, pres-ences.

[1] A great difficulty one encounters time and again when interpreting and analyzing Heidegger (as well as Aristotle or medieval thinkers) in English is the formal impossibility to distinguish between *das Sein* (τὸ εἶναι, *esse*), on the one hand, and *das Seiende* (τὸ ὄν, *ens*), on the other hand. According to the tradition, which has already taken shape, I render as a general rule Heidegger's *Seiendes* as "entity," and *das Sein* as "being." But sometimes, when it seems to be justified, and especially in this section, I write "a being," or even "what-is," for *ein Seiendes*, or "(the) beings" for *das Seiende* in the sense of τὰ ὄντα, *entia*. In the latter case Heidegger's expressions referring to the *ontological difference*: "das Sein des Seienden," "das Sein ist kein Seiendes" are translated respectively as "the being of beings," "being is not a being." Parmenides' τὸ ἐόν is rendered in sect. 2 of this chapter as "being."

Heidegger has taught us not to neglect the opacity of the hidden backgrounds of this revelation, which means to be attentive to the erased traces of the fateful (*das Geschickliche*), to learn to abide by the *totality* of the historical (*das Geschichtliche*),[2] to stop searching within it for a kind of central point, a kind of spiritual ἀκμή where the closest proximity to the a-historical "eternal truth" has been achieved, in order to position oneself in that point and to take up residence there.

Hegel's philosophical task, "to comprehend and to express the truth not as a substance only, but as a subject as well,"[3] implies all the same that the "living substance," in the movement of self-assertion, may outlive itself, become closed in itself, fulfill its cycle; restore, through a sequence of necessary steps of speculative thinking, the original unity with itself and, after absorbing everything, become, in its totality, Parmenides' *well-rounded globe* (fr. 8DK, l. 43). This unity is "becoming itself (*das Werden seiner selbst*), a closed cycle which implies its own end as its goal and has it as its beginning, and which is actual only in its completion and its end."[4] The word "end," repeated twice, refers to *spirit* or *mind*[5] as to a form existing in itself and which has to become, through a kind of internal movement, existing for itself. Parmenides' globe, the perfectly contented *sphairos*, is exactly a prototype of such a spirit, which exists within and for itself.

The self-willingness of being just mentioned implies that the spirit "breathes where it wills" and gives itself in the "occurrence" (*Ereignis*) of a gift when it wills, not because it has been compelled (or has compelled itself) and forced to regulate its steps according to the "necessity of the matter which proceeds coolly on her way,"[6] i.e., to subject itself once and forever to itself,[7] to its own pre-eternal morphology. For Heidegger, there exists a "giver" (*das Es, das gibt*)[8] behind all "givenness" (*es gibt*), and this "giver" eludes our vision and our desire to name it; like a really generous giver, it hides its face from us.

---

[2] As we have mentioned, history, according to Heidegger, is not only and perhaps not so much the history of disclosure as the history of concealing and hiding; history is *der Irrtum*. The main formula of being's concealment reads as follows: "Das Sein entzieht sich, indem es sich in das Seiende entbirgt." See "Der Spruch des Anaximander," in *Holzwege* (Frankfurt a. M.: V. Klostermann, 1957), p. 310. "Dergestalt beirrt das Sein, es lichtend, das Seiende mit der Irre. Das Seiende ist in die Irre ereignet, in der es das Sein umirrt und so den Irrtum (zu sagen wie Fürsten- und Dichtertum) stiftet" (*ibid*). The concealment is at the same time the *oblivion* of being, and "the oblivion of being is oblivion of the distinction between being and beings." It follows that the restoration of the un-concealment, of the *a-lētheia*, the *truth* of being, is connected with unveiling and thematizing the *ontological difference*. The ontological difference, the distinction between being and beings, will be one of the most important subjects of our research.

[3] G.W.F. Hegel, *Phänomenologie des Geistes*, hrsg. v. J. Hoffmeister (Berlin: Akademie Verlag, 1971), p. 19. Hereafter cited as PhG.

[4] *Ibid.*

[5] Neither word is a wholly satisfactory translation of Hegel's *Geist*. Cf. E. Craig, *The Mind of God and the Works of Man* (Oxford: Clarendon Press, 1987), p.174, ftn. 2: "Mind is not theological enough in its overtones, spirit not intellectual enough..."

[6] "kalt fortschreitende Notwendigkeit der Sache" (PhG 13).

[7] Like the god of the Stoics. As Seneca says: "Ipse creator et conditor mundi semel iussit, semper paret."

[8] M. Heidegger, "Zeit und Sein" in *Zur Sache des Denkens* (Tübingen: Max Niemeyer Verlag, 1969), pp. 1–25.

Yet the "self-willingness" of the truth cannot by any means justify our arbitrariness regarding the truth; Hegel's demand to bear the burden of the concept cannot be ignored — but do we understand what a concept is and *how* it is? Has Hegel laid down the ultimate rules concerning the concept (*der Begriff*) as such?

In one of Heidegger's later works, *The Anaximander Fragment* (1946),[9] "eschatology" of thought is the subject matter. Such is the self-perception of today's epoch of philosophizing: this epoch is τὸ ἔσχατον, a term, an ending, a temporal limit, a sunset before the coming night.[10] Of course, a limit does not exclude the existence of something "beyond." "The later philosophy" is just a term testifying to the long distance separating us from the Greek starting point, if not to an expectation (for expectation implies a different mood), at least to the acceptance of the possibility of a new start.[11]

It is impossible not to take into account such a self-definition of philosophy. Perhaps such an eschatological departure from classical philosophy allows us, "the late philosophers" (*den Spätlingen der Philosophie*), to better hear the message of its dawn. In some surprising way the latest comes up with the earliest. "The early philosopher" Heraclitus says (fr. 103): ξυνὸν γὰρ ἀρχὴ καὶ πέρας ἐπὶ κύκλου περιφερείας — "on the border of a circle the beginning and the end (limit) coincide (literally: 'are common')." *Qua* "the late philosopher" Heidegger defines the eschato-logical philosophical thinking as coming together, merging, gathering together (λέγεσθαι, λόγος) its limits (τὰ ἔσχατα).

The attempt to perceive the earliest message, a movement towards the philosophical dawning and into the depth of chronology is, undoubtedly, a movement "away from the light and towards the darkness."[12] Yet this movement is not a lapse into a confusion of notions and indistinctness. It is rather an attempt to trace back the origin and motive of distinction, the *morphogenesis* of the distinct. This task of tracing is precisely what is termed "late philosophical thinking." It differs drastically from the intelligent tracing in contemplation of mutual participation or communication of the *ideas* or genera. Such is Plato's way of philosophizing, called "*di-*

[9] See ftn. 2.

[10] Heidegger alludes to Hölderlin's "holy night," where the last poets, the "priests of Dionysus" are wandering: "Aber sie sind, sagst du, wie Weingottes heilige Priester, / welche von Lande zu Land zogen in heiliger Nacht." (*Brot und Wein*, 7)

[11] I would like to quote one passage from J. Derrida as evidence of such a self-consciousness of the "later philosophy." "That philosophy died yesterday, since Hegel or Marx, Nietzsche, or Heidegger — and philosophy should still wander toward the meaning of its death — [...]; that philosophy died *one day, within* history, or that it has always fed on its own agony, on the violent way it opens to history by opposing itself to non-philosophy, which is its past and its concern, its death and wellspring; that beyond the death, or dying nature, of philosophy, perhaps even because of it, thought still has a future, or even, as is said today, is still entirely to come because of what philosophy has held in store; or, more strangely still, that future itself has future — all these are unanswerable questions. By right of birth, and for one time at least, these are problems put to philosophy as problems philosophy cannot resolve." See J. Derrida, "Violence and Metaphysics. An Essay on the Thought of Emanuel Levinas" in *Writing and Difference*, trans. A. Bass (London: Routledge, 1990), p. 79.

[12] M. Heidegger, *Platon: Sophistes* (Marburger Vorlesung Wintersemester 1924/25), Gesamtausgabe, Bd. 19 (Frankfurt a. M.: V. Klostermann, 1992), p. 10.

alectics" in *The Sophist.* The late philosophical thinking means: *Das bisherige Wesen des Seins geht in seine noch verhüllte Wahrheit unter.* The setting sun is here the old "essence" of being which was secretly prevailing, the being-ness of being which was disclosing itself as presence, as a permission-to-be-exposed to intelligent vision. The setting sun is Plato's idea of the good that allows the other ideas — i. e., εἴδη, the intelligible *species* — to be present to in-spection and speculation of the intellect. This essence goes down into its as yet hidden truth.

"The midday philosopher" Hegel says: "In spirit (*im Geiste*) it is otherwise [as opposed to the 'realm of nature' — *A.Ch.*]; it (sc. *der Geist*) is consciousness, it is free because in it *the beginning and the end coincide.*"[13] Hegel understands the eschatology of thinking as a *fulfillment* of time, as a coming of spirit to itself — from its embodied state in history to its disembodiment in its own element of speculative thinking. The history of philosophy, the existing, self-writing historical text of philosophy, is just an absolute form of movement, i. e., correlation of facets of the "one and self-identical idea," which recognizes itself in the "dim mirror" of accidental historical circumstances while yet transgressing them (in the sense of Hegel's *aufheben*). This "achronous" movement, which only takes on a shape of self-explication within time for a finite "individual" consciousness, is realized in its own element by speculative dialectics. "Philosophy has now become for itself the apprehension of this development and as conceiving thought, is itself this development in thought. The more progress made in this development, the more perfect is the philosophy."[14] Thus the perfection, the complete fulfillment of the form of philosophical thinking (the midday sun of the Idea, one could say, standing still at its zenith) removes the historical and frees the spirit from "lapsing into history." To be more precise, according to this picture, it is genuine history itself that lapses into time, for the spirit's pre-eternal history is indistinguishable from its morphology.[15] For Hegel, "gathering together the limits" meant nothing but identifying the beginning of the speculative development of the Concept with its end, the fulfillment of "being-for-itself" what it already was in itself. This Gnostic myth, turned into a project of philosophical speculation, was worded as follows: "This being-at-home-with self, or coming-to-self of spirit may be described as its complete and highest end: it is this alone that it desires and nothing else. Everything that from eternity has happened in heaven and earth, the life of God and all the deeds of time are simply the struggles of spirit to know itself, to make itself objective to itself, to find itself, be for itself, and finally unite itself to itself. It is alienated and divided, but only so as to be able to find itself and *return to itself.*"[16] Thus instead of searching for the erased traces of original thinking, to which we all are heirs in philosophy's historical text, a mind loyal to this goal must rather be able to see in itself (such as it was created from

[13] G.W.F. Hegel, *Lectures on the History of Philosophy*, Book I, trans. E. S. Haldane (Lincoln/London: University of Nebraska Press, 1995), p. 22. (The translation is modified.) "Im Geiste ist es anders. Er ist Bewußtsein, frei, darum, daß in ihm Anfang und Ende zusammenfällt."

[14] *Ibid.*, p. 27.

[15] "...die Entwicklung der Geschichte fällt in die Zeit." G. W. F. Hegel, *Die Vernunft in der Geschichte*, Sämtliche Werke, hrsg. v. G. Lasson, Bd. 8 (Leipzig: F. Meiner Verlag, 1923), p. 133.

[16] G. W. F. Hegel, *Lectures on the History of Philosophy*, p. 23, italics mine.

the very beginning, not "made" by the historical *poiesis*), footprints of the "necessity of the matter which proceeds coolly on her way" and to fix what is happening by means of the only fitting word (the proper name of the thing itself — *der Sache selbst*), provided to thought by the ready-to-hand variety of the (German) language. Yet the "late philosophy" sees itself differently...

ὁ ἄναξ οὗ τὸ μαντεῖόν ἐστι τὸ ἐν Δελφοῖς οὔτε λέγει οὔτε κρύπτει ἀλλὰ σημαίνει.
The lord whose oracle is in Delphi does not speak nor conceal, but signifies.[17]

In these words of Heraclitus, Heidegger sees a reference to the nature of philosophical discourse. Even when trying to be as clear and explicit as possible, philosophy does not tell all. It does not hide or conceal, for it is possible to conceal only what is already in one's possession, what has been somehow discovered and is (or can be) in the well-lit circle of the meaning-ful. Unsaid remains what is by no means known, but borders on what is. Unsaid remains what is not questioned, what has not become (and this means in most cases: has not *yet* become) a matter of inquiry. This area of silence and non-manifestation surrounds and delimits philosophical discourse. The phenomenon of opening and disclosing meaning, which takes place in discourse, always remains within a limited, finite horizon, though the boundaries are not perceived from within the scope of already acquired and stored meanings. There is no external prohibition and no pious self-discipline, not even recognition of the inability to cope with what has *already* become an object to work on. The matter is rather the self-consciousness of "late philosophy" and the acceptance of *precisely such* a vision of the truth.

In a way late philosophy has no claims of its own, since it is heir to the tradition, and so accepts to bear the burden of unfulfilled promises. For it, the true is τὸ ἀ-ληθές, i. e., un-concealed, un-forgotten, un-forgettable. Un-concealed does not mean manifest, open, publicly available. The true, as un-concealed, responds to the effort of discourse, to the attempt at telling, but remains unsaid, is retained (ἐπέχει) in expectation, postpones its revelation, limits the discourse, locates the discourse in the sense of Aristotelian physics where "location" or "place" (τόπος) is defined as "the limiting surface of the containing [or surrounding] body."[18] A late philosopher understands that an "epoch" (i.e., a retention of time, a retardation) is always defined not so much by what is expressed (for every new attempt at and every new outline of philosophical thinking appears unlimited to itself) as by what remains unsaid, by what will be articulated later. The logic of such a *logos*, which has renounced claim to the totality of meaning, which picks up and collects (λέγει) what has been given by being itself, while never turning it into a completed system, presupposes *sigetic*[19] thinking.

[17] Heraclitus, fr. 93DK.

[18] *Phys.* IV 4, 212a6f: πέρας τοῦ περιέχοντος σώματος.

[19] The adjective is derived from the Greek σιγάω — "I do not say," "I am silent." Cf. M. Heidegger, *Beiträge zur Philosophie*, GA 65 (Frankfurt a. M.: V. Klostermann, 1994), p. 58: "das *anfängliche* Denken ist in sich *sigetisch*."

"Leaving unsaid, being silent is the 'logic' of philosophy insofar as it asks its main question[20] on another basis," Heidegger says. "On another basis" means on a basis different from the starting point of the Greek metaphysics we know, which defines essence and being-ness as *idea*, *eidos*, an intelligible species exposed to speculation, an "intelligible face" of the existent, as "whatness" (*quidditas*), capable of being thought of and expressed in its proper definition — λόγος οἰκεῖος. Yet all the same the new catches up with (*überholt*) and repeats (*wiederholt*) the old. Only what is unique and happens only once allows a real repetition and restitution.[21] Such a repetition cannot be a simple reproduction, a mechanical replica (since the time has changed and the text's acoustics has changed along with it, so that the same pronunciation produces a different sound). A repetition is always a "deed" of thought, and real deeds cannot be reproduced mechanically.

Repetition as a renewal is not guessing at the universal hidden behind particular historical instances of philosophical thinking. Late philosophy is aware that the historical is not merely a kind of a corrupted, estranged text of thinking, out of which the midday philosophical consciousness reads "authentic" speculative meaning; the historical cannot be removed from the logical as something accidental. The acceptance of the wholeness of truth must lead to a process of learning to live in a multi-dimensional space of meanings that includes the synchronous logical development as well as the diachronic generation of meanings. Philosophical concepts, topics and motives always contain, concealed in them, the traces of their development; they are interrelated not only on the synchronous plane of formal logical operations, but also (though in a non-manifest way) along all the depth of their chronology and genealogy (including linguistic genealogy). One more reason why the historical cannot be transgressed within the logical medium is that the philosophical *logos* cannot be ultimately freed from the element of discourse, out of historically given language, and turned into a "soundless" "intelligible" correlation of pure concepts within some kind of "eidetic" element, which would then be effortlessly "made sounding," i. e., expressed and preserved, by language. This impossibility is one of the major points of disagreement between Heidegger and Husserl, who insisted on the transparent and "unproductive" (i.e., purely reproductive) character of the "layer of expression."[22] Here we come to grips with the willfulness of language, with its primarily creative and poetical dimension, which constitutes an obstacle to an "eidetic" disembodiment of philosophical discourse. This obstacle cannot be surmounted either by "heroism of the spirit" (Hegel) or by "the heroism of reason" (Husserl).

---

[20] This question is: "*wie west das Seyn?*" *Ibid.*, p. 54.

[21] "Nur das einmalige wieder-holbar." *Ibid.*, p. 55.

[22] "Die Schicht des Ausdrucks ist — das macht ihre Eigentümlichkeit aus — abgesehen davon, daß sie allen anderen Intentionalien eben Ausdruck verleiht, nicht produktiv. Oder wenn man will: *Ihre Produktivität, ihre noematische Leistung, erschöpft sich im Ausdrücken ....*" E. Husserl, *Ideen zu einer reinen Phänomenologie und phänomenologischen Philosophie*, 5. Aufl. (Tübingen: Max Niemeyer Verlag, 1993), p. 258. Hereafter cited as *Ideen I*.

When we begin (time begins) to conquer the area of silence, it immediately becomes evident that the philosophical text, which strives to avoid incompleteness and ambiguity, is actually full of unexplained allusions. The attempts to attain absolute clarity and absolute unambiguity always appear incomplete, unsuccessful and hopeless. What was intended as a plastic form of language, perfect, sufficient and satisfying to itself and to the contemplating mind (attracting the gaze of the intellect and consequently compelling the latter), becomes the pretext of numerous and often contradictory interpretations, the object of a greedy understanding which adds to the text and rewrites it time and again and multiplies its copies. The interpretation, trying to repeat the movement of the (classical) text, takes new routes. The road map gets more and more complicated; new roads split on new bifurcations, go farther or get lost in oblivion, so that new commentators, tired of the well-trodden paths, may explore them, asking themselves: "Is there no other way of thinking?"

In *The Sophist* of Plato, the Stranger intends to commit "patricide" (i. e., a rebellion against Parmenides' way of comprehending being) because the "necessity of the matter which proceeds coolly on her way" compels him to do it. But did not Necessity, 'Ανάγκη, herself lead the pupil (κοῦρος) Parmenides, through the gates of day and night, to where the revelation concerning being was given to him?[23] Does the Stranger believe that he knows better what the real necessity is, that 'Ανάγκη looks with more favour on him than on the Father and is more truthful when talking to him? Yet it is not possible to ascribe inconstancy to Necessity herself, who holds being "in the inescapable bondage of a limit" (fr. 8DK, ll. 29f.). Or, perhaps, the god, who is never mentioned in the text because he is associated with the forbidden, invisible and silent path (fr. 2DK, ll. 6–8), the god Chronos, proves to be mightier and more cunning than powerful (κρατερή) Necessity? It might happen that Plato's Stranger hides behind "logical necessity" another motive to commit patricide.[24]

Yet loyalty to ancestors and the "goodwill" of the interpreter, his desire to accept the authority of the text — these, too, make him ask the question: "Is it not possible to think differently?" The philosophical text (especially in the form of a "philosophical system") itself lays claim to an absolute and universal validity and *totality*. If this claim is to be taken seriously, it is in itself a good reason to ask time and again: "Is it not possible to think differently?"

---

[23] Parmenides does not call his divine preceptress by name. Some ancient commentators believed it was 'Ανάγκη, others thought that Parmenides received the ontological revelation from Justice (Δίκη). It is Justice who opposes Chronos, because "she did not loosen the being in her fetters and move it either to come to be or to be perishing but holds it fast" (fr. 8, ll. 14f.). Heidegger in his lectures on Parmenides (1942–1943) says that his goddess is Truth herself, 'Αλήθεια. Cf. M. Heidegger, *Parmenides*, GA 54 (Frankfurt a. M.: V. Klostermann, 1982), pp. 1–9.

[24] Nietzsche says in a fragment entitled "Homer's contest" (1872): "The greater and more sublime a Greek is, the brighter the flame of ambition that glares out of him, consuming everybody who runs on the same course... every great Hellene hands on the torch of contest; every great virtue kindles a new greatness..." And in another fragment "We, the philologists" (1875) Nietzsche continues: "The greatest fact remains always the precociously panhellenic Homer. All good things derive from him; yet at the same time he remained the mightiest obstacle of all. He made everyone else superficial, and this is why the really serious spirits struggled against him. But to no avail. Homer always won. The destructive element in great spiritual forces is also visible here..."

The notion of "differently" is naturally defined in relation to what is already present and understood in a certain way. "Different" does not mean "contrary." A difference is not necessarily a negation (or negation of negation). That is why the development of thought is much more chaotic, complicated and arbitrary than Hegel liked to believe.

It is all the more difficult to make the classical text sound in the new acoustics created by time. Yet without a correlation and an interaction with what is old nothing new can begin. That is why late philosophy has to try to reach the impossible goal attempting to repeat what is unique, in the hope that the impossible is, paradoxically, inevitable: for 'Ἀνάγκη is the boundary of the "well-rounded sphere" of being, and consequently the necessity of an eternal return. "It is indifferent to me whence I begin," Parmenides says, "for to that place I shall come back again" (fr. 5DK).

2. ONTOLOGY AND CHRONOLOGY

Πῶς οὖν ἂν εἴη τὶ ἐκεῖνο ὃ μηδέποτε ὡσαύτως ἔχει;

Plato, *Cratylus* 439e1f.

Being and time, time and being... These words are more than just titles of Heidegger's most important works. A major part of philosophical effort in the 20th century has been spent on topics related to the conjunction "and" (the conjunction *par excellence*), in attempts to understand the meaning of the relation it establishes. The results are striking. Chronos, the primary metaphysical antipode of "what truly is" (τὸ ὄντως ὄν), the antipode whose very being is dubious, who contaminates with ambivalence even *being* as such, who turns it into "becoming" and "ceasing to be," who transforms beyond recognition the intelligible image of beings and who (unless the mind resists stubbornly) makes thinking itself lapse into the realm of *doxa*,— it is this antipode who becomes the origin of thinking about *what-is* and the meaning of the being of beings. Pure consciousness is "time" (the absolute temporal flux). Such is the ultimate conclusion of Husserl's phenomenology. The being of beings as distinct from beings (i.e., in its enigmatic difference from beings) temporalizes itself as the original temporality. Such is the ultimate conclusion of Heidegger's "fundamental ontology." This strange metamorphosis becomes the emblem of "late philosophy." *Late onto-logy is chrono-logy.*

The Greek origin of philosophy (the "first origin" as Heidegger terms it in order to distinguish it from the new origin, coming to maturity, temporizing itself in the form of eschatology of thinking) is, essentially, a warning against Chronos. Long before the corpus of Aristotle's *Metaphysics* was compiled, and of course long before the term *ontology* made its appearance on the philosophical stage, philosophy received a metaphysical or ontological revelation — the Poem by Parmenides, to which the Greek doxographers later gave the traditional title of Περὶ φύσεως —

On nature.[25] It follows from the speech of the nameless goddess who instructs Parmenides, that time is the master of the way (or of the way-wardness) taken by the meaninglessly wandering "double-headed mortals" (fr. 6DK), for whom "to be" and "not to be" is and is not the same. And time is nothing but the conjunction "and" in this "to be and not to be," "is and is not." Time is inseparably connected with change (movement). In Aristotle's theory of time, as we shall see later, this connection forms the basis of all reasoning. And movement, as Aristotle testifies, has been at all times an ontologically dubious subject: "some people consider movement to be otherness, inequality and non-being."[26] Moira, Parmenides says, forces being to be "entire and changeless" (fr. 8DK, ll. 37f.), she "has bound fast" (ἐπέδησεν) all beings, and strong Necessity holds it "in the bondage of a limit," so that it remains equal to itself, remains the same, rests (literally, "is lying") by itself (καθ' ἑαυτό) and in this way perpetually dwells "in this very same place" (αὖθι μένει).[27] Yet the mortals live among words devoid of real meaning — "to be born," "to perish," "to be and not to be," "to change place," "to change the appearance" (8, 38–41). And most devoid of meaning is the word "time." Time contaminates language, poisoning it with words devoid of meaning; time contaminates grammar, making us say "it was," "it will be..." It is in a different way that being should be spoken of (8, 5f.):

οὐδέ ποτ' ἦν οὐδ' ἔσται, ἐπεὶ νῦν ἐστιν ὁμοῦ πᾶν, ἕν, συνεχές...

It never was nor will be, since it is now, all together, one, indivisible...

Yet, just as "being" and "to be" can be rescued from meaninglessness and from the thoughtlessness of the "double-headed mortals" by deleting "non-being" and "not to be" from the language of thought, so time, too, can in a certain way be purified and justified, if it is turned into an immutable, persistent "now," "all at once," if it assumes the function of *tempus praesens*, the tense of the verb "is." If time is to be allowed at all, then as the *prae-sens* only, by turning time into eternity, into an ever-present "now." Being remains steadfastly here and is always in the same place (αὖθι μένει), it is pres-ent in the present, and what is abs-ent (ἀπεόντα) is to be steadily viewed by the mind as pres-ent (παρεόντα).[28] "*Is*" — *that* is the right word and the major sign to be discovered by those who follow the right way of inquiry. Since being is perfect in its immutability and persistence, since it abides in its totality without interruption, there is no room for another time alongside (πάρεξ) of being. Παρουσία, the presence in the present, the present as the presence — this is the

[25] I have used an excellent critical edition: "The Fragments of Parmenides. A critical text with introduction, translation, the ancient testimonia and commentary by A. H. Coxon" in *Phronesis, Supplementary Volumes*, vol. 3 (Assen, etc.: Van Gorcum, 1986). Still in the references I preserve the *numbering* of the fragments according to Diels–Kranz (adding the indication "DK" to the number). Coxon suggests another order of the fragments but the first one has been for a long time generally accepted and, I believe, is immediately recognizable.

[26] *Phys.* III 2, 201b20.

[27] Fr. 8, 29f.

[28] Fr. 4.

only meaning of time, all the rest has no sense and is a mere invention of the mortals, whose minds are full of disorder and perplexity. Then, Parmenides says, "time is not nor will be another thing alongside being" (8, 36f.).[29] *Chronos* is a redundant, misplaced, void and dangerous name.

We are quick to pronounce the word "being"; we have become used to the fact that it is the key word of Western metaphysics. Tradition has taught us to bring to light, and to immobilize (to posit) being as the main (essentially the only) object of thought. Yet we must not forget that this name itself, τὸ ἐόν, was "introduced" by Parmenides in his logical prophecy, was discovered by him, was encountered by him on his philosophical journey beyond day and night.

Parmenides speaks about the way of searching or questioning. The movement along this way is referred to by a special verbal derivative δίζησις. The verb δίζημαι itself can be found in Heraclitus (e. g., fr. 101DK), who tells us that he carefully studied *himself*, and searched for the truth that he had by or in *himself*. "I questioned (studied, searched) myself and learned everything from myself." Parmenides wants to oppose his way of searching to the procedure employed by the Ionians, which was designated by the word ἱστορίη — *gathering information*; Heraclitus himself says that knowing many things (ἱστορίη) does not make one intelligent (fr. 40). Logical revelation responds to the ultimate effort of asking. In Parmenides' revelation there must be finally no room for unexpressed latent elements: a logical prophet's piety lies not in abstaining from questioning, but in questioning perseveringly.

What does Parmenides ask about? How does he go about asking it? What does the nameless goddess, Parmenides' preceptress, allow and order to be asked about, and how? Here is the proper question:

ὅπως ἐστίν τε καὶ ὡς οὐκ ἔστι μὴ εἶναι (fr. 2DK, l. 3)[30]

An undefined nameless object (if an object at all), that which is discovered on the way of searching, is introduced at first with a special figure of *aposiopesis*; the sentence is abruptly broken off, as it were, in a silent pause and then continued:

Come now, I will tell you — and do you preserve my story (μῦθος), when you have heard it — about those ways of inquiry which are alone conceivable. The one: that... [an area of silence, as yet nameless "subject" of the sentence, defined by its namelessness — outlining the space for the name yet to come]... *is* and [for it, that is for the object occupying the area of name-giving silence] there is no non-being. (fr. 2DK, ll. 1-3)

Yet, as it turns out later, there is only one word which can force the *topos* of silence to speak, there is only one "logic" which can replace the "sigetic": the predicate "is" can have only one unique subject. This is the precept received by Parmenides from his guide. To understand this is to understand the meaning of the verb "to be," or to be more precise, of its personal form "is," which is, essentially, the only meaningful form, unlike the others, invented by sightless mortals. This requires the ulti-

---

[29] I accept Coxon's reading: οὐδὲ χρόνος ἐστὶν ἢ ἔσται ἄλλο πάρεξ τοῦ ἐόντος. A cogent argument in favour of this reading can be found on pp. 210–211 of Coxon's book.

[30] Fr. 3 according to Coxon's numbering. Coxon translates: "that a thing is, and that it is not for not being."

mate effort of intelligent contemplation, the perseverance implied by the verb δίζημαι, the resoluteness of one who has *taken on his shoulders the yoke of necessity*, as Aeschylus said speaking about Agamemnon (*Agamemnon* 1071).

The goddess says:

I will tell you, and you, on hearing my words, preserve, keep (κόμισαι) them, make them grow, resisting vagueness, oblivion, fading away, and carry them with you.

There is an unusual intensity in the way this is said and heard: an essential meaning is here laid bare and brought to light in words.[31] The name is pronounced later on (6, 1):

It is necessary to assert and conceive that this [i.e., that for which the area of silence has been reserved] is *being* (τὸ ἐόν).[32]

As we have seen already, much will be said about being (τὸ ἐόν) further on (fr. 8): it is ungenerated and imperishable, entire, unique, unmoved and perfect... Yet despite all its importance, this speech (which seems to attribute something to being), is only a collection of metaphors. Just as Parmenides allows only one subject to accompany the predicate "is" (if its true meaning is to be understood), so only one predicate can be connected with the subject "being" (τὸ ἐόν): it *is*. The perseverance of intelligent contemplation (νοεῖν) must not allow (so the goddess says) "into the contemplation," and must remove and put off,[33] anything related to difference, plurality, change, coming to be and ceasing to be — anything, in a word, which belongs to Chronos' realm,— as the mortals believe. Everything removed in contemplation, everything the mind keeps at a distance, everything negated in the goddess's speech with α *privativum* and οὐδέ "is incommensurable with being, its measure is different."[34]

On the other hand the perseverance of intelligent contemplation of what exists must see with the mind's eye that which is not present, without letting it out of the sphere of presence, out of the boundaries and limits of what-is.

Gaze on even absent things with your mind as present and do so steadily. For it will not severe being from cleaving to being... (fr. 4DK)

In this case the mind renounces the habit to follow "the sightless eye, the noisy ear and tongue" (7, 4ff.), bases its judgment on the *logos* (7, 5) and becomes an ally of the mighty 'Ανάγκη, who holds it *in the bondage of a limit*, of the de-limitation,

[31] A close connection between Parmenides, the poet, and the Panhellenic poet Homer is well established and amply discussed. Homer says somewhere, using the same verb κομίζω: ἔγχος ἐν χροῖ κομίσασθαι — "carry away a spear in one's body," i. e., "to be wounded by a spear." This "intertextual consonance" is only one of many non-accidental occurrences within the poetical space: what has been said by the goddess is a stab of meaning, what has been heard by Parmenides is a scar of meaning.

[32] I understand ἐόν in the first line predicatively (cf. Coxon *The Fragments...*, pp. 54, 181f.) and τό, which precedes λέγειν, as a demonstrative pronoun pointing to that, which was discussed earlier and for which a silent place had been reserved.

[33] "Weg-sehen, fort-sehen" — says Heidegger commenting upon fr. 8 in the lectues *Einführung in die Metaphysik*, GA 40 (Frankfurt a. M.: V. Klostermann, 1983), p. 104.

[34] *Ibid.*

de-finiteness of what-is as such. To gaze even on what is absent as being present, means to grasp what-is in its being as something uninterrupted and indivisible (συνεχές). A division into what is present and what is absent cannot, according to the goddess's precepts, belong to the being-ness[35] of what Parmenides calls τὸ ἐόν. Any "this and that" or "was and will be" statement, any difference, any multiplicity implies a boundary and a division — a split, a fissure in the uninterrupted compact solidity of Parmenides' sphere. Yet what-is cannot be split, divided or taken apart in its being (8, 22). The mind sees the being of beings or entity in its being, and then it becomes clear that "being is incapable of being more than being in one regard and less in another... for it is equal with itself from every view and encounters determination all alike" (8, 47–49), and "everything is full of being" (8, 24). Being is uninterrupted and continuous (ξυνεχὲς πᾶν), for what-is leans on what-is (8, 25). For what could mark a boundary or an interruption in being, what could separate what-is (as such) from what-is (as such)? Something entirely different from what-is. But what differs from what-is is not: "neither has non-being any being which could halt the coming together of being" (8, 46f.).

What-is-not, non-being, is under a prohibition, the goal and meaning of which are not quite clear. The second way of questioning is mentioned by the goddess only to be immediately prohibited as being beyond the scope of thought and speech:

> The other [path]: that [an area of silence again, which, as we shall see, cannot be made to sound and, essentially, cannot name anything] is not, and it must needs not be, this I tell you is a path wholly without report, for you can neither know what it is (for it is impossible), nor tell of it... (2, 5–8)

This is a wild, sightless, incomprehensible (ἀνόητον), nameless (ἀνώνυμον), false (οὐκ ἀληθής) path (8, 16ff.). Going along it, we find no true name and no sign-posts. The "name-giving area" in fr. 8, l. 5 remains silent and void. What is not cannot become a subject of thinking or a subject of speech. This means that what *has been said* here about a spacing, a gap, a split in being, cancels itself, goes beyond itself as a non-thought, as unthinkable: "neither has non-being any being which could halt the coming together of being." Strictly speaking, *non-being*, τὸ μὴ ἐόν, is not a name at all. This is a sound marking an impossibility to name, a *designation* of the area of silence, which should contain the name of the impossible subject of the sham predicate *"is not."* But nothing is sub-ject to such a predicate, nothing underlies the "*is not*"; it is impossible here to go down to a firm basis and to lay or posit (*sub-jicere*) a "positive" foundation. An attempt to grasp the subject gets lost in a void. Much later the tradition will dare to pronounce the forbidden name: the infinity of this void, this is *time* (time as a negation, the *nihil originarium*,[36] as an absolute temporal flow lacking any "positivity").

---

[35] Heidegger uses the verb "wesen" in order to coin a term which would allow us to express the invariant core of the pres-ent and the abs-ent, of *das An-wesende* and *das Ab-wesende*. "Das Seiende ist, das Sein aber *west*."

[36] Cf. M. Heidegger, *Metaphysische Anfangsgründe der Logik im Ausgang von Leibniz*, GA 26 (Frankfurt a. M.: V. Klostermann, 1978), p. 252.

Aristotle remarks in his *Metaphysics* (XIV 2, 1088b35–1089a6) that if one accepts, following Parmenides, the thesis asserting homogeneity and uninterruptedness of being, it becomes impossible to explain the plurality of *entities*, unless non-being is somehow allowed. The Platonists, he says, "thought that all things that are (τὰ ὄντα) would be one, (to wit, *being* itself) if one did not come to issue with and refute the saying of Parmenides: 'for this principle will never be proved, so as to allow things to be that are not.'[37] Thus it seemed to be necessary to demonstrate that non-being somehow is; so that the things that are, in order to be multiple, would consist of being and something other [in relation to being]." And indeed the patricide committed by the Stranger in *The Sophist* consists in proving that "non-being is in some way, and, vice versa, being in some way is not" (241d). Yet in order to make such a proof possible, a latent re-interpretation (latent since the Platonists understood the related problems in the old way – ἀρχαϊκῶς, and consequently were not aware of the substitution) of the *meaning of being* was necessary. The crucial transformation consists in reinterpreting *being* as one genus among the other "most universal genera" (μέγιστα γένη) – "movement," "rest," "the same" and "the other." The multiplicity is posited in *The Sophist* from the very beginning, since in the Stranger's argument "being," "movement" and "rest" are enumerated in a sequence one after another. Only one more step ("each [one of these three genera] is something other than the [remaining] two, but the same as itself") is enough to bring out the genera of "the other" and "the same." For Plato the assertion "X *is*" means "X *participates in (or partakes of) the genus 'being'*." Of course, *being is*, for the genus of "being" participates in itself, like any genus (that is why in *The Sophist* movement moves, and rest rests). But besides that, rest *is*, like movement, and this means that both partake of *being*. The *"is"* mediates and articulates the mutual participation of ideas, the *"is not"* mediates and articulates their *difference*: "X is not Y."

Yet for Parmenides himself this way of thinking is impossible: there is no gap in being, it is not possible to escape from the uninterrupted compactness of being. Time, which is the primary difference, the difference *par excellence*, is declared a void name: there is no room for time alongside of *what-is*, there is no room for anything alongside of being; alongside or "outside of" (πάρεξ) *being* there is no room at all – being has no *alibi*.

What is not cannot be a subject of thinking, since

τὸ γὰρ αὐτὸ νοεῖν ἐστίν τε καὶ εἶναι. (fr. 3DK)

Clement of Alexandria, Plotinus and Proclus, the authors who made known to us this famous fragment, were unanimous in interpreting it as an expression of the notorious "identity of being and thinking" and, to a certain extent, ascribed to this thesis a soteriological significance. In other words, they believed that this identity bears on what is *expressed* by the infinitives νοεῖν (*to see in the mind, to grasp mentally, to conceive*) and εἶναι (*to be*). It immediately follows from such an interpreta-

[37] Fr. 7, 1.

tion that Parmenides' *what-is* is the same as *what-thinks*, i.e., the mind or intellect (the neo-Platonists, of course, meant *Nous*).

I believe, all the same, like Zeller and Coxon, that this fragment is to be read as follows: "the same thing is for conceiving as is for being," i.e., one can contemplate or grasp mentally exactly that which is, and only that which can be seen in the mind, is.[38] The meaning of the verbs νοεῖν and εἶναι must be investigated simultaneously. And since, strictly speaking, there is only one thing-which-*is* — unwavering, never born, never perishing, immobile, compact, uninterrupted, similar to a solid mass of a well-rounded globe, there is only one thing to be thought: *being-is-and-is-not-for-not-being*. Thinking always takes place "for the sake of" (8, 34) this unique thing to be conceived.

The same thing is for conceiving as is that for the sake of which the thought takes place; for not without being, when something has been said about it, will you find thought (conceiving). (8, 34ff.)

When the question is raised whether Parmenides, while introducing the word τὸ ἐόν to designate the only one thing which is, understood this "is" as an "essence" or as an "existence,"[39] the fact is disregarded that the *distinctio essentiae et existentiae* was invented at a much later time in a quite different context of thought. For Parmenides' "being" this distinction has no sense. As we have seen, *what-is* has no other determinations than that of being, its de-finition lies in the "bondage of limit" in which Necessity holds it. To this inescapable bondage there corresponds the perseverance of contemplation, which, dismissing the multiplicity of its objects — ever and ever becoming different from what they are, appearing and perishing — sees only their "is." Of course, the "is," i. e., the presence, παρουσία, of the varied (and this means for Parmenides — inauthentic) objects of thought, is impossible without a certain *quidditas*, a delimitation of meaning. Yet the presence as such neither depends on a given "what," nor coincides with the *existence* (in the sense of medieval metaphysics) of a particular "what" (a particular *essence*). What is important here is not the semantic appearance of this and that (any) thing, nor the names of the manifold beings, which mortals, "confident that they are real" (8, 39), have established and preserved in their language, but the manifest character, the manifestation of the meaning itself, the fact that it is intended to be thought, and this means — to be *tout court*. The temporality (so far as we can mention Chronos' name at all as being inside Parmenides' globe) of such a presentation of meaning is the present, an ever-present "now." "Was," "will be" are words invented by mortals, and they mean something. Yet precisely because they mean something, and in every one of them

[38] This interpretation was first proposed by Zeller in his *Philosophie der Griechen*. The phrase recurs in frr. 3DK; 5, l. 1; 8, l. 34 with the same meaning in all places. Coxon remarks that "the use of tenses of εἶναι followed by a transitive infinitive, the object of which is understood from the subject of the finite verb, is idiomatic in the fifth century and later, e.g., Aeschylus, *Pers.* 419, θάλασσα δ' οὐκέτ' ἦν ἰδεῖν; Eupolis, fr. 139, 2K, ὁ δὲ Γνήσιππός ἐστιν ἀκούειν" (*The Fragments...*, p. 174).

[39] Cf. Coxon, p. 175: "Parmenides' use of the phrase shows that it is mistaken to understand his concept of being as 'existential.' His introduction of 'is' and 'is not' in isolation from either subject or further predicate does not mark an 'absolute' use of the verb..."

the always identical phenomenon of the revelation of meaning is reproduced, there hides behind "was" and "will be" — for the persevering mind — the "is." It was said about being:

It never was nor will be, since it is now all together, one, indivisible. (8, 5-6)

I would like to conclude this essay on the pre-history of onto-chrono-logy with a long quotation from Heidegger, which could form a bridge to the following investigation.

What is the significance of the fact that ancient metaphysics determines the ὄντως ὄν — the being which is in the highest degree[40] — as ἀεὶ ὄν? The being of beings is understood here as permanence and persistence. What project lies at the basis of this comprehension of being? The project relative to time (*der Entwurf auf die Zeit*), for even eternity, taken as the *nunc stans*, for example, is thoroughly conceivable as "now" and "persistent" only on the basis of time.

What is the significance of the fact that a being in the proper sense of the term *(das eigentlich Seiend) is understood as* οὐσία, παρουσία, i.e., basically as "presence," an immediate and always present possession (*gegenwärtiger Besitz*), a "having" (*Habe*)?

This project reveals that *being* (*Sein*) is synonymous with *permanence in presence*.

Are not temporal determinations accumulated in this way, i.e., in the spontaneous comprehension of being? Is not the immediate comprehension of being developed entirely from a primordial, but self-evident projection of being on (*auf*) time? [...]

Where lies the ground for such a spontaneous comprehension of being out of time? [...]

The essence of time as it was fixed — and, as it turned out, decisively — for the subsequent history of metaphysics by Aristotle does not provide an answer to this question. On the contrary, it would be easy to show that it is precisely Aristotle's conception of time that is inspired by a comprehension of being which — without being aware of its action (*Tun*) — interprets being as permanent and present, and consequently determines the "being" of time from the point of view of the "now," i. e., from that character of time which is always and permanently present in it, and thus properly *is*, in the ancient sense of the term. (KPM 233f.)

[40] "Das Seiende, das so seiend ist, wie Seiendes nur seiend sein kann."

CHAPTER TWO

TIME AS NUMBER AND CALCULATING SOUL

(*Aristotle's theory of time. The prototype of the ontological difference*)

> Vielleicht, daß man einige Jahrhunderte später urteilen wird, daß alles deutsche Philosophieren darin seine eigentliche Würde habe, ein schrittweise Wiedergewinnen des antiken Bodens zu sein, und daß jeder Anspruch auf „Originalität" kleinlich und lächerlich klinge im Verhältnis zu jenem höheren Anspruche der Deutschen, das Band, das zerrissen schien, neu gebunden zu haben, das Band mit den Griechen, dem bisher höchst gearteten Typus „Mensch."
>
> Nietzsche, *Der Wille zur Macht*, Aph. 419.

1. ENERGEIA AND ITS INTERNAL FORM

1.1. *The Definition of Movement*

Aristotle says:[2] "For time is just this — number of movement in respect of 'before' and 'after'."[3]

Time is inseparably connected with movement; it is, according to Aristotle, κινήσεώς τι. Time is a definite formal moment of movement. In its turn movement is the main topic of Aristotle's physics. We have seen in the previous chapter that according to Parmenides "change," as well as "time," is a redundant, misplaced, and

[1] These words preface Heidegger's lectures on *Metaph.* Θ 1–3: *Von Wesen und Wirklichkeit der Kraft* (1931), published as vol. 35 of *Gesamtausgabe* (2. Aufl. Frankfurt a. M.: V. Klostermann, 1990).

[2] As a general rule, citing Aristotle in English, I make use of the translations included in the two volumes of *The Complete Works of Aristotle. The Revised Oxford Translation*, ed. J. Barnes (Princeton NJ: Princeton Univ. Press, 1985). *Physics* is translated by R. P. Hardie and R. K. Gaye, *Metaphysics* and *Nicomachean Ethics* – by W. D. Ross, *On the Soul* – by J. A. Smith.

[3] τοῦτο γάρ ἐστιν ὁ χρόνος, ἀριθμὸς κινήσεως κατὰ τὸ πρότερον καὶ ὕστερον. (*Phys.* IV 11, 219b1f.) We shall see later on that the expression refers to a definite relation – the "before-and-afterness," so to speak. Everywhere in my book I render Aristotle's κίνησις as 'movement' (Hardie and Gaye: "motion," Ross: "movement"); κίνησις is a very broad concept in Aristotle, which includes not only locomotion, but alteration and increase/decrease as well.

ontologically void name: there is no room for time and change (movement) alongside (πάρεξ) of being. The *Physics* is actually nothing but an *ontological justification* of movement.

Aristotle gives a definition of movement, and is the first to do so.[4] The two main versions of this definition are:

Phys. III 1, 201a10f.· ἡ τοῦ δυνάμει ὄντος ἐντελέχεια, ᾗ τοιοῦτον, κίνησίς ἐστιν...
The *entelecheia* [5] of what is potentially, as such, is movement...

Metaph. XI 9, 1065b16: τὴν τοῦ δυνάμει ᾗ τοιοῦτόν ἐστιν ἐνέργειαν λέγω κίνησιν
I call the *energeia* of [a being] in potentiality insofar as it is in potentiality, movement.[6]

This definition allows us to include motion within being, contrary to Parmenides' prohibition. Indeed, because potentiality is a genuine mode of being, as we shall see in section 3, change and motion are rightfully beings.

Aristotle's definition, paradoxical at first sight (indeed, its structural skeleton is: *the actuality of potentiality qua potentiality*), has been a constant object of attention for commentators since antiquity.[7] Without discussing it in detail, I shall nonetheless indicate my understanding of it.

The definition speaks of the actuality of a *definite* possibility as, precisely, possibility. The *entelecheia* (or *energeia*) meant here carries out to perfection, makes actual, and constitutes a *definite* possibility *qua* this *definite* possibility, e.g., this piece of copper as a *possible* statue *and not* as a possible shield. In movement a certain tendency becomes manifest, and this presence of the meaning of a definite possibility must be connected to an *eidos* or form different from the form of copper *qua* copper as well as from the form of the finished statue. This quasi-form (I shall later explain why I choose such a cautious term), which forms the possible as possible, which al-

---

4 According to a testimony of Simplicius (397, 15) nobody before Aristotle attempted to give a definition of movement.

5 In what follows I leave this most important term of Aristotle's metaphysics without translation. The generally adopted rendering "actuality" does not allow it to be distinguished from ἐνέργεια. Of course, the two terms can be used as complete synonyms, but nevertheless the subtle differences in meaning are always retained. ἐντελέχεια signifies, according to the internal structure of the term (ἐν-τελής + ἔχειν), "being-at-the-goal," "completeness" or "perfection," and has nothing to do with *action* and *activity*. The common Latin translation of the term is "perfectio." In the *Revised Oxford Translation* the English equivalent for ἐντελέχεια is "fulfilment." However the same word is generally used as the translation of Husserl's *terminus technicus Erfüllung* (see chap. 5, sect. 2). The closed interrelation between ἐντελέχεια (ἐνέργεια) and *Erfüllung* will be discussed further on in connection with my "phenomenological interpretation" of Aristotle and "Aristotelian interpretation" of Husserl (cf. chap. 5; 7).

6 Ross: "I call the actuality of the potential as such, movement." I either leave Aristotle's term ἐνέργεια untranslated or render it as "actuality."

7 On various ways to interpret the Aristotelian definition of movement in the most important historical commentaries cf. F. Brentano, *Von der mannigfachen Bedeutung des Seienden nach Aristoteles* (Freiburg im Breisgau: Herder'sche Verlagshandlung, 1862), Kap. IV, § 2. See also another excellent work taking into account the most important contemporary studies: R. Brague, *Aristote et la question du monde. Essais sur le contexte cosmologique et anthropologique de l'ontologie* (Paris: PUF, 1988), ch. IX, § 58.

lows the possible to have the meaning (λόγος) of the possible, *is* movement;[8] this quasi-*eidos* allows the statue-of-Hermes-in-possibility to be *present* during casting as distinguished from the shield-in-possibility.

Thus movement constitutes the possible (the potential) as a *definite* possibility (potentiality). Yet how must we understand the way of being of this definiteness? Aristotle says: as *energeia* or *entelecheia.* And if we want to avoid a vicious circle in the definition, *energeia* cannot be interpreted here as the *energeia of movement* (ἐνέργεια κατὰ κίνησιν). The clue to solving the problem is given by the fact that the noun δύναμις (*possibility, force, ability, power*), as well as the noun ἐνέργεια, are used equivocally.

The following passage from *Metaph.* IX 6 is extremely important,[9] because here we find a fundamental distinction between the two main meanings of ἐνέργεια:

> Not all [entities] are said to be actual (ἐνεργείᾳ) in the same way (univocally), they are called so only by analogy: as A is in B or relates to B, so also C is in D or relates to D. In one sense actuality (ἐνέργεια) relates to potentiality (δύναμις) as movement to the ability [to move], in another sense as substance to some sort of matter. (1048b6–9)

Thus the equivocality specified in this passage is not accidental homonymy: although, strictly speaking, there is no general term in Aristotle's language designating the two meanings of *energeia* just mentioned, because there is no common genus in relation to which they could be considered as species, these meanings are connected by the unity of analogy. Aristotle says (cf. *Metaph.* IX 6, 1048a36ff.) that the meaning of δύναμις and ἐνέργεια, like the meaning of τὸ ὄν or τὸ ἕν, is "one" only in the sense of analogy, κατ' ἀναλογίαν, and we must be content not to demand a definition (given by *genus* and *differentia*) but to grasp the analogy and see the nature of the underlying anonymous universal unity (which is not and cannot be in this case a common genus) by studying the instances of it. In one sense *energeia* is activity, actuality of change, and in relation to such *energeia* δύναμις is ability as the source or "principle" of change.[10] In the second sense *energeia* is fulfilment, perfection of *eidos*, of the form of the thing embodied in matter;[11] and in relation to such *energeia* δύναμις is matter as something indeterminate (and therefore not present),

---

[8] Cf. Brentano, *op. cit.*, p. 58: "Die κίνησις ist die Aktualität des Potenziellen als solchen, [...] d.h. sie ist die Aktualität (ἐνέργεια), die ein in Möglichkeit Seiendes (τοῦ δυνάμει ὄντος) zu dem macht, was es ist (ᾗ τοιοῦτόν ἐστιν), nämlich zu diesem in Möglichkeit Seienden, oder mit anderen Worten, die ein Mögliches als Mögliches (ein im Zustande der Möglichkeit Befindliches als in diesem Zustande befindlich) konstituiert oder formiert."

[9] In what follows we *retell*, not actually *translate* it.

[10] It is defined as "the principle of change or transition, which lies in something other or in the thing itself insofar as it can be considered something other" (1046a11), – for example, when a physician treats himself for an illness, being at the same time his own patient.

[11] *Metaph.* 1048a30-33: "Actuality means the presence of a thing (ἔστι δὴ ἐνέργεια τὸ ὑπάρχειν τὸ πρᾶγμα) not in a way which we express by 'potentially,' we say that potentially, for instance, a statue of Hermes is in the block of wood and the half-line is in the whole, because it might be separated out, and even a man who is not contemplating [truth] we call a man of science if he is capable of contemplating." (Ross' translation is slightly modified.) ὑπάρχειν means among other things "to be already there," "to be available." Heidegger's "vorhanden-sein" is a very fitting translation of this Greek verb.

subject to determination, capable of embodying the form and so allowing it to be present within the *compositum*. Only what has been formed can be present, and that is why *energeia* in the second sense is almost synonymous with *eidos* and *presence* as such. *Energeia* means that the thing is present or available: τὸ ὑπάρχειν τὸ πρᾶγμα (1048a30f.). And it is only in this sense that *energeia* becomes synonymous with *entelecheia*, for the completeness or fulfilment of the form is the *goal* of all coming into being, and *entelecheia* signifies "being-at-the-goal." Matter is the indefinite possibility of attaining the goal, that is to say, to be determined by the form and to embody it completely.

1.2. *The Concept of Energeia*

After distinguishing the two meanings of *energeia*, we can speak of *energeia in contradistinction* to movement. The way Aristotle posits and articulates this distinction deserves most serious attention. I am here referring to the famous fragment of *Metaph.* IX 6, 1048b18–35.[12] The text is very much corrupted, and the author's manner extremely laconic and hasty;[13] that is why the efforts of the editors to understand the passage in question and to render it understandable resulted in a multitude of variants. In what follows I cite the passage in my own translation (the comments in brackets inserted in the text seem to be inevitable, for they make the meaning I intend to convey manifest). I accept the reading suggested by R. Brague[14] and, in general features, follow his interpretation of the fragment. Ross' translation makes the text mute in some nuances,[15] which are of crucial importance for the further development of my interpretation of Aristotle's concept of *energeia.*

Since of the actions (πράξεις) which have a limit none is an end in itself, but they all belong among the means aiming at an end [other than themselves],— e. g., losing weight [by means of a prescribed treatment] is, indeed, of this sort in relation to the aim of weight to be lost,— and since the things involved in the process[16] of making somebody lose weight are in movement in this way that the result itself for the sake of which the movement takes place is not yet readily present, this is not an action proper or at least not a perfect one: for it is not an end in itself. *But in a [proper or perfect] action the end must already be present, and [still the action must remain] an action.*[17]

[12] The history of transmission of this fragment is outlined in R. Brague, *Aristote et la question du monde*, pp. 454–456. The author calls it "un aérolithe aristotélicien."

[13] W. Jaeger in his edition of the *Metaphysics* (10th impression, Oxford 1989, p.184) remarks in connection with this fragment: "oratio est admodum dura et obscura."

[14] R. Brague, *op. cit.*, pp. 456–461.

[15] A cogent argument against Ross' reading can be found in Brague, *ibid.*

[16] Ross considers αὐτά in 1048b20 as referring to "the parts of the body themselves when one is reducing their bulk" (cf. *Aristotle's Metaphysics. A Revised Text with Introduction and Commentary by W.D. Ross.* Oxford: Clarendon Press, 1988, vol. II, p. 253), while Brague believes that here the "means aiming at and end" (τὰ περὶ τὸ τέλος in line 19) are meant. Since this matter is of no importance for my further analysis I choose a "neutral" way to interpret αὐτά as "the things involved in the process."

[17] In ll. 1048b22f. I accept the correction proposed by R. Brague, who reads: ἀλλ' ἐκείνῃ (ἐκείνη codd.) ἐνυπάρχει τὸ τέλος καὶ ἡ πρᾶξις. Here ἐκείνῃ refers to πρᾶξις τελεία. Thus, the sentence says that the following two features belong to the perfect action simultaneously (are at once inherent in it):

Thus, somebody is seeing and at the same time has already seen, is understanding and has understood, and also is grasping mentally and has already grasped. But it is not true that at the same time somebody is learning and has learnt, or is being cured and has been cured. At the same time somebody is living well and, indeed, has already achieved a good life, is experiencing happiness and is already happy. If not, the activity would have had sometime to cease, just as the process of losing weight ceases [when the goal is achieved]; yet this is not what takes place but, on the contrary, somebody is both living and already alive. Of these activities, then, we must call the one set *movements*, and the other — *energeiai*. For every movement is incomplete: losing weight, learning, walking, building; these are movements, and incomplete. For it is not true that at the same time somebody is walking and has already walked or is building and has already built, or is coming to be and has come to be — in a word, it is not true that something is being moved and has already completed its movement: these are distinct [states of affairs]. But simultaneously somebody has seen and is seeing the same thing,[18] and is grasping mentally and has already grasped. The latter sort [of actions], then, I call *energeia*, and the one mentioned earlier, movement.

From this difficult text we can gather at least the following: Aristotle divides activities or actions in the broadest sense into perfect, completed (πρᾶξις τελεία) and imperfect, incomplete, intermediate (ἀτελής).

The incomplete or imperfect actions must be called movements, and the completed ones, *energeiai*.

The latter are characterized by the end being inherent in them. In this case it is impossible to distinguish the action from that for the sake of which the action is per-

---

a certain state of affairs which has come to be as the result of the action and its end (this meaning is expressed by Greek perfect tense) and, on the other hand, the continuing activity (expressed by the present tense). Brague's translation of the passage 1048b18–23 goes as follows: "Puisque, parmi les actions, aucune de celles dont il y a une limite n'est une fin, mais qu'elles relèvent des (moyens) qui concernent ce but — par exemple, par rapport au fait de faire maigrir, la cure d'amaigrissement est justement cela —, et puisque ces moyens, chaque fois que l'on fait maigrir, sont en mouvement de façon telle qu'ils ne sont pas en eux-même les résultats en vue de quoi le mouvement (se produit), ces (moyens) ne sont pas une action, ou en tout cas pas une action parfaite. En effet (cette action) n'est pas une fin; en revanche, en celle-là (sc. dans cette action parfaite) est inhérente la fin et (du coup) l'action (qui y mène)." (*Op. cit.*, p. 458 f.)

[18] I follow Jaeger's reading and excise καὶ κινεῖ καὶ κεκίνηκεν (1048b33). Ross preserves these words and considers ἕτερον in l. 32 and τὸ αὐτό (ll. 33 f.) as forming an opposition; hence his rendering of the passage: "...it is a different thing that is being moved and that has been moved and that is moving and has moved; but it is the same thing that at the same time has seen and is seeing, or is thinking and has thought." Ross' summary of this fragment reads as follows: "It is not the case that a thing at the same time is being moved and has been moved; that which has been moved is different from that which is being moved, and that which has moved from that which is moving." (*Aristotle's Metaphysic. A revised Text with Introduction and Commentary*, p. 254.) This latter interpretation seems to be doubtful for two reasons. First, because Aristotle himself states *ex professo* that one always can say correctly that something is moving and has already moved ἀεὶ γὰρ ἅμα κινεῖ καὶ κεκίνηκεν (*Phys.* VII 5, 249b29f.). Second, because it seems to be obvious that τὸ αὐτό in ll. 33f. is not the subject but a direct object as in *Soph. El.* 178a9ff: ἆρ' ἐνδέχεται τὸ αὐτὸ ἅμα ποεῖν τε καὶ πεποιηκέναι; οὔ. ἀλλὰ μὴν ὁρᾶν γέ τι ἅμα καὶ ἑωρακέναι τὸ αὐτὸ καὶ κατὰ ταὐτὸ ἐνδέχεται. Brague translates: "Est-il possible de faire et d'avoir fait en même temps la *même chose*? Non. Il est pourtant possible de voir et d'avoir vu *la même chose* sous le même rapport" (*op. cit.*, pp. 461f.). The opposition Aristotle intends to articulate in *Metaph.* 104818–35 is the opposition between two kinds of actions with different internal structures: it is not true that a thing at the same time is being moved and has completed its movement – here Greek *praesens* and Greek *perfectum* express two distinct meanings; but it is true that at the same time someone is seeing and has seen the same thing in the same aspect, and it is true, of course, that someone or something is moving a thing and (thus) has set the thing in motion (*praesens* = *perfectum*).

formed. In other words the goal is already achieved by the very fact of activity. Yet all the same the activity remains activity; the end achieved does not mean cessation of activity. On the contrary, imperfect activities (*movements*), while still actually being carried out, have not yet achieved the goal. The end is external to them, the achievement of the goal, the "being-at-the-goal"; i. e., the *entelecheia*, sets a limit to the action, and vice versa, any action which has a limit (stops at a certain result external to the activity itself) is imperfect. All such activities are just *media*, they are intermediate, and have a limit set to them by the fact that they aim at an end other than themselves, with the attainment of which they come to stop. Aristotle's formula οὐδὲ γίγνεται καὶ γέγονε (ll. 31f.) reveals a fundamental character of the actions called "movements": each of them is *coming to be of a certain result*, a certain state of affairs; and it is not the case that "something is coming to be and at the same time has come to be." Thus Aristotle's οὐδὲ refers here to the cleft of "before and after": it is movement that reveals the "before-and-after" relation, not *energeia* — here only the ἅμα of the "now" is present in its presence.

To the criterion allowing us to distinguish between *energeia* and movement which Aristotle offers here, *viz.* that of an activity we are exercising and have exercised at the same time, he adds in the *Nicomachean Ethics* X 2, 1173b2 another: we cannot be said to ἐνεργεῖν quickly or slowly (though we can move quickly or slowly, and we may be quick or slow in passing into the state of *energeia*).

At first the concept of *energeia* is introduced in the *Metaphysics* by a simple reference to the fact of movement: "For it is thought that *energeia* is first of all movement."[19] Now *energeia* and movements are distinguished as different *species* of activity.

It is to be remarked that in the passage of interest, the examples cited play a major role. Yet, surprisingly, all the "meaning-forming power" of these examples lies in the comparison, in the collation of two forms of the same verb, of the present and the perfect tenses, linked by the vocable ἅμα — "at the same time," "at once," "simultaneously."[20] While dealing with movements, the activities expressed by the present and the perfect are distinguished, and while dealing with *energeiai*, they are identified with each other. The latter alternative means that we have before us, at the same time and without any temporal interval in between, the activity being carried on now *(tempus praesens)* and "the state of things following the action and existing at the present time." The phrase between parentheses is the generally received definition of the grammatical category of *perfectum indicativi*.[21] The paradox is that Aristotle's ἅμα destroys the temporal gap: the result does not *follow* the action, is not generated when the action ceases, but is achieved in the very action. The action is its

---

[19] *Metaph.* IX 6, 1047a31f.

[20] ὁρᾷ ἅμα καὶ ἑώρακε, καὶ φρονεῖ καὶ πεφρόνηκε, καὶ νοεῖ καὶ νενόηκεν, εὖ ζῇ καὶ εὖ ἔζηκεν ἅμα, καὶ εὐδαιμονεῖ καὶ εὐδαιμόνηκεν. ἀλλ᾽ οὐ μανθάνει καὶ μεμάθηκεν οὐδ᾽ ὑγιάζεται καὶ ὑγίασται.

[21] In his book R. Brague discusses in detail the function of the Greek perfect tense in connection with the fragment in question. See *op. cit.*, ch. IX, § 52 "Présent et parfait," pp. 461–466.

result, and that is why it never exhausts itself (has no limit), because it is *at once* completed in itself by the very fact of being carried on. Such an activity is *at once* whole.[22]

1.3. *Sensations as Energeiai*

Let us look more closely at the first example: ὁρᾷ ἅμα καὶ ἑώρακε – "[someone] is seeing and at the same time already has seen." It should be mentioned that in itself this formula seems to be one of those logical and semantic curiosities which were very popular in their time:

> "Is it possible to be doing and to have done the same thing at the same time?" – "No." "But, you see, it is surely possible to be seeing and to have seen the same thing at the same time, and in the same aspect." (*On Sophistical Refutations*, chap. 22, 178a9ff.)

And yet the formula "is seeing and at the same time has seen" in the *Metaphysics* is not just an example of eristic weapons or sophistic paradoxes, but (alongside of the other examples) a major way of articulating the difference between completed and incomplete activities.

What exactly does Aristotle want to explain to us? It is certainly not true that somebody *is looking at the object and thereby already has perfectly perceived it by sight.* If what I see is a spatial thing, then looking at it I understand that it has sides hidden from my glance which *I have not seen yet*, and when I shall have seen them, I shall get a more complete notion of what is before my eyes. This means that my activity of seeing is not at all completed from its very beginning. Using Kant's and Husserl's language (not in order to disclose some "deeper layers" of Aristotle's text, but to explain what I have in mind while commenting upon it) we could say that the act of visual perception (the act of perceptual synthesis) is not all completed at once, but has duration; it develops temporally: not all sides of a spatial thing are disclosed to the spectator's glance "at once" (ἅμα).[23] Aristotle must have meant something else.

First of all, the spatial thing must not be held for the proper object of sight. We read in *De anima* II 7, 418a26f: "The object of sight is the visible, and what is visible is *colour...*" Of course one may say: "I see the son of Diares "or "What I see is the son of Diares." Yet strictly speaking the object of sight is not the son of Diares (cf. *De anima* II 6). What I see is something white, and it is only indirectly, *per accidens*,

---

[22] ὅλον γάρ τι ἐστί, as it is said elsewhere (*Eth. Nic.* X 3 (iv), 1174a17). This chapter is devoted to the nature of pleasure. According to Aristotle, pleasures are neither qualities, nor movements in the soul, but *energeiai*. In order to provide a firm ground for this thesis Aristotle needs to "start from the very beginning," *viz.* from the metaphysical distinction of *energeiai* and movements. That is why this chapter of *Ethics* is most illuminating in relation to Aristotle's understanding of "*energeia.*" We shall come back to this text time and again.

[23] Cf. Husserl's story about the "the call of a thing" in the lecture course *Grundprobleme der Logik*: "Tritt näher und immer näher, sieh mich dann unter Änderung deiner Stellung, deiner Augenhaltung usw. fixierend an, du wirst an mir selbst noch vieles neu zu sehen bekommen, immer neue Partialfärbungen usw., vorhin unsichtige Strukturen des nur vordem unbestimmt allgemein gesehenen Holzes usw." See *Analyse zur passiven Synthesis*, hrsg. v. M. Fleischer, *Hua* XI (Den Haag, 1966), p. 7.

since this white thing happens to be the son of Diares, that I see him: "... because that which is perceived — the son of Diares — only belongs incidentally to the whiteness."[24]

Even this description simplifies the situation. *Seeing* is always *seeing something*, but this *something*, the primordial object of sight, is, strictly speaking, unnameable. It can only be referred to in speech when the name is used as a pointing gesture. When I say "the seen is white," I express in words only the intellectually comprehensible form of the visible (εἶδος τὸ νοητόν). For "sensation in actuality is always sensation of a particular thing" (417b22), and "it is only accidentally that vision sees colour, since this particular colour it sees is a colour" (*Metaph.* XIII 10, 1087a19f.).

When we say that someone is looking at a thing but has not yet perfectly perceived it by sight, the "not yet" refers to the object understood in a certain way and is connected to the meaning of thingness selected beforehand (for instance, this may be "a spatial thing"). But "am seeing" always means "am seeing and have already seen something." The primordial object of vision cannot be excluded from the act of vision just as it cannot be named. Actually it does not matter *what* it is. I am seeing and am already in the complete actuality (*energeia*) and perfection (*entelecheia*) of vision. The aspects of the visible added later can "complete" the form of the thing, but not the *form of vision or vision-as-form.*

In this respect vision does not differ from the other senses. In a certain way one can say that an object of hearing as a whole may be complex, made up of elements: the syllable *ba* consists of *b* and *a*, and it is not just a sum of certain sounds: it has an internal form which makes it one syllable (*a follows b)*, this and not that: for "*b* and *a* are not the same as *ba*, just as flesh is not the same as fire and earth" (*Metaph.* VII 17, 1041b13f.). We cannot say, somebody is hearing and already has heard *the syllable ba*, for the sounds *b* and *a* reach the ear consecutively, one after the other (cf. *De sensu* 6, 446b6). And yet the hearer is all the time hearing *what is heard* and is in the complete actuality of hearing. Strictly speaking, what we are hearing is a sound, and the syllable is heard only accidentally. The incomplete character of the form of syllable does not mean the incompleteness of the form of hearing.

> Now, even if one always is hearing and has heard — and, in general, is perceiving and has perceived — at the same time, and these acts do not come into being but occur without coming into being — nevertheless, though the stroke which causes the sound has been already struck, the sound is not yet at the ear (and, then, we need to admit that the sound moves, and some interval of time must elapse in which the movement takes place — *A. Ch.*).[25]

When Aristotle speaks about sensation as perfect activity, he is interested not in the form of what is sensed, but in sensation itself as a form.

---

[24] 418a22f.

[25] *De sensu* 6, 446b2–6 (trans. J. I. Beare).

1.4. *Internal Form of Energeia*

In the *Nicomachean Ethics* vision is again cited as an example of perfect activity in the chapter devoted to pleasure:

Seeing seems to be at any moment complete, for it does not lack anything which coming into being later will complete its form (*eidos*); and pleasure also seems to be of this nature. For it is a whole, and at no time can one find a pleasure whose form will be completed if the pleasure lasts longer. For this reason, too, it is not a movement. For every movement (e.g. that of building) takes time and is for the sake of an end, and is complete when it has made what it aims at. It is perfect (τελεία), therefore, only in the whole time or in the final moment. The several movements occupying portions of time of the whole are imperfect, and are different in form (*eidos*) from the whole movement and from each other.[26]

According to Aristotle it is the activity that gives pleasure, and each activity is directed towards an object, one way or another. These activities can be imperfect, i.e., aiming at a certain end and limited. Yet pleasure is always completed and perfect, whatever activity it is connected with. Pleasure is *energeia*, i.e., a perfect activity; in a way it makes any activity perfect: "pleasure gives perfection to the activities, and therefore to life." (Cf. *Eth. Nic.* 1175a11–17.)

A very important idea, which Aristotle develops in *Eth. Nic.* X 3, is that every activity in a broad sense (be it movement or *energeia*) has, precisely as an activity, its internal form.

In particular, movements are distinguished by their internal form, and besides it is impossible, while a movement lasts, to point to a portion of the time of the whole within which the movement would be perfect as to its form.[27] A movement is directed towards its end and is "made up" of stages of reaching it — τὰ περὶ τὸ τέλος (as it is said in *Metaph.* IX 6, 1048b19).

The movement of moulding a statue defines two actualities — the actuality of copper as copper and the actuality of the statue — as something preceding in relation to something succeeding, constituting in this way the relation of before-and-after-ness. Establishing the connection between the two, the movement "exhibits" the piece of copper as a being-in-potentiality (δυνάμει ὄν) in relation to the actuality of the statue; it is exactly this connection that allows a being in potentiality to be *actually disclosed* as such, as this particular possibility of the statue and not a shield. This disclosure is nothing but "presence" (*energeia*) of a potentiality *qua* potentiality. The difficulty lies here in *the way* of "exhibiting" or "being present." The metaphor of intellectual *contemplation*, of an instant grasping of the intelligible species (*eidos*) in the act of intelligent vision cannot be applied here. The form of movement or movement as form (if it may be spoken of that way) does not fit into the "now," is never disclosed in the winking of an eye; more than that, it can be thought only as *extended* temporally. This is a form, which is not "seen," but rather "calculated" and articulated (λέγεται), and time as the number of movement is one of its aspects.

---

[26] *Eth. Nic.* X 3, 1174a14–24. Ross' translation modified.

[27] *Ibid.*, ll. 21–24.

The form of movement is perfect when the movement is completed and its end achieved, when no missing stages, not yet gone through, can be pointed to; when the *for-the-sake-of-which* (= *that-towards-which*) of the movement has been reached. In such a situation the Greek language may use the *perfectum indicativi* to designate the corresponding action. Yet in the case of movements, as soon as it becomes possible (and necessary) to articulate the meaning of what is going on by means of the perfect tense, use of the verb in the present is prohibited. We have one more paradox here: the form of movement becomes completed when there is no more movement.

Yet this is not so in the case of *energeiai*: the present tense may and must be used alongside of the perfect tense. It is so that the "grammatical positing of meaning" is realized in *Metaph.* 1048b18–35.

The form (*eidos*) of *energeia*, in contradistinction to that of movement, is perfect "at once"; it does not require duration to be completed, or, as Aristotle put it, it is perfect and whole not in time (μὴ ἐν χρόνῳ), but in every moment, in every "now" (τὸ γὰρ ἐν τῷ νῦν ὅλον τι).[28]

1.5. *Identity of Energeia and Form*

We shall go back now to the discussion of sensation as *energeia*. The famous passage of *De anima* III 8, 432 a1–3 reads:

Thus the soul is, as it were, a hand: as the hand is the instrument of instruments, so the intellect is the form of forms (the *eidos* of *eidē*), *and sensation is the form* (*eidos*) of the sensible.

It follows that vision is in a sense the "form of the visible." This elliptical formula requires a detailed explanation.

To this end, we shall introduce the term (missing in Aristotle's works) of "*event (occurrence) of seeing.*" In a certain sense (and we know already in what sense) one can say that the event of seeing is completed "every moment," though the last phrase is not quite fitting since seeing as *energeia* is not "in time" (μὴ ἐν χρόνῳ), but all of it is in the "now," and the "now" is not a part of time.[29] *Seeing* always means *seeing something*, and this "something" must be called *the seen* (τὸ ὁρατόν). The expression "the seen" (as well as "the sensed," τὸ αἰσθητόν, in general) is ambiguous. On the one hand, the "sensed" or the "perceived-by-senses" designates a thing or a sensible quality of a thing. For instance, "the seen" is the colour, and the colour is a quality of a thing, which "is able to produce movement in the actually transparent (sc. in the transparent medium)" (*De anima* II 7, 419a8). On the other hand, if we are to distinguish the meanings more accurately, we must say that the colour and the seen are not the same (*Phys.* III 1, 201b4), just as copper and the statue-in-possibility are not the same. Τὸ ὁρατόν as that which *can* be seen is defined in relation to

[28] *Ibid.*, b9. Aristotle asserts this in relation to pleasure, but precisely in so far as the latter is considered as a particular case of *energeia*: "It is impossible to move otherwise than in time, but it is possible to be pleased, for the experiencing of pleasure is in every moment (alreay in the 'now') a perfect whole."

[29] *Phys.* IV 10, 218a6. This thesis will be a subject of our closer consideration.

another ability, the ability of the spectator to see (vision as a faculty of his soul); τὸ ὁρατόν as something actually seen is defined in relation to the actuality of vision (the actual exercising of this faculty). The sensed (or the perceived-by-senses) is not a proper definition of the thing. The *energeia* (actuality) of the sensed *as sensed* and the *energeia* (activity, action) of sensation are one and the same *energeia* (425b26).

Generally, about all perceptions, we can say that perceptions is what has the power of receiving into itself the sensible form of a thing without the matter. (*De anima* II 12, 424a17)

The event of seeing is the actual perception of such a sensible form without the matter, and the form (*eidos*) itself is what is actually seen. Yet being a form without the matter this actually-seen serves, nevertheless, as matter in relation to another form, to *seeing as form*, the form (*eidos*) of *all* that is seen.

The *act* of vision, seeing as *energeia* (perfect activity), is at the same time a complete form (*eidos*). This form is invariable and self-identical in its completeness in every event of seeing (irrespective of what is seen). Nonetheless we can speak of *different* and distinct events of seeing. Therefore also we have to consider some "matter" as the cause of distinction of what is formally identical (εἴδει ἕν). Of course, it is the "object of sight" (= the "actually seen") that serves as the matter of the event of seeing. But the object of sight must be understood here as a sensible form-without-matter, "received into the sense." The Aristotelian matter/form distinction generates a hierarchy of forms: that which is designated as a sensible form without matter serves as the matter for the higher level form — the form of seeing as such (independent of what is actually seen). Vision thus understood is *energeia and* form, a form that is a certain *energeia* or an *energeia* possessing its own internal "energetic" form (seeing, and not hearing; pleasure, and not pain). This is the *energeia* (= form) of sensation as of perfect activity *(am sensing and have already sensed)* and at the same time the *energeia* (= actuality; actual presence in sensation) of the sensible object as such (i.e., of the seen as seen and not as red; of the heard as heard and not as loud). Seeing as a perfect activity (*energeia*) of the spectator has *its own intrinsic determination*; determination of activity is expressed in Greek by the verb in the *praesens indicativi* (in English we use *present continuous tense* to express the same meaning): *am seeing*, and not *am hearing*, not *am grasping intellectually*, not *am being happy*. And at the same time this *energeia* is a presence (in *energeia* = in actuality), a "manifest-ness" of the seen as seen.

Aristotle says: "The *energeia* of the sensible object and that of the sense is one and the same *energeia*, and yet the distinction between their being remains."[30] It follows that the formula ἅμα... αἰσθάνεται καὶ ᾔσθηται, "one is always perceiving (sensing) and has perceived (sensed) at the same time" (*De sensu*, 446b4–6) means the identity of activity and passivity. The presence in the soul (in sensation) of the sensi-

[30] *De anima* III 2, 425b26f: ἡ δὲ τοῦ αἰσθητοῦ ἐνέργεια καὶ τῆς αἰσθήσεως ἡ αὐτὴ μέν ἐστι καὶ μία, τὸ δ' εἶναι οὐ τὸ αὐτὸ αὐταῖς. The expression τὸ εἶναι τῷ X refers to the *essence* of X. The essence, the *meaning* of the "sensation" and of the "sensed" is different, although the actuality of the sensation is at the same time the actuality of the sensed, and vice versa. And this is one and the same *energeia*.

ble form of a thing is at the same time the *energeia* of the sensed and the *energeia* of the sensation, a kind of "letting-be-present." This topos of indistinctness or indifference, the place of encounter, the location of purely "energetic" character must become the starting point of all *phenomenological* analysis — and it is to such an analysis that the present book is tending. Of course, this does not hold for *first philosophy* as Aristotle projected it. Here a universal ontological configuration of first principles and foundations has been constructed, which has the entity belonging to the first category, *the substance*, for its centre.[31] According to this onto-logic, activity and passivity serve as determinations (*accidentia*) of a substance. And this means that within the framework of this scheme we can and must distinguish between the *energeia* of sensation, the activity (an *accidens*) of a living body endowed with sense, on the one hand, and the *energeia* (in-fluence) of the object, which affects the sense organ. This latter *energeia* in its turn is to be interpreted as an accidental determination of the thing-as-substance (the external object of sensation), so that we must consider "one and the same *energeia*" as being a replica of two different determinations of two different substances and call them by different names. Aristotle writes:

The *energeia* of the sensible object and that of the sensation is one and the same *energeia*, though their essence is not the same; in saying that they are not the same, I mean actual sound and actual hearing: a man may have hearing and yet not be hearing, and that which has a sound is not always sounding. But when that which can hear is actively hearing and that which can sound is sounding, then the actual hearing and the actual sound come about at the same time (these one might call respectively hearkening and sounding).[32]

2. NUMBER AS "ARTICULATION" OF A FINITE SET

2.1. *Magnitude and Number*

"Time is just this — number of movement in respect of 'before' and 'after'" (*Phys.* IV 11, 219b1f.). If we want to really understand what Aristotle means, not to read our own notions into his text, we need to reconsider not only Aristotle's concept of movement, but his understanding of number (ἀριθμός) also.[33]

Number is that by means of which something can be numerically estimated, yet the ways of this estimation (measurement) are essentially different. *Number* (ἀριθμός) is to be distinguished from *magnitude* (μέγεθος). By means of number we estimate what is more and what is less numerous; number is the measure of what is discrete, the measure of an aggregate of separate and distinct elements-units. Aris-

[31] In the next section we shall discuss this ontological scheme in more detail.

[32] *De anima* III 2, 425b26–426a15. Smith's translation modified.

[33] R. Brague remarks in his book *Du temps chez Platon et Aristote. Quatre études.* (Paris: PUF, 1982) that the chapters of the *Physics* devoted to the problem of time must serve as the main source for our interpretation of the concept of number as it is used by Aristotle, and not as a domain where we can apply our modern idea of number: "Le concept de nombre ne devra donc pas être présupposé. Le sens précis que le mot *arithmos* a pour Aristote dans le traité du temps devra au contraire être déduit du texte même." See the section entitled *L'"Arithmos,"* pp. 134–144.

totle calls such a discrete set πλῆθος. Magnitude is the measure of what is continuous, of what Aristotle calls τὸ συνεχές (*continuum*), of an internally coherent and connected manifold which can be divided into parts in potentiality only (but is not actually divided), and in which for any two "points," however near to each other, there always can be found a third one lying between them. Sometimes such a manifold itself is called μέγεθος.

A quantum (ποσόν τι) is a plurality (πλῆθος) if it is numerable, a magnitude (μέγεθος) if it is measurable. (*Metaph.* V 13, 1020b8f.)

Number, defined elsewhere as a "measured *plurality*,"[34] and magnitude, a "measured continuum," are two *distinct* kinds of the same genus "quantum" (τὸ ποσόν).

Aristotle says: "Number is perceived by the negation of continuity" (*De anima* III 1, 425a19), whereas magnitude is continuous by definition and serves as the measure (or metrics) of the continuous. Here we are dealing with one of the riddles of Aristotle's definition of time, for the author of the *Physics* insists on the *continuity* of time. Time is continuous *because* it is one of the formal aspects of movement.[35] And movement, Aristotle says, *follows magnitude* (or *is dependent on magnitude*).[36] And since magnitude is continuous, movement is continuous too (*Phys.* IV 11, 219a10–13); and time is also continuous because it inherits in a way the structure and therefore continuity of movement. Via this chain of mediation time itself "imitates" magnitude and is *analogous* to it (cf. 219a17f.).

And yet, it seems to follow immediately from Aristotle's definition that time cannot be continuous (as we could expect on the ground of the previous arguments), but is the *number* of movement, "negating" continuity. This apparent contradiction must become the first object of our analysis.

2.2. *Number as Numbering and as Numbered*

Aristotle distinguishes several senses in which the term ἀριθμός may be used: on the one hand, we speak of the numbers that are counted in the things in question, and, on the other hand, of the numbers *by which* we count them and *in which* we calculate. Imitating Aristotle's (and the corresponding Latin) terminology we shall call number in the first sense "numbered number" (τὸ ἀριθμούμενον), and that by means of which we count, ᾧ ἀριθμοῦμεν,— "numbering number." But because we do not always *actually* count something countable, we need to be as precise as possible and to distinguish within the notion of the *numbered number* (1) something

[34] Cf. *Metaph.* XIV 1, 1088a4ff.: ἀριθμός = πλῆθος μεμετρημένον.

[35] Indeed, the starting point of the Aristotelian theory of time is the statement, "we perceive movement and time together [...]. Hence time is either movement or something that belongs to movement. Since, then, time is not movement, it must be the other [sc. it must pertain to movement in some way]" (*Phys.* IV 11, 219a6–10).

[36] *Ibid.*, 219a11f.

which is counted in actuality and also (2) the "numerable number" (τὸ ἀριθμητόν), i.e., something which *can* be counted.[37]

Number, we must note, is used in two ways — both of what is counted or countable and also of that with which we count. Time, then, is what is counted, not that with which we count: these are different kinds of things. (219b6ff.)

The number by means of which we count is, in today's terms, an element of the natural number system: 2, 3, 4, ..., "one" excluded, for it cannot, strictly speaking, be called an ἀριθμός, since "the smallest possible number (if we speak of it without any further qualifications) is the dyad" (220a26). The "numbering number" can be regarded as a universal form (εἶδος τὸ καθόλου) of a finite discrete set, while the numbered number is a form embodied in matter, a form inherent in a concrete complex whole (εἶδος ἐνόν), composed of matter and form.

And the number of a hundred horses and a hundred men is the same, but the things numbered (literally: that of which it is the number) are different — the horses for the men. (220b10ff.)

This means that the set consisting of a hundred horses and the set consisting of a hundred men are identical in relation to number as to the universal numerical form (εἴδει ἕν), they are iso-morphic in this respect, or, as the mathematicians say, these sets are "equal" as far as their *cardinality* is concerned, but nevertheless they are different sets. The numbered or numerable number must not ignore this difference ("matter"); these concepts are sensitive, so to speak, to the nature of counted things. Let us repeat once more: the number of elements of a concrete set is thought either as something which can be counted (*numerabile*), and in this case Aristotle uses the term τὸ ἀριθμητόν (just as τὸ αἰσθητόν is that which can be perceived by the senses, and τὸ νοητόν that which can be grasped by the intellect), or as what is actually being counted (*actu numeratur*), what becomes the object of counting, is manifested, shown in the act of counting, as a numeric form, a numeric appearance-eidos of this particular discrete set.

The numbered number is, in its most general form, the structure of a discrete (finite) set implying a singling-out of its elements as indivisible units (or rather unities),[38] their distinction from one another and collocation in one whole a set, which allows us to speak of a pair, a hundred etc. of such and such things. It is not without reason that Aristotle cites as an example two groups consisting of creatures belonging to the same kind (e.g. a hundred men, a set in which all elements are identical in kind): we must be concerned not with the character and the meaning of the differ-

---

[37] Cf. 219b6f. Thomas Aquinas in his commentary explains that while speaking of number one needs to distinguish the two following meanings: 1) numerus *numeratus*, which can be understood either as the result of actual counting or as something that can be possibly counted (id, quod numeratur actu, vel quod est numerabile); and 2) the number by means of which we count or calculate (numerus *quo numeramus*). Thomas Aquinas, *In physicorum Aristotelis lib.* IV, cap. XI, lect. XVII.

[38] W. D. Ross in his edition of *Physics* remarks referring to Iamblichus, *In Nicom. arithm. introd.* (see p. 10, ll. 8f.), that according to the definition ascribed to Thales the number is σύστημα μονάδων (a system of units). "Some Pythagoreans called the unit ἀριθμοῦ καὶ μορίων μεθόριον, the boundary between number and fractions."

ences between single elements, but only with the distinctness of that which has no "face" of its own within the procedure of "calculation" described, *viz.* with the singling-out of units.

We have already referred to the very important passage of *Metaphysics* VII 17 in which Aristotle speaks about the compound (τὸ σύνολον) of matter and form, and declares form to be the foundation of its unity: form is that which makes matter (in particular, the aggregate of elements) one thing. The whole thus formed is not just a heap (σωρός) of elements: the syllable *ba* is not the same as *b* and *a*, but in addition to that at least a certain *order* of sounds following one another. Using modern language, one might say that the syllable *ba* is not described by the set of phonemes {b, a} = {a, b}, but rather by the ordered set (b, a). Nonetheless if the "heap" is thought of as *one*, as one aggregate, we must ask about the foundation of this unity: what makes the whole *one* and *united*? And the foundation is nothing but the "numbered/numerable number." The number in this sense is the form of a discrete set of elements as such; the very first "layer" of being-formed on which other formal structures can be superimposed. In the hierarchy of forms the discrete set as something "numerable" occupies the lowest rank. For instance, Aristotle calls "stones" matter of the house, yet when in the very beginning of the process of building the stones are gathered in one place, they are already a united whole, though the form of this whole can be thought of in a very weak sense only (the "heap" of stones is a totality as numerable, i.e., possessing the form of number).[39] The totality of stones may serve as material, which in the process of construction turns into a complex whole having the form of a house. Put together, *b* and *a* may become the syllable *ba* or *ab*, yet the set {a, b} already has by itself the form of the number two (the corresponding *cardinality*).

Time, then, is what is counted [sc. a countable or numbered number], not that with which we count. (219b7f.)[40]

2.3. *The Numbering Soul*

Numbered number here may refer to something *actually being counted* or to something, which *can be counted, counted in potentiality*. Yet the potential as such is defined in relation to some actuality. The potential that cannot become something actual is an absurdity: "evidently it cannot be true to say 'this is capable of being [this

[39] Cf. R. Brague, *op. cit.*, p. 137: "Un *arithmos* est avant tout une structure, un assemblage."

[40] Thomas Aquinas leans upon this thesis of Aristotle in order to explain the apparent contradiction between the characteristic of time as dependent on magnitude, and therefore continuous, and its definition as number, and therefore discrete: "Et ideo, licet numerus sit quantitas discreta, tempus tamen est quantitas continua, propter rem numeratam, sicut decem mensurae panni quoddam continuum est, quamvis denarius numerus sit quantitas discreta." This interpretation implies of course that the unit of measure (μέτρον) is chosen and fixed. Indeed, Aristotle in *Metaph.* XIV 1 declares magnitude to be a "sum of many units (of measure)." I believe nevertheless that Thomas' commentary does not clarify why Aristotle should use the term ἀριθμός at the strategic point of his discourse, i.e., in the key definition of time.

or that thing], but will not be'."[41] It follows that the numerable number, that which can be counted, is defined in relation to an actuality — to the activity (operation) of counting.

We read in *Phys.* IV 14 (223a16–26):

It is also worth considering how time can be related to the soul [...]. Whether, if soul did not exist, time would exist or not, is a question that may fairly be asked; for if there cannot be someone to count there cannot be anything that can be counted either, so that evidently there cannot be number; for number is either what has been, or what can be, counted. But if nothing but soul, or the intellect of the soul, is qualified to count, it is impossible for there to be time unless there is soul [...].[42]

It is true that this is followed by an important reservation, which does not allow us to interpret this passage as Aristotle's final judgment on the connection between the intellect of the soul (ψυχῆς νοῦς) and time. But we are now interested in the meaning of the premise.

If that which is capable of counting is not and will not be present, there is nothing countable either. For that is countable (in potentiality) which can be actually counted, and the actuality of counting and the actuality of number is one and the same actuality. The actuality of counting is an activity of the soul counting (or of the intellect of the soul). Yet what does "to count" or "to calculate" mean here? "To calculate," if we stick to our interpretation of the term, means *to recognize/constitute the structure of the finite set*, i.e., to distinguish and to put together the elements-units. The simplest operation of counting or calculating thus understood is distinguishing and putting together the one and the other, of units which admit no more of internal distinction and separation within the counting procedure. The simplest act or step of counting is adding a unit to what is already thought of as one united totality. In the most fundamental sense every step of counting is the generation of a pair, a dyad: 1 + 1.

"One" is not yet a number, but rather it is a unit of measure (μέτρον), something indivisible either in kind (all men in a gathering of men are identical in kind) or for sensation (*Metaph.* XIV 1, 1088a2f.). *One* is the measure of a plurality (a discrete set), and number is the plurality measured-counted (by single units) or a sum of several units (1088a4ff.).

That is why it is correct to say that one is not a number, just as a unit of measure is not a sum of such units, but the unit and the measure are a principle (ἀρχή). (1088a6f.)

Number as a measured-counted set is the result of distinguishing and putting together what is indistinguishable individually (indivisible in kind). A unit has no internal determinations, which would allow us to distinguish it from another unit, and that is why distinction necessarily implies com-position; distinction (taking apart) and composition (putting together) are here one and the same thing. The smallest possible number is *two* (*Phys.* 220a26), the result of the simplest counting operation of distinguishing in and by composition.

[41] *Metaph.* IX 4, 1047b4.

[42] The translation by Hardie and Gaye slightly modified.

Following a witty suggestion of R. Brague, we could call this operation *articulation*. One could say also that the countable (counted) number itself is "an articulated set." The simplest *articulus* (a joint, a chain link) is here the number *two*.

The simplest *articulus* of time[43] is what is distinguished and put together (i.e., calculated in the simplest act of counting) "now" and "now," the preceding and the succeeding, the *before* and *after*. The "now" articulates time just as the unit articulates, i.e., counts, the set, for the simplest act (or step) of counting, as was already said, is the addition of one unit to what has already been understood as one (united via the form of number). That is why Aristotle, while speaking about time as about number, likens the "now" to the *unit*: although the "now" *is not a part* of time, it is the source and principle (ἀρχή) of the temporal form, as the unit (the one) is the principle of the numeric form.

On the other hand, according to an important formula of Aristotle:

ᾗ ἀριθμητὸν γὰρ τὸ πρότερον καὶ ὕστερον τὸ νῦν ἔστιν. (219b25, 28)

The "now" is (exists) insofar as the *before* and the *after* are countable.

[An alternative interpretation (reading ἐστιν instead of ἔστιν): "the 'now' is the before and after, *qua* countable."][44]

The counted, articulated "now" *always appears to us*[45] as a dyad.

It is true that we recognize a lapse of time, when we have already determined movement defining the before-and-after in it; for it is when we have had the sensation of something prior and something posterior in movement that we say that some time has elapsed. And we delimit [the prior and the posterior] due to the fact that we grasp them as the one and the other with something different from both in between. For when we have thought of the extremes as distinct from the middle, and the soul pronounces the "now" to be **two**: one prior (*the before*) and one posterior (*the after*),— this is when we say that a certain time has passed, and that is what we call time. For it is precisely what is delimited by means of the "now" that seems to be what we mean by time. (219a22–29)

The "now" is (exists) when and as the soul "pronounces" (ὅταν εἴπῃ) the "now" to be two, when it articulates the prior and the posterior. In this articulation two "now"s (or the "now" as a dyad) appear to the soul, the preceding in relation to the succeeding and the succeeding in relation to the preceding. Aristotle does not mention here any isolated "properly present now." He deals with the "now" counted in

[43] Thanks to a happy coincidence (or perhaps due to a concealed correlation of meanings) *articulus temporis* signifies (for example, in Cicero's writings) "exact, critical, crucial moment of time," in a word, has exactly the same meaning as the Greek καιρός (the moment of decision and resoluteness) or German "Augenblick" as it is used by Heidegger (following Kierkegaard). This concept will be of crucial importance for us in the next chapters.

[44] The first variant of the translation is mine. Hardie and Gaye (l. 25): "... and if we regard these (sc. the 'before and after' in motion) as countable we get the now." P. H. Wicksteed and F. M. Cornford (*in Loeb Classical Library*, vol. 228): "...and it is in virtue of the countableness of its (?!) before-and-after, that the 'now' exists." A footnote gives the alternative translation: "the 'now' is the before and after, *qua* countable." This possible interpretation is very important for our analysis in this section and also in sect. 5. Cf. Simplicius (732, 29f.): καθόσον δὲ ἀριθμητὸν τὸ πρότερον τοῦτο καὶ ὕστερον, κατὰ τοσοῦτον, φησί, νῦν δείκνυται.

[45] Cf. "νῦν δείκνυται" in Simplicius.

movement, and there is no movement in the isolated "now," insofar as it is possible to speak of it (*Phys.* VI 3).

In the simplest joint of time the "now" always appears as a dyad, the preceding and the succeeding, with a gap, an interval (διάστημα), an "empty" difference between them. In this "phenomenological" context (for here we are dealing with our understanding, grasping, sensing of time) time is opposed to the dyad of "now" and appears as otherness itself, a difference indifferent to events. Only the extreme "now"-points are the "now" of some occurrence, as linked to some events: "before" (in the preceding "now") Coriscus was in the point A, and "after" (in the succeeding "now") he is (or will be) in the point B. There is an event-less, indifferent, empty interval between these two *eventful* "now"s. Time is delimited by the "now," but every "now" is constituted as such and articulated as a dyad by this event-less gap. Thanks to this gap the "now" is counted as the prior and the posterior, and the "now" *is* (exists) insofar only as it is counted. It follows that the counted and event-linked "now" is "the other in relation to something other" (to use an expression taken from Plato).[46]

The "now" in one sense is the same, in another it is not the same. In so far as it is in something other again and again, it is different (which is just what its being "now" was supposed to mean)... (219b13f.)[47]

3. BEING AND ENTITY

3.1. *Aristotle on the Manifold Meanings of Being*

Aristotle says: "Being is said in many ways (τὸ ὂν λέγεται πολλαχῶς)" (*Metaph.* VII 1, 1028a10). And this statement is one of the most important in the whole of the *Metaphysics.*[48]

"Being (τὸ ὄν) is said (1) on the one hand, in an accidental way (*per accidens*), (2) on the other hand, by itself. (1017a7f.)."[49] "Further, (3) 'being' and 'is' mean that [a statement] is true, 'non being' that it is not true but false,— and this alike in affirmation and negation" (1017a31–33). "Again, (4) 'being' (τὸ εἶναι) and 'that which is'

[46] *Soph.* 255d1: ἕτερον πρὸς ἕτερον. Plato argues here that the genus "being" coincides neither with "the same" nor with "the other," but partakes of both, since it is *the same* with regard to itself and *other with regard to other,* and *other* is always said in a relative sense and to be what it is [sc. *the other*] it needs to be other for something other. We shall return to the discussion of this passage in chap. 6, sect 3.

[47] Trans. Hardie and Gaye (slightly modified).

[48] Heidegger remarks in a lecture course delivered at Freiburg University in 1931: "Dieser Satz: τὸ ὂν λέγεται πολλαχῶς begegnet bei Aristoteles wie eine stehende Formel. Aber es ist keine bloße Formel, sondern in diesem kurzen Satz prägt sich aus die ganze fundamental neue Stellung, die Aristoteles in der Philosophie gegenüber seiner ganzen Vorzeit, auch Plato gegenüber, sich erarbeitet, nicht im Sinne eines Systems, sondern im Sinne einer Aufgabe." M. Heidegger, *Aristoteles. Metaphysik Θ 1–3. Von Wesen und Wirklichkeit der Kraft*, GA 33 (Frankfurt a. M.: V. Klostermann, 1990), p. 12f. F. Brentano writes in his famous treatise *Von der Mannigfachen Bedeutung des Seienden nach Aristoteles* (Freiburg i. B., 1862, p. 5.): "So bildet die Erörterung der mehrfachen Bedeutung des Seienden die Schwelle der aristotelischen Metaphysik."

[49] My translation. I would like to remind the reader once more of the impossibility of distinguishing in English, without resorting to the help of artificial construction, between τὸ ὂν and τὸ εἶναι. Ross: "Things are said to be (1) in accidental sense, (2) by their own nature."

(τὸ ὄν), in these cases we have mentioned, sometimes mean being-in-potentiality, and sometimes being-in-*entelecheia*" (1017a35–b2).[50]

Such are the four primordial meanings of being according to Aristotle. And one of them must in its turn undergo an internal division: "there are as many ways to say about things that they are by themselves as the schemata of category designate" (1017a22f.).[51]

In the universally received scholarly Latin σχήματα τῆς κατηγορίας is rendered as *figures of predication*. They refer to the different ways of predicating (i.e., of saying *something about something else*) distinguished by Aristotle. And as we shall see, the discussion of categories belongs not so much to the domain of logic or semantics as to ontology. "First philosophy" asks about being as being (τὸ ὂν ᾗ ὄν), asking *what* a being (entity) is insofar as it possesses being.[52] Yet the question "what is?" refers to meaning. The clarification of the meaning of being as being (entity as entity) requires a systematisation of the ways of speaking of entity, of the ways of pronouncing the word "is" while bearing witness to being[53] "logically." Aristotle's creation, which we call *logic* today, was for its creator himself the science dealing with the way entity exhibits itself (its being, its beingness-essence), bears witness to itself in the logos. Logic is always onto-logic, an aspect of ontology.

The most important feature of entity as entity is its ability to be said. Being (that because of which entity is called entity = a being) can always disclose, exhibit itself in words. The central theme of onto-logy is *being (or entity, or the being of entity) as expressed in the logos*. Being-in-actuality can be thought of only as having-a-form. As we have seen in the first section of this chapter actuality (*energeia*) in its *metaphysical* sense is almost synonymous with completeness (*entelecheia)* of the form-*eidos*, and the *eidos* is already *designed* to be sounded in words, essentially capable of being said.[54] In this respect Aristotle's position does not at all differ from Plato's.

Yet for Plato the manifestedness of being, its inherent possibility of being expressed in logos, are connected with the concept of ἰδέα, and this implies an invariable, always self-identical and self-sufficient manifestedness, clarity, internal luminosity of meaning. For Aristotle κατηγορία (*witness to..., saying something about something else*) becomes the central onto-logical term, which comes to replace or to displace ἰδέα. *That-about-which* (τὸ ὑποκείμενον) of the statement is shown through, and by means of, something else (τὸ κατηγορούμενον); it always discloses its kind-*eidos* in a certain logical perspective depending on the "schema of category," yet always remains itself and the same in its *many various* ways of being shown.

---

[50] Ross: "...sometimes being potentially and sometimes being actually."

[51] Ross: "Those things are said in their own right to be that are indicated by the figures of predication: for the senses of 'being' are just as many as these figures."

[52] *Metaph.* IV 1, 1003a21.

[53] Κατηγορέω meant originally "to speak against somebody before the judges, to bear witness to something in order to accuse someone," and in general – "to bear witness, to show, to indicate, to prove" (cf. εὖ γὰρ φρονοῦντος ὄμμα σοῦ κατηγορεῖ, Aeschylus, *Agamemnon*, 271).

[54] Cf. *Metaph.* IX 1, 1045b29–31, where the equality sign is put between what-is and what is (or can be) said (τὰ ὄντα = τὰ λεγόμενα).

Conversely, the same meaning-*eidos* can be revealed in a multiplicity of individual things. *That-about-which* (τὸ καθ' οὗ) of the statement, *that about which something else is predicated* (1028b36), is present at hand and available. It is present before us (ὑπάρχει) in its facticity, it can be pointed to and the gesture of pointing can be imitated by means of speech, saying *this here* (τόδε) or *something here* (τόδε τι). Aristotle says about the followers of Plato that they are blind to the factually present.[55] The disclosedness of the meaning in *this here* amounts to factual presence. In the last sentence both terms *factual* and *presence* are important. It is impossible to be "present and available" without being definite, that is to say, without possessing form-*eidos*. The eidetic determination is a necessary condition of factual presence, but not the sufficient one. We have to ask: "the determination of what?" The *eidos*, too, can be thought of, according to Aristotle, first of all as the definition of something factually present at hand, as the definition of a *this-here*.

Thus the term "category" implies the difference between the presence and what is present, and means the presence (manifestedness) of the one *through* or *in* the other. Accordingly the schemata of category provide a list or typology of the ways of being so present, of the meanings of being. *Being* means, since Parmenides, a simple and manifest presence, comprehensible by the intellect in one simple act. Yet the "meaning" of being, according to Aristotle, is defined by the sense in which the "is" is said; i.e., in which *sense* the one [subject] *is* (which means, is shown as) the other [predicate]: the person asking about being must take into account the dual nature of being-present: it is always the presence of what is present.

About the subject itself we can say: "it *is*," only having in mind: "it is an *eidos* X"; the former is an elliptic form of the latter. From the onto-logical point of view, such "is" (designated as *copula* according to its logical function) does not connect one entity with another,[56] nor one concept with another, but shows the internal duality of being itself; it articulates the fact that *being* always is *something's-being-something*. Using the technical terminology of Aristotle we could say: matter (undetermined, but having to be determined) *is* only as formed by means of something else (form), and form *is* (strictly speaking) only as having formed something else (matter). It is true that certain exceptional extreme ontic situations are possible, which lie outside this general ontological scheme, but we cannot discuss them here.

The being of entity manifests itself in the logos, and first of all in the manifesting, "indicative" (as ἀποφαίνεσθαι) way of speaking which is called λόγος ἀποφαντικός in Greek,— in other words, in an apophantic assertion. Its structure is: *S is P*. If we ask Heidegger's question "what is the meaning of being?" remaining within the framework of Aristotle's ontology, that is to say, if we ask: "what does the expression 'to be' mean?" — this question must be detailed further. Its final (complete and correct) form is: "what does it mean for S to *be* P?" For instance, "what does it mean for Socrates to be a man?" — τί ἐστι τῷ Σωκράτει ἀνθρώπῳ εἶναι;

[55] ἀθεώρητοι τῶν ὑπαρχόντων (*De gen. et corr.* I 2, 316a8f).

[56] Though the traces of such a point of view can be found in the treatise on *Categories*.

The question just cited as an example is fundamental, since the answer to it must express the *being-a-man* proper and essential to Socrates. The question aims at the ultimate ontological foundation of Socrates' being-what-he-is. The correct answer to this question is called *definition* (the proper logos of the entity). This ontological ground pointed by the strategy of searching, the form itself of the question asked ("what does it mean [in the most essential sense] for a thing to be [what it is]?"), is codified by the famous Aristotelian formula τὸ τί ἦν εἶναι.[57]

3.2. *The Being of the Copula*

A being (τὸ ὄν) in the sense, which is fundamental for this ontology is to be understood as a participle, derived from the copula "is" of the proposition: *S is P.* That is, "a being," generally speaking, is not a self-sufficient term, it needs to be completed and (thereby!) refined, it is "open" towards the subject and predicate. When you say simply "being" by itself, explains Aristotle, this does not mean anything and only refers to a connection which cannot, however, be thought of without what is being connected.[58] To acquire its full meaning, the participle *being* must be put between an implied subject and an implied predicate, i.e., must be included into the following construction:

S — being — (*what* or *as what*) P

A being (ens) is said about S (and in this case implies a certain P) or about P (and implies a certain S).

In this construction the subject acquires its semantic determination (the form expressed in the language) thanks to the predicate without which it can only be pointed to in speech by means of an imitation of gesture, the demonstrative pronoun τόδε τι. Yet this gesture itself (not just τόδε, but τόδε τι) is already an empty possibility of disclosing meaning in predication.

---

[57] The analysis of the syntactic structure and the meaning of this formula has a long and intricate history. I think it is difficult to find a more or less serious investigation of Aristotle's metaphysics, which lacks a section devoted to this topic. A thorough and profound investigation of the subject can be found in the article by Friedrich Bassenge "Das τὸ ἑνὶ εἶναι, τὸ ἀγαθῷ εἶναι etc. etc. und das τὸ τί ἦν εἶναι bei Aristoteles." *Philologus* **104** (1960), pp. 14–47, 201–222. See also: P. Aubenque, *Le problème d'être chez Aristote* (Paris : PUF, 1962); Françoise Coujolle-Zaslawsky, "Aristote: Sur quelques traductions récents du TO TI HN EINAI." *Revue de Théologie et de Philosophie* **113** (1981), pp. 61–75; H. Schmitz, *Die Ideenlehre des Aristoteles*, Bd. I, 1. Teil: *Kommentar zum 7. Buch der Metaphysik* (Bonn: Bouvier Verlag, 1985), pp. 13–22.

[58] *De int.* 3, 16b23ff: πράγματος, οὐδ' ἐὰν τὸ ὂν εἴπῃς ψιλόν. αὐτὸ μὲν γὰρ οὐδέν ἐστιν, προσσημαίνει δὲ σύνθεσίν τινα, ἣν ἄνευ τῶν συγκειμένων οὐκ ἔστι νοῆσαι. Cf. H. Schmitz, *Die Ideenlehre des Aristoteles*, Bd. I, 2. Teil: *Ontologie. Noologie. Theologie* (Bonn: Bouvier Verlag, 1985), p. 66: "'Seiendes' ist hier wie nach *De interpretatione* vielmehr ein Ausdruck ohne eigene, isolierte Bedeutung, der bloß mitbezeichnend (*synkategorematisch*) eine Zusammensetzung aufzeigt (16b23–25), aber Farbe und Charakter gewinnt durch die verschiedene Perspektiven des Ansprechens, die die 'Figuren der Prädikation,' die Kategorien, dieser prädikativen Synthesis von Subjekt und Prädikat verleihen."

Of course, a name pronounced without any connection with other names refers to a meaning, but not to a *being* possessing this meaning (cf. *De int.* 4, 16b28–30), not to being, since according to Aristotle being must be understood as being of a meaning (form) *in* what has the meaning (matter). This holds true for verbs also. Aristotle writes in *De int.* 3, 16b19–25:

When uttered just by itself a verb is a name and signifies something — the speaker arrests his thought and the hearer pauses — but it does not yet signify whether it *is* or not. For not even "to be" or "not to be" is a sign of the actual thing (nor if you say simply "that which is" ["a being"]); for by itself it is nothing, but it additionally signifies some combination, which cannot be thought of without the components.[59]

The entity exhibits itself in the logos depending on the logical perspective corresponding to the way of asking about the entity. The question is precisely the *direction* of the "logical glance" looking attentively at the entity and disclosing it in its being. The expression of being, its logos, the witness (κατηγορία) to the being of entity are the answer to the questions: *What is the entity? What is the entity like? Where, how, when... is it?*

The schemata of category (figures of predication) correspond to the ways of answering, and these correspond in their turn to the ways of asking, to the typology of questions.[60] Before we get an answer, the entity is for us only *this-here,* that to which the gesture points, that at which the question aims and that from which the answer must come.

For when we intend to say of what quality a thing is (sc. to answer the question "what is... like?"), we say that it is good or beautiful, but not that it is three cubits long (for we do not answer the question "how long ... ?" which requires another interrogative pronoun in Greek — *A. Ch.*) or that it is a man; but when we say *what* it is (answer the question "what is...?" — *A. Ch.*), we do not say "white" or "hot" or "three cubits long," but "man" or "god."[61]

So the entity bears witness to its being in many ways. The ways of witnessing (categories) are certain meanings of being and refer to the ways of being. There is a first fundamental witness among them, Aristotle says, *viz.* the *first* category. This is the answer to the question "*What is* [this thing]?" In answer to this question the entity must show itself as substance. All the rest (*accidentia*) are called beings only because they are in some way connected to this first entity and represent, in relation to it, either quality, or quantity, or location, or state, etc., i.e., because they can take up the position of the predicate in the open (*syncategorematic*) figure

" ... — being — ...."

The white *is* (is an entity) because a substance, i.e., an ontically autonomous and self-sufficient — "separable" (χωριστός), as Aristotle sometimes puts it,— thing can be found, named S, about which we can say: "S is white." In every such asser-

---

[59] Trans. by J. L. Ackrill.

[60] Charles Kahn writes in the article entitled "Questions and Categories" (in *Questions*, ed. by H. Hiż. Dordrecht, 1978, p. 227): "Aristotle's classification of the categories is derived from a list of interrogative forms or questions asked in reference to a given subject." Kahn ascribes this observation to William Ockham.

[61] *Metaph.* VII 1, 1028a15–18.

tion (predication) we always refer to a *substance*. Substance as the entity of the first category is ontically independent of the other and is not said about the other.

The formal distinctive characteristic of substance itself is that "it is that which is not said about (predicated of) any subject, but that about which all the rest is said" (*Metaph.* VII 3, 1029a8f.). One could suppose that *"substance is said about matter"* (a23f.). Yet this is true only in a certain conventional sense when that which serves as matter in relation to the form we are concerned with (or a totality of forms) is in itself a definite entity (for instance, copper). In the absolute sense substance cannot be said about matter, since there is *nothing* to be said about: for matter in the ultimate sense (as the *first matter*) is not "something," but a pure possibility to become something. Yet for the entity of the first category, also the internal duality of being, the splitting expressed in the formula τὶ κατά τινος does not disappear without leaving a trace; it is still there as the distinction of matter and form; only matter as such shifts to the region of the anonymous and is now a pure possibility to be (or not to be) *this-here* entity.

We have treated of that which *is* primarily (τὸ πρώτως ὄν), and to which all other categories of being are referred — i.e., of substance. For it is in virtue of the *logos* of substance that the others are said to be — quantity and quality and the like; for all will be found to participate in the *logos* of substance... [62]

All the other meanings of being can be traced back to the entity as substance. So these meanings are united not by the unity of common genus, but by the unity of analogy, of which substance serves as the "centre." In the Middle Ages this analogy was termed *analogia entis*, and we shall go back to discuss it in the following chapters of this work.

Yet if the entities of other categories *are* (*exist*) only depending on the substance, it is not an easy task to decide whether such things as, for example, *walking, sitting, or being-healthy* really are entities or not (1028a20f.).

Aristotle's point of view in the *Metaphysics* differs essentially from the one expounded in the *Categories*.[63] In *Metaph.* VII 1 Aristotle is already careful not to speak of such an individual entity as this here individual white which is in this here individual man as in the subject (ἐν ὑποκειμένῳ, 3a7). The "white" is rather one of the ways in which the being of the subject is exhibited in the logos. It is absurd to consider as a separate entity some individual act of sitting or walking *in* this-here sitting or walking man. Of course, it is possible to speak of such an entity as "walking," but "walking" is (exists) because this is a walking man, because a man *is* and "walking" can be predicated of him: *the man is walking*. "Is" and "being" (*ens*) as designations of *white* and *walking* are neither synonyms nor homonyms, but *paronyms* (*Cat.* 1, 1a12) in relation to "is" and "being" said of, for example, Socrates. An accidental determination (P) can be called *a being* since something else, that about which as a subject (καθ' ὑποκείμενον) P is said (that of which P is predicated), is inevitably co-present in its being-presence.

[62] *Metaph.* IX 1, 1045b27–31. Ross' translation slightly modified.

[63] Cf. E. Tugendhat, TI KATA TINOΣ. *Eine Untersuchung zu Struktur und Ursprung aristotelischer Grundbegriffe* (Freiburg/München: Verlag Karl Alber, 1968), § 7, pp. 44–49.

We could say that "white" is *Socrates*, an autonomous and self-sufficient entity; and that the "white" is as far as it is manifested in Socrates' being according to a special logical perspective, prescribed by the question "what like?"[64]

We have already remarked that in Aristotle's ontology the participle "being" (*ens*) must be taken as a natural elliptical form of the unnatural (from the linguistic point of view) construction "S *being* P," i.e., the participle "being" is in some way derived from the copula "is." Yet if S names an entity of the first category (substance), then among all the variants of true predication there is a special one: "S is S." For the fact, says Aristotle, that a thing *is* identical to itself (αὐτὸ δὲ ὅτι αὐτό) is the single reason and the single cause in all cases, "because each thing *is* inseparable from itself, and its being one just meant this."[65] In my opinion, it is to the fundamental meaning of *such* "is," ἐστί, transformed into an infinitive noun τὸ εἶναι, that the Aristotelian formula τὸ τί ἦν εἶναι refers; in this case it must be expanded into: τὸ τί ἦν τῷ S (subject) S (noun of the predicate) εἶναι, and this is nothing but the designation of the essence = internal form (εἶδος ἐνόν) as the foundation of the thing's being.

The fundamental meaning of the participle "being" is also connected with this fundamental meaning of the verb "to be." The Schoolmen were distinguishing between "being" as the participle proper (expressing participation in an action) and "being" as a noun, between *ens sumptum participaliter* and *ens sumptum nominaliter.*[66] When we simply consider S as a *being (sumptum nominaliter)* without any further qualifications, we mean a participle derived from the copula of the assertion "S is S."

The entity defined by this "closed" expression of identity is nonetheless *open* to further ontological definitions: it can be shown (in the sense of the Greek ἀποφαίνεσθαι) in various ways in its being depending on the logical perspective chosen. This fact allows us to detect the topos of the "ontological difference" already predelineated within Aristotelian ontology. The *ontological difference* (i.e., the difference between *being* and *what-is* which will be the subject of close attention in chapter 7) understood in accordance with this onto-logical scheme is founded on the difference of the identity: "S is S" (which is closed in itself), on the one hand, and the openness to the further ontic determinations ("questions and answers," *witnessing* of the category) expressed by the formula: "S is –... "

To put it differently, the implicit prototype of the ontological difference in Aristotle's ontology, or rather its "projection" onto the purely logical plane, must be understood as the difference between the function of copula in the closed identity formula "*S is S*," and the openness of the "*S is* –" for varied predicates. Of course, the "is" as identity (*S is S*) preserves its copulative function and indicates being in its internal duality, discovered by Aristotle. This duality means generally, as we have seen,

[64] The language allows us to say: "this white is Socrates," and this is true, for Socrates indeed is [something] white. But here the grammatical relationship between the subject and the predicate inverts the proper ontological relationship; that is why it is impossible in this case to call the phrase an assertion proper (*An. Post.* 83a15–17).

[65] Cf. *Metaph.* VII 17, 1041a15–20: αὐτὸ δὲ ὅτι αὐτό, εἷς λόγος καὶ μία αἰτία ἐπὶ πάντων, [...] τοῦτο δ' ἦν τὸ ἑνὶ εἶναι.

[66] We shall come back to this distinction in the next chapter.

that *that-about-which* (the subject, τὸ ὑποκείμενον) of an assertion is "shown" in its being through and by means of something else (the predicate, τὸ κατηγορούμενον). The identity assertion "S is S" is only a limit case of logical manifestation (ἀπόφανσις). Nevertheless, if we had only one and unique way of manifestation, only one and unique mode of being we could never even pose the question about the distinction of entity and (its) being. Indeed, entity (*ens*) in general is, by definition, exactly *that-what-is*, and the "is" in its "proper philosophical sense" would always signify one and the same thing, for example, Parmenides' identity. It was Aristotle who first discovered the manifold meaning of being by discovering (among other things) different figures of predication, different functions of the "is" as copula.

In the next two sections I intend to demonstrate how this *logical* distinction is connected with the *phenomenological* distinction of the *noetic* and *dianoetic* energies of the intellect, and the *chrono-logical* distinction of the two characters of the "now" indicated by Aristotle: the "now" as one and the "now" as multiple. We shall consider this latter (*chronological*) distinction first.

4. THE "NOW" AS ONE AND AS A MULTIPLE

After the onto-logical detour undertaken in the previous section we return to our analysis of the "now" as *articulus temporis*. Aristotle says in Physics that in a sense the "now" is identical to itself and invariable, and always different in another sense.

The "now" as identical [67]	The "now" as different
τὸ νῦν ὅ ποτ' ἦν	τὸ εἶναι αὐτῷ (219b11)
ὅ ποτε ὂν ἐστι τὸ νῦν	ᾗ μὲν γὰρ ἐν ἄλλῳ καὶ ἄλλῳ (219b13–15)
ὅ ποτε ὄν	τῷ λόγῳ (219b18–20, 220a8)
ὅ ποτε ὂν νῦν ἐστι	τὸ εἶναι (219b26f.)

ᾗ μὲν γὰρ ἐν ἄλλῳ καὶ ἄλλῳ, ἕτερον — τοῦτο δ' ἦν αὐτῷ τὸ νῦν εἶναι — ὅ ποτε ὂν ἐστι τὸ νῦν, τὸ αὐτό. (219b13–15)

In so far as it (sc. the "now") is in something other again and again, it is different (which is just what its being "now" was supposed to mean) — yet that being which the "now" every time is [what it is], is the same. [68]

---

[67] Cf. P. F. Conen, *Die Zeittheorie des Aristoteles* (München: C. H. Beck'sche Verlagsbuchhandlung, 1964), p. 47.

[68] The awkward construction in the second part of the phrase imitates Aristotle's syntax and serves as a raw material for the further interpretation. It allows us at least to avoid too hasty decisions concerning the meaning of this important thesis of Aristotle. I shall quote the "translation" of P. H. Wicksteed and F. M. Cornford (though the relation of this text to that of Aristotle remains a great puzzle to me): "...for inasmuch as the point in the flux of time which it (sc. the 'now') marks is changing (and so to mark it is its essential function) the 'now' too differs perpetually, but inasmuch as at every moment it is performing its essential function of dividing the past and the future it retains its identity" (*Loeb Classical Library*, vol. 228, p. 389). Hardie and Gaye: "The 'now' in one sense is the same, in another it is not the same. In so far as it is in succession, it is different (which is just what its being 'now' was supposed to mean), but its substratum is the same." Heidegger "retells" a very similar passage (219a10f.) in the lectures on *The Basic Problems of Phenomenology* as follows: "...das Jetzt ist dasselbe hinsichtlich dessen, was es je schon war, –

In section 2 we partly discussed the thesis that to be (what?) "now" means to be always *in* something else, to be other in relation to the other (the "now" is perceived always as an element of a dyad, says Aristotle), and therefore the "now" is in some way "*otherness" itself* as such; we shall go back to this subject later.

Dealing with the self-identity (invariability) of the "now" we encounter, among other things, an extremely serious obstacle of linguistic character, the phrase ὅ ποτε ὄν, which seems to be artificial and invented by Aristotle *ad hoc* (though it can be found in other texts also) while speaking of time. To be more precise, we have to do here with an opposition or a distinction which is of major importance for the text discussed: ὅ ποτε ὄν ἐστι τὸ X / τὸ εἶναι τῷ X. The phrase ὅ ποτε ὄν was called "the pitfall of Aristotle's terminology" very long ago.[69] I shall try to elucidate the meaning of this formula, but only insofar as I need it for the present investigation, and refer the reader to more detailed works devoted to this topic.[70]

As we have established in the preceding paragraph, not all that is *called* entity "by itself" (τὸ ὂν καθ' αὑτὸ λεγόμενον), and not *per accidens*, is an ontologically self-sufficient entity, for it may depend in its being on something else. The "by itself" spoken of in *Metaph.* V 7 implies a necessity of additional specification. If we say about the "white as such" that it is a *being*, we imply (though we do not express in words) that *there is* some unexpressed, anonymous entity of the first category (a substance) and that it *is* white: there is a substance *being* white. It follows that we must look for the ontic foundation of "white" not in "the white itself," but in something else. We can ask: *What is that, being which, the white is [what it is]?*

Sometimes this question can be answered in a straightforward way by referring to (i.e., naming one way or another) the subject (an entity of the first category, a substance) as having such and such a property. Yet in the case when the answer depends every time (ποτέ) on the situation within which the question is being posed and changes as it changes, or in the case when the subject is unknown or its being is dubious or undefined, we can, as Aristotle does, refer to this quasi-subject by means of a phrase reproducing the form of the question. Here is one example. The question

---

d.h. in jedem Jetzt ist es Jetzt; seine *essentia*, sein Was, ist immer *dasselbe* –, und gleichwohl ist jedes Jetzt in jedem Jetzt seinem Wesen nach ein anderes, das Jetztsein ist je *Anderssein* (Wiesein – *existentia* – ἕτερον)" (GP 350). In my view this interpretation, which leans on the *distinctio essentiae et existeniae*, is lame in some respects, but it shows, at least, that Heidegger himself felt the connection of the distinction specified in this fragment of Aristotle's text with the problem of *ontological difference*.

[69] H. Schmitz, *op. cit.*, p. 168: "Das zeigt sich an [...] Verwendung einer auffallend künstlich klingender Formulierung, auf die schon Wilamowitz den Finger legte, als er schaudernd vom angeblichen 'Höllenzwang' der Terminologie des Aristoteles sprach."

[70] The first elaborated philological commentary dedicated to this subject is by A. Torstrik: "ὅ ποτε ὄν: Ein Beitrag zur Kenntnis des aristotelischen Sprachgebrauch." *Rheinisches Museum für Philologie*, **12**, 1857, 161–173. The history of interpretations of this expression (the so called 'OPO formula') and the critique of the solutions proposed by different authors can be found in the book by P. Conen *Die Zeittheorie des Aristoteles*, Kap.III. An elegant interpretation of the OPO construction is given in Chap. III ("Sur la formule Aristotélicienne ὅ ποτε ὄν.") of R. Brague's treatise *Du temps chez Platon et Aristote*. Cf. also H. Schmitz, *op. cit.*, pp. 167–171. I am grateful to Prof. A. P. Bos who has drawn my attention to the recently published book by Elena Cavagnaro *Aristotele e il tempo*, Leeuwarden, 1995, where the OPO expression is circumstantially discussed.

is: "From what is the thing made or generated?" The "empty," or tautological, answer is": From that which the thing is made or generated from"; or, in a more artificial form,— from the thing's "that-from-which" (τὸ ἐξ οὗ). It is thus that Aristotle designates matter; this is not a void name or a metaphor ("wood" — ὕλη), but a substantival construction reflecting the form of the question.

In a similar way the Aristotelian ὅ ποτε ὄν, in ontologically undetermined situations (e.g. when discussing time as entity or the "now" as entity), refers to the universal answer to the question "What is that, being which, X is [what it is]?" The tautological answer — "That, being which, X every time is [what it is]" — is empty in the sense that the thing looked for remains anonymous. Nonetheless the Aristotelian formula ὅ ποτε ὄν ἐστι τὸ X generates a kind of *semantic operator* pointing to the meaning searched for. Torstrik, who was the first to investigate seriously Aristotle's expression we are dealing with, proposed to give the name "substrate"[71] to the result of applying this operator. This rendering sacrifices the fine details of meaning, of course, but grasps its basic core.

What, then, remains invariable in various events taking place "now"? Aristotle's answer is: *that, being which, the* now *every time is [what it is]* (according to Torstrik, *the substrate of the* "now"). The fact alone that Aristotle uses here his semantic operator ὅ ποτε ὄν, shows that the thing searched for cannot be named and that its being is doubtful. Nonetheless we shall try to gain some understanding of it, following the advice of Aristotle himself.

In *Physics* IV 11 it is said that movement follows magnitude (ἀκολουθεῖ τῷ μεγέθει) and time follows movement. In the case of (spatial) locomotion the fact that *movement follows magnitude* means that *the moving body follows a point*; the magnitude is represented here by an (e.g., straight) line on which a starting point is marked and an orientation fixed, so that all points are ordered depending on their position on the line. In this sense the relation of *before-and-after-ness* is already defined in an obvious way (219a16): e.g., for two points A and B situated on the right-hand half-line, A precedes B if it is nearer (as regards the distance along the line) to the starting point. In this case point A is *the prior* (τὸ πρότερον) and point B is *the posterior* (τὸ ὕστερον). The body, moving along the straight line, moves away from the starting point; it *follows a point* in the sense that it goes past the points of the line one by one (if this expression in relation to a continuous point-manifold still may be tolerated), changes its location, and it is by this that we recognize spatial movement. And since the points are ordered, as mentioned, we can learn what is prior and what is posterior in the movement by the position of the moving point (217b17f.).

The next premise of Aristotle is: the "now" accompanies the moving body as time accompanies movement (219b22f.).

---

[71] This solution follows essentially the reading of Thomas Aquinas, who used the word "subjectum" to designate the meaning in question. To be sure, terms like "substratum," "subjectum" refer back to the Greek ὑποκείμενον. Such a "translation" from Greek into Greek can be found in the commentaries of Torstrik himself and also of W. D. Ross. The same "solution," as we have seen, is accepted by Hardie and Gaye.

Thomas Aquinas, expert in analogy, notes this correspondence as the following "proportions" in his commentary:

(1) the "now" : time = moving body : movement

(2) time : movement = the "now" : moving body[72]

That-which-moves as such *is* only in a secondary sense, for *that-which-moves* (like *that-which-walks* discussed in the example analysed above) is a determination of a substance. That being which that-which-moves is every time [what it is], is a constantly self-identical entity of the first category, "be it a point, or a stone, or something else of the kind, though it may be different as regards its *definition* (τῷ λόγῳ)" (b19f.). Of course, what is meant here is not the definition of stone as stone or of point as point (the answer to the question "what is this thing?"), but the definition related to the category of location (the answer to the question "where?"). Coriscus going on a journey from Lyceum to Agora, remains himself (Coriscus, a man) yet is, while moving, in a way, becoming other and other, since he attains various definitions relating to the category of location, i.e., he is now in one place and now in another.[73] The distinction between two different functions of the copula "is," mentioned above, is crucial for expressing this state of affairs. On the one hand, Coriscus remains one and the same; Coriscus *is* [now, and always, and essentially] Coriscus; and the copula signifies here just what his *being Coriscus was supposed to mean*, — τὸ τί ἦν Κορίσκῳ εἶναι, the *esse essentiae*. On the other hand, one can say: "Coriscus is now at the point A, now at the point B, etc." In the latter case the "is" expresses openness to different accidental definitions (συμβεβηκότα) belonging to the category of location.

Thus *that-which-moves* remains invariable in one sense (as a substance, as the entity, which one can call, imitating Aristotle's language, "that being which that-which-moves *is* as such") and changes in another sense (attains various definitions answering the question "where?"). In a similar way the "now" remains identical to itself in one sense and becomes other and other again in another.[74]

We distinguish between the prior (the *before*) and the posterior (the *after*) in *movement* because we recognize identity in diversity. Actually point A precedes point B on the way from Lyceum to Agora simply because it is nearer to Lyceum, whether there is something moving along this way or not. For what precedes and what succeeds are already contained in the "magnitude." If Coriscus is at point A, and Aristotle is at point B (thereby marking them), we recognize what is prior and what is posterior in *magnitude*, not in movement. It is only if now Coriscus is at point A, and *the same* Coriscus is now at point B, and the soul notes (counts) the two

[72] See *Commentarium in physicorum Aristotelis libros*, cap. XI, lect. XVII: "Sic igitur se habet nunc ad tempus, sicut mobile ad motum: ergo secundum commutatam proportionem, sicut tempus ad motum, ita et nunc ad mobile."

[73] "Sic igitur patet, quod id quod movetur est alterum secundum rationem, in eo quod est alibi et alibi, licet sit idem subjecto" (*ibid.*).

[74] For Thomas Aquinas this observation is a good reason to assert that the "now," as well as the moving body, is identical as to its subject – *idem subjecto*. But the question remains, *how should we understand* this enigmatic *subject* of the "now."

"now"s and the interval between them, that we recognize movement and what is prior and what is posterior in it. Thanks to the unity of what moves we know what precedes and what succeeds in movement, and the "now" is (exists) because the prior and the posterior are countable (can be articulated as the before-and-after) (219b25, 28).

Yet the "now" is said to "accompany" the moving body, and this probably means that it is analogous to it as regards its self-identity and as regards its otherness = its internal difference. Aristotle speaks about this as follows (219b26f.):

> In them (sc. in the "before and after") that, being which, the "now" is every time precisely "now," is the same. For this is the prior-and-posterior in movement. Yet to be the "now" means in each case quite a different thing (for the "now" is [exists] to the extent to which the prior-and-posterior are countable).[75]

At least the following can be gathered from this difficult text: it is the moving body that enables us to recognize and distinguish what precedes and what succeeds in movement. For the moving body is an entity of the first category, a *this-here*, it can be pointed at, and it is most comprehensible (b29). It is thanks to the unity, to the self-identity of the moving body in its changing *where*-determinations that the prior and the posterior in the movement can be recognized. Counting and articulating the prior-and-posterior, as we have seen, mean distinguishing in composition. The moving body as a body, by remaining the same, allows us to connect what has been distinguished, to posit the identity within the diversity or, which means the same, to posit the difference in the determinations of what is identical to itself.

The fact that the "now" accompanies the moving body allows us to transfer this reasoning from the topological to the temporal order. Let us do it by means of the first proportion of Thomas Aquinas — the "now" : time = the moving body : movement.

The articulation of what is prior and what is posterior in time, i.e., the positing of the priority-posteriority relation (the relation of *before-and-after-ness*), implies an aspect of identity and one of difference. What "was before" and what "will be after" must be understood as, correspondingly, the preceding "now" and the succeeding "now." This tacit "now"-origin or "now"-substrate of any event constitutes the underlying identity principle. On the other hand, the preceding "now" and the succeeding "now" must be linked to the corresponding states of things; i.e., they must mark different occurrences, must have *different* "event content": *now Coriscus is at point A, and now he is at point B.* Since Coriscus in Lyceum, in the sense mentioned above, is different from Coriscus in Agora, the "now"s corresponding to these events must be distinguished in some way also.

Yet the "now"-definiteness of an event (independent of concrete event content) is always identical to itself, according to Aristotle. It is to this "now," always identical to itself and not included in the time order, i.e., in the time sequence, that the formula [τοῦτο] ὅ ποτε ὂν [τὸ νῦν] ἐστι νῦν refers.

[75] 219b26f. The phrase is in fact once more an imitation of Aristotle's syntax. Another possible reading: "The 'now' is the 'before-and-after' *qua* countable."

Since the meaning of the "now" as the countable *prior-and-posterior*, as the fundamental operation of generating a dyad or as the simplest articulation, is acquiring its shape (by means of Thomas' proportions) on the basis of a *difference* of the corresponding events, we must, apparently, search for the (quasi-) "now" identical to itself (the "subject" or "substrate" of the "now") on the basis of the indifference or identity in relation to the changing content of events. And this is nothing but the "now" of the actual (energetic) presence, abstracted from the factually present, the "now" as the temporal determination of any *energeia*; for, *energeia*, according to Aristotle, is perfect and whole not in time (μὴ ἐν χρόνῳ), but in every moment, in every "now" *(Eth. Nic.* X 3). This is precisely the presence expressed by the present tense of the verb "to be" in the statement "Coriscus is (what he is), i.e., Coriscus is Coriscus." The operation of abstracting[76] referred to by the phrase ὅ ποτε ὄν[77] discards the predicate related to the category of location. Following this impulse of meaning, we abstract our attention from the varied logical perspectives, from varied ways of bearing witness to being (categories) and concentrate on the centre pointed to by the *analogia entis*, on the entity of the first category, on the identity: *Coriscus is Coriscus;* for the identity of the "now" imitates the self-identity of the moving substance as substance. And then we see that the "now" identical to itself is nothing but the "now" of the energetic presence of the substance, the "now" of its identity to itself, the present tense of the verb "to be" in the formula "S is S." And if, according to Aristotle's intention, the calculating intellect has been chosen as the starting point of interpretation, then the "now" as invariable must be interpreted as the *energeia* of the presence of the intellectually contemplated form, independently of the form present, i.e., as pure noetic *energeia*: *Someone is grasping mentally and has already grasped* (νοεῖ καὶ νενόηκεν ἅμα).

Thomas Aquinas says in his commentary that the changing event content of the "now" (temporal flow) is, as it were, the movement of the invariable substrate or subject of the "now" itself. The "now" in its movement-flow forms time as a point traces a line. It is easy, Thomas says, to gather from this analogy some understanding of eternity. The "now" as a substrate the "now"-event, always identical to itself, is the constant "now" (*nunc stans*) or the "now" of eternity (*nunc aeternitatis*), and it is to be distinguished from the "now" of time (*nunc temporis*). The latter is numbered (articulated) in movement, and the former is the number (or rather the unit-unity) of the self-identical moving thing.[78]

---

[76] I use this term as it is commonly used in the contemporary scientific language, without any allusion to Aristotel's term ἀφαίρεσις.

[77] This operation of abstracting is implied by the phrase under consideration. Aeschylus says (*Agamemnon* 160): Ζεύς, ὅστις ποτ᾽ ἐστίν... ("Zeus, whoever he might be...").

[78] *Commentarium in physicorum Aristotelis libros*, cap. IX, lect. XVIII, n. 5: "Sicut igitur nunc temporis intelligitur ut numerus mobilis, ita nunc aeternitatis intelligitur ut numerus, vel potius ut unitas rei semper eodem modo se habens."

5. TIME AND THE INTELLECT OF THE SOUL

In this paragraph we shall go deeper into the context, extremely important for Aristotelian chrono-logy, which must be called phenomenological (in the broadest possible sense), since Aristotle searches here for the answer to the question about the "nature of time" (217b32), analysing how movement and time *are given to the soul*, how we *perceive, distinguish, and recognize temporal determinations*.[79]

5.1. *Movement and Number As Objects of the Sensus Communis*[80]

According to Aristotle we possess a sense of movement, we perceive movement sensually, as it is expressed by the Greek verb αἰσθάνεσθαι. And movement, rest, number, figure (σχῆμα) and magnitude are objects of the so called *sensus communis* (cf. *De anima* II 6, 418a17f.).

We cannot discuss in detail here the complicated theory of *sensus communis* developed in the treatise *On the Soul*, yet some remarks seem necessary. In *Metaph.* XII 9 it is said that a sensation is always a sensation of something else, but incidentally it is also a sensation of itself.[81] In my opinion *sensus communis* is connected precisely with this structure of ἐν παρέργῳ (1074b36), with the structure of co-sensation. *Sensus communis* is, if you like, a particular sense which every time grasps itself ἐν παρέργῳ.[82] *Sensus communis* is the particular sense *plus ultra*. And this *plus ultra* is nothing but the difference of the sensible form without matter (object of any one of the five senses), on the one hand, and the *energeia* of the sensual presence ("being present") itself, on the other. Due to *sensus communis* this difference is given to the soul without mediation (that is why Aristotle speaks of a *sense*). The dif-

---

79 Of course, this "phenomenological" approach is not the only one strategy of Aristotle, and perhaps not even the most important. Yet in order to understand its significance in relation to the question of time, it suffices to look at the passage 219a22–219b2, where the main premises of the time definition are accumulated: we are aware of the lapse of time, while perceiving, grasping, thinking, recognizing, determining, etc. something in movement.

80 *Sensus communis* is the generally accepted Latin equivalent for Aristotle's αἴσθησις κοινή. I use this Latin term because the English expression "common sense" has (as a result of a long chain of semantic transformations) a rather specific meaning and the English translators of *De anima* try to avoid it. J. A. Smith coins for this occasion the expression "common sensibility," W. S. Hett (*Loeb Classical Library*, vol. 288) – "common faculty."

81 *Metaph.* XII 9, 1074b35f.: φαίνεται δ' ἀεὶ ἄλλου ἡ ἐπιστήμη καὶ ἡ αἴσθησις καὶ ἡ δόξα καὶ ἡ διάνοια, αὑτῆς δ' ἐν παρέργῳ.

82 In the Scholastics *sensus communis* was connected with *conscientia concomitans*. In Descartes' writings sensation as such was a species of *cogitatio* (cf. *Principia philosophiae* I, 9: "...etiam sentire idem est, quod cogitare"), and that is so because we always have a *concomitant awareness* of sensation ("quatenus eorum in nobis conscientia est"). Sensation possesses the formal character which marks the *cogitationes*: *cogito me cogitare*. In Kant's *Critique* Aristotle's *sensus communis* (via Scholastic mediation) becomes the "inner sense" (*der innere Sinn*). And according to Kant the inner sense has the same objects as the outer senses, only they are "reinterpreted" sensually as determinations of the mind, not of the external things. That is to say, these determinations are removed from the spatial order (which is the form of the external senses) and posited within the order of time. Indeed, time as we know is the form of the inner sense, according to Kant. Everything except time is external, *alien*, as Seneca says in one of his letters: "Omnia, Lucili, aliena sunt, tempus tantum nostrum est" (*Epistulae morales ad Lucilium*, I, 1).

ference in question has been already defined and discussed in section 1 of this chapter: in the numerically one event of perception (which is the energetic manifestedness of sensible form), *to be the form* and *to be the energeia* mean different things, as it is not the same: *to be perceived* and *to be perceiving*.

Sensus communis, being related in a specific way to all particular senses and their objects, is nevertheless *one* definite faculty of the soul. It is *sensus communis*, according to Aristotle, that allows the soul to distinguish between the proper objects of particular senses: e.g., to distinguish between sweet and white and hot. It would be impossible unless the soul could somehow combine heterogeneous sensible objects in one "part" (by means of one faculty). In his work *On Sense and Sensible Objects* (7, 449a14–17) Aristotle explains the unity of *sensus communis* by means of an analogy: the same thing can be white and sweet and have many other sensible qualities. Yet this does not mean that these qualities are actually separated one from another, but for one and the same thing its *being-white* and *being-sweet* is different. For *sensus communis* as the *one* faculty of the soul, to-be-the-sense-of-vision (vision given to itself) and to-be-the-sense-of-taste (taste given to itself) is not the same either.

5.2. *How the Understanding of Difference Is Possible*

Let us now go back to our main problem: *how does sensus communis perceive movement?* And this means: *how can it make distinctions?* The distinction *qua* distinction is not given to vision, it is colour that is given to vision, for instance the colour white. Yet *sensus communis* (= vision *plus ultra*) lets the soul escape the "blind" identity of the white to itself, to pose the distinction of the white as white and the white as seen, i.e., the difference between the white colour as a visible form and the visibility itself as its *energeia* (a form of sensible forms). We have discussed this difference in detail in the first section of this chapter. It is important that by means of *sensus communis* this difference becomes perceived, it is "given" to the senses aware of themselves (according to Aristotle, of course, aware of themselves alongside with other things). Thanks to the invariable *identity* of the *energeia* of sensation (sensation as *energeia* = form of sensible forms), the difference in sensible contents is given to the soul.

Yet this is just a preliminary answer. The fundamental difficulty is still awaiting solution: the difference between white and sweet is simultaneous (*synchronus*); the differences in an object changing its colour are successive (*diachronous*). If the "now" is the form of any *energeia* (actuality), it is not clear how a diachronous difference can be *actual* (be actually present *in* or *for* the *sensus communis*).

Aristotle says: "what is intelligible is contained in the sensible forms" (*De anima*, 432a4f.).[83] The noetic form of movement must be contained in its sensible form, and therefore also the aspect of the noetic form called *number* by Aristotle, i.e., time. The question, how the soul is capable of distinguishing and connecting what has been distinguished, must be posed and solved simultaneously regarding *sensus communis* and the intellect, since the difficulty is of the same nature in both cases.

[83] ἐν εἴδεσι τῶν αἰσθητῶν τὰ νοητά ἐστι. Smith: "The objects of thought are in the sensible forms."

It is necessary, Aristotle remarks, for the distinguishing faculty to be one, and for the contents to be distinguished, to be manifested in this one faculty: for if I perceive the white, and you perceive the black (or the sweet), it is impossible to recognise the distinction between the two since there is nobody who could draw it (*De anima* III 2, 426b16–17). And it is said of this one distinguishing faculty that it both grasps mentally (intellectually) and senses (καὶ νοεῖ καὶ αἰσθάνεται) the different determinations in their difference (b22).

How are sensation and intellectual comprehension (*noesis*) of what is different in its difference possible? The difficulty lies in this:[84] what does it mean, generally speaking, to grasp a thing intellectually? In Aristotle's *noetics* we find quite a definite answer to the question: in the event of grasping intellectually (*noesis*) the intellect becomes the intellectually comprehensible form of the thing without matter. The *energeia* (perfect activity) of the intellect and the *energeia* (actuality, perfect presence) of the intellectually comprehensible form are here the same *energeia*. Yet how can the intellect, distinguishing concrete definitions A and B, *become at once A* and *B*? Aristotle words this difficulty as follows:

But, it may be objected, it is impossible that what is self-identical (sc. the self-identical faculty of the *sensus communis* or the *intellect of the soul*) should be moved at one and the same time (ἅμα) with contrary movements in so far as it is undivided and in an undivided moment of time.[85]

Yet the solution could perhaps be as follows: as a thing, being one, admits of various definitions belonging to the category of quality (it may be, say, white and sweet), so the intellect also, remaining one and self-identical, may be divisible "as to its being" (427a5). This means that, at the same time, being-the-form-of-the-white-colour and being-the-form-of-the-sweet-taste may be proper to the intellect.

Still, this explanation cannot be accepted in the case that interests us most. Let us imagine that something white is getting black. We are dealing here with movement (to be more precise, with one kind of movement — the change of quality, *alteration*); to grasp movement, as we have said, the intellect must grasp the difference in the determinations of the same thing. What does it mean in this case *to grasp the differences*? What differences can be meant here? Clearly — the difference of light and dark. The intellect must at the same time (ἅμα) contemplate, in the object, determinations which are not just different, but contrary. This means that the intellect must at once become the intellectual forms of both *contrary (or even contradictory)* determinations. And the solution proposed above is no good here, since contrary determinations (even *accidentia*) cannot be proper *at the same time* to any one self-identical substance. The same thing cannot be at the same time white and red, still less can it be white and black.

Although the same indivisible thing can be both contraries potentially, it is not so in the sense of its *being* [these contraries] (one and the same thing cannot *be* both contraries simultaneously in actuality — *A. Ch.*). [...] And as something one and indivisible cannot be at the same time white and black, so the

[84] We concentrate upon the noetic side of the problem, because in connection with *sensus communis* the difficulties have precisely the same nature. Cf. 427a9 where the intellect and *sensus communis* are once more mentioned together as equally relevant objects of the discussion.

[85] 426b29–31.

distinguishing faculty, remaining one, cannot perceive at the same time the form of the white colour and the form of the black colour; while to think and to sense means to become such forms. (427a6 – 9)

5.3. *The "Now" and the "Point"*

Aristotle tries to find another solution. The situation is the same, he says, as when we consider what the geometricians call a point. A point of a line, being by itself utterly one and indivisible, divides the line into two parts and can therefore be regarded at the same time (ἅμα) as the end of the left half and as the beginning of the right half.

To the extent that the distinguishing faculty is indivisible, it is one and distinguishes simultaneously, but being divisible [in some other sense] it cannot be considered something one. For it (as it happens in our example) uses one point as two points, as it were. Since it uses the same limit twice, it recognizes a pair of distinct [determinations of a thing] by becoming divided in itself, as it were. Yet since, on the other hand, the distinction takes place by means of what is one, it takes place at once (ἅμα).[86]

The ability to distinguish is at the same time a unit and a dyad, because it is like the limit; in its unity, it is simultaneously and at once, a dyad.

In connection with a similar difficulty the point concept is mentioned also in another passage of the treatise *On the Soul*, the sixth chapter of the third book. While the second chapter is devoted mainly to *sensus communis* (the distinguishing intellect is mentioned there only incidentally, insofar as the analysis of distinguishing in thinking and in sensing encounters obstacles, which have the same structure, according to Aristotle's opinion), the sixth chapter treats thinking (*noesis*) as the main subject. Its central topic is the indivisible (ἀδιαίρετον) grasped by the intellect. Aristotle distinguishes two meanings of "indivisible": 1) that which *cannot be* divided (e.g., a point), and 2) that which *is not* divided *actually* (e.g., a part of a straight line as a whole),— and he says that all that is indivisible in the first (strong) sense can be manifested to the intellect only as a privation (στέρησις). A point is defined in a negative way as that which has no parts (Euclid, *El.* 1, def. 1). It is in a similar way that we recognize blackness and evil: the first as a privation of colour, the second as a privation of good. Such objects of thinking, Aristotle says, are grasped by the intellect by means of their opposites. For instance, to think of a point (as a privation, a "negation" of extension) the intellect must think of extension. The intellect must think of X and of non-X. The paradox is not that, thinking of X, the intellect must be, in some way, non-X in possibility (for thinking of X the intellect remains all thinkable things in possibility); the paradox is that the intellect must be actually and at the same time X and non-X, while *to think* (to grasp intellectually in the sense of the Greek verb νοεῖν) means to *become the intelligible form without matter*, and the

86 *De anima* III 2, 427a10–14, my translation. The passage is extremely difficult as the ancient commentators of *De anima* already considered it to be. The translation by J. A. Smith (with corrections by W. D. Ross) reads as follows: "Just as what is called a point is, as being at once one and two, properly said to be divisible and indivisible, so here, that which discriminates is *qua* undivided one, and active in a single moment of time, while *qua* divisible it twice over uses the same dot at one and the same time. So far then as it takes the limit as two, it with what in a sense is separated; while so far as it takes it as one, it does so with what is one and occupies in its activity a single moment of time."

act of thinking is precisely the energetic being of such a form. How is this possible? How is it possible for intellect *to be* an *actual contradiction*?

5.4. *Two Energeiai of Thought*

The answer comes to a definite outline when we compare these passages of the treatise *On the Soul* to the discussion of the double-faced "now" in the fourth book of the *Physics*. Actually, the point dividing the line and marking at the same time a beginning and an end is a universal model of the "now" (which is the end of the past and the beginning of the future) which appears again and again in the *Physics*, although Aristotle warns against an unrestrained application of this analogy.

The "now," as we have seen, is a form (or an *energeia* of presence) always equal to itself. But on the other hand, the "now" must be understood as a universal "form" of distinction, a boundary, a limit, an *articulus*, the simplest act of counting which generates a dyad. Here the "now" appears as the *generation* itself of the difference, as the *differentia differens*.[87] It is in this "now" that the activity of articulation is performed. The fact that this operation (analogous to the addition of a unit), which posits a dyad and, as a result, the multitude, remains one and indivisible in itself, should not seem paradoxical to us any more. This operation is a perfect activity ("I am distinguishing and have already distinguished"), an *energeia*, to which corresponds the actuality (*energeia*) of the limit, the boundary. Intellectual activity can thus be considered in two ways: on the one hand, it is the *energeia* of presence of an intellectual form (we have called such an *energeia noetic*), and on the other hand it is the *energeia* (perfect activity) of "counting," articulating, setting a limit. The intellect then is not only the form of forms, but also the differentiating difference (*differentia differens*, differential). We shall call this second perfect activity or *energeia* of the intellect *dianoetic*.

Dianoetic *energeia* is expressed in the logos: it plays the role of the distinguishing-connecting "is" in the statement "S is P."

We have pointed above to the context in Aristotle's works within which, as I believe, it is possible to detect the distinction drawn between being and entity. Aristotle does not directly discuss the ontological difference, and yet he refers to it in an indirect way, in particular by means of the opposition ὅ ποτε ὄν ἐστι τὸ X / τὸ εἶναι τῷ X. Since the entity manifests its being in an apophantic (ex-pressive, in-dicative) logos, since there are various modes of such manifestation, not belonging to the same common kind, we can say (and have said already) that from the point of view of logical optics and "schemes of category" (different ways of logical manifestation of a being in its being), being (*esse*), as opposed to entity (*ens*) and *distinguished* from it, is represented as the "open clause" "*X is* –," not as the "closed" identity "X is X."

In the horizon of *energeiai* of the intellect this formal distinction has quite a definite counterpart, a *foundation* as it were, because according to Aristotle an

[87] In chap. 7, sect. 2 we shall discuss the concept of the *differentia differens* within the context of Heidegger's philosophy of time.

apophantic statement is a *connection* of *noemata*, and it is the intellect that connects them.

The formula "X is –" expresses a certain situation of the activity of the intellect, certain facts of the matter, and these are as follows: X *is now* as one and the same thing; the intellect grasps the noetic form of X, *becomes* this form, becomes the energetic presence of the form of X. Yet at the same time the intellect posits the connection of X with something else, as yet defined only negatively, setting a boundary between X and non-X. The actuality of this boundary cannot be distinguished from the perfect activity of the intellect that we have called *dianoetic energeia*.

X *is* now as an openness to the other, i.e., to the various ways of logical *apophansis*: "X is Y, Z, T..." This openness posits X in its being, and this being is sounded and designated by the copula.

Thus it is possible to grasp the being of X in two ways:

1. X *is* now *(and always, and essentially)* X. The being, expressed by this "is," refers to the *essence* of X, to its internal form (εἶδος ἐνόν),[88] to that which it means (and always has meant) to be this X. The Scholastics used to speak in this connection of *esse essentiae*.
2. X *is* now –...; X *is* now as openness to the various ways of logical manifestation: *X is Y, Z, T...* This openness posits X in its being and this being is articulated and designated by the copula. If X is the name of a thing composed of matter and form (a *compositum*), the "is" here implies also openness to the accidental (the "matter-of-fact," the "here-and-now") determinations of the thing, and thus points to its *existence*, to *esse existentiae*. (We shall come back to the medieval distinction of essence and existence in the next chapter.)

The "now" must be understood differently here. In the first case it is the permanent presence of the persistent substantial form, which is always self-identical. In the second case it is open to the factual event-content. These "now"s are the two temporal forms of the noetic and of the dianoetic *energeiai* respectively. "Phenomenologically" speaking (i.e., from within the context of the active intellect of the soul), the distinction of these two *energieiai* is the distinction of the two aspects of the "now."

Within the context of Aristotle's ontology this distinction serves as the condition for distinguishing (though never accomplished explicitly) between a being (ὄν) and (its) being (τὸ εἶναι) to the extent that the latter is disclosed in thought and expressed in words, i.e., in the *logos*. And this same distinction is the condition for the possibility of varied event content of the "now" or, to use Thomas Aquinas' phrase, for the possibility of distinguishing between the *nunc stans* and the *nunc fluens*, the initial *fluere* of the *nunc fluens*. This distinction generates the totality of time, perceived by the *sensus communis* and "counted" ("articulated") by the soul, as a moving point generates a line or as a pen's point generates a text.

[88] Cf. *Metaph.* VII 11, 1037a29f.: ἡ γὰρ οὐσία ἐστὶ τὸ εἶδος τὸ ἐνόν, ἐξ οὗ καὶ τῆς ὕλης ἡ σύνολος λέγεται οὐσία.

CHAPTER THREE

DISTINCTIO ET COMPOSITIO ESSENTIAE ET EXISTENTIAE AS INTERPRETED BY MARTIN HEIDEGGER

1. MEDIEVAL DISCUSSION OF THE ONTOLOGICAL DIFFERENCE

Diversum est esse et id quod est.

Boethius, *Quomodo substantiae in eo quod sint bonae sint...*

1.1. *Medieval Ontology and "The Basic Problems of Phenomenology"*

During the years subsequent to the publication of *Being and Time* Heidegger's project to restate explicitly the question of being, the question that "has been forgotten" and lost some of its power but has never ceased to be at work in the history of philosophy, becomes more and more clearly outlined. *How* must philosophy question being? *What* does it *mean* to "ask in a primordial manner," to ask *de profundis*? In the second half of the 20's Heidegger already has an answer: "Only as phenomenology, is ontology possible"(SZ 35). The reverse is true also: "with regard to its subject-matter, phenomenology is the science of the being of beings – ontology" (SZ 37). Being of beings, being of the entity insofar as it *differs* from the entity, this enigmatic difference of being (*esse*) and entity (*ens*) – such is, according to Heidegger, the main problem of phenomenology. In a series of lectures entitled precisely *The Basic Problems of Phenomenology* (*Die Grundprobleme der Phänomenologie*), delivered at Marburg University in summer 1927, while pursuing his usual hermeneutical strategy, Heidegger tries to introduce his audience to phenomenological ontology (or ontology-as-phenomenology) by explaining phenomenologically several important statements, or "theses" concerning being (*Thesen über das Sein*), "which have been advocated in the course of Western philosophy since Antiquity" (GP 20). One of these statements requiring a phenomenological elucidation is "the thesis of medieval ontology (Scholasticism) which goes back to Aristotle [and asserts that] to the constitution of being of a being belong (a) whatness (*Wassein, essentia*), and (b) being-at-hand or extantness (*Vorhandensein, existentia*) (*ibid.*).

Francisco Suarez's XXXI *Metaphysical Disputation*,[1] included in his great treatise *Disputationes metaphysicae* and entitled "De essentia entis finiti ut tale est, et de illius esse, eorumque distinctione" ("On the essence of the finite entity as such and on its being, and on the distinction between them"), is dedicated precisely to this topic. It is therefore not surprising that Heidegger's attention focuses on this text, in which Suarez not only expounds his own views concerning the distinction and composition of *essence* and *existence* in created entities, but also analyses carefully the problem's history, the tradition of its solution and the arguments of his precursors.

1.2. *Essence, Existence and Ontological Difference*

Why does the "thesis of medieval metaphysics" mentioned above seem so important to Heidegger? Why is it designed to serve as an introduction to phenomenological ontology — i. e., to ontology as such?

In the Marburg lecture course of 1927 mentioned above the notion of *ontological difference* is discussed for the first time. The term itself refers to a concept basic to Heidegger's ontology — to the difference between entity and being. Language allows us at least to articulate this difference by means of different suffixes: *das Seiende — das Sein*,[2] and so to initiate our understanding of it — which, however, may later reveal itself as an illusion. For "we do not always [actually] possess," as Leibniz wrote, "the idea of the subject we are aware of thinking about."

Strictly speaking, the scholastic distinction between essence and existence does not coincide with the ontological difference, as Heidegger defines it. For even before any detailed analysis of these concepts one can say that existence, *existentia*, signifies *presence*,[3] or as we can formulate it now,— the *energeia* of being-present, the actuality of an entity, i. e., it signifies "*the way* in which something actual or existent (*Existierendes*) *is*" (GP 109); whereas essence, *essentia*, refers to the *what*-character belonging to the ontological constitution of an entity, to its *being-something*, "somethingness" or "thingness" (*realitas, Realität, Sachheit*), understood as "whatness" (*quidditas, Washeit*). Thus, Heidegger says, the distinction between essence (reality) and existence belongs on the side of one member distinguished (or to be distinguished) in the ontological difference: "neither *realitas* nor *existentia* is an entity; rather it is precisely two of them that make up the structure of being. The distinction between *realitas* and *existentia articulates* being more particularly in its essential constitution" (GP 109). Therefore the "distinction and composition" of *essentia* and *existentia*, as the basic articulation of being, namely, each single entity's being, are rather one of the *moments* or ways of positing the ontological difference.

---

[1] F. Suarez, *Opera omnia*, ed. C. Berton, T. XXVI (Paris, 1861).

[2] It is the German language, of course, that allows Heidegger to express immediately the difference in question. We already referred to the difficulties one encounters while discussing the same topic in English.

[3] *Presence* in the sense of *extantness, being-ready-to-hand*. In Kant's *Critique of Pure Reason* the terms *Existenz* and *Dasein* are complete synonyms.

1.3. *Semantic Distinctions*

Our starting point must be the assumption that the ontological difference — whether it proves to be intelligible or unintelligible, actual or imaginary — is already implied and expressed in language, and for the time being the way it is expressed has to become the matter of our analysis.

First philosophy takes as its subject τὸ ὄν ᾗ ὄν,[4] i. e., entity insofar it is entity (*ens, in quantum ens*), entity as such. But what can we say about an entity, a being as such? We say, "a being (an entity) *is*." The last sentence would be little more than a tautology if the meaning of its subject and predicate were equally clear to us. Nonetheless it is its being that defines the entity as such. Suarez writes:

> the entity, insofar as it is entity, receives its name from being (*esse*) and because of being; in other words, it has the meaning of entity (*rationem entis*) because of its destination to be...[5]

Thus strictly speaking *ens* = *aliquid habens realem actum essendi*:[6] the entity is something which "has" — that is, *accomplishes or performs*,— *the act of being*. Consequently, the main problem of first philosophy — τί τὸ ὄν; "what is entity?" — seems to be identical with the question: τί τὸ εἶναι; "what is being?" Yet if the word ὄν, *ens, entity* were capturing *only this meaning* of the participle, the meaning of participating in an act (action, actuality), i.e. of participating in being, the distinction we are interested in would be just a ghost, an unimportant difference between two ways of expressing the same content.

Those scholars who discussed the problem of *the distinction and composition of essence and existence in created entities*[7] in the Middle Ages faced a similar difficulty. It was discovered that the possibility of stating the distinction or difference between *ens* and *esse* or (in another form) between *essentia* and *existentia* as a proper subject of inquiry (that is, according to Heidegger, the possibility of the *basic articulation of being*), is determined by the remarkable ambiguity of the little word ὄν, *ens.*[8] While discussing, with the thoroughness peculiar to him, the concept of entity, *conceptus objectivus entis*, Suarez says that the meaning of the word (*vocis*) *ens* is twofold.

1. As a participle of the verb *sum*, which means, taken absolutely (not as a copula), nothing else but the act of being, *actus essendi*, *ens* is to be understood as *that which is*. It follows that in this sense *entity* refers to being as an *act* only, no matter

---

[4] Aristotle, *Metaph.* IV 1, 1003a21. The traditional Latin translation of this formula (which Suarez also refers to) is: *ens in quantum ens*.

[5] F. Suarez, *Disputationes metaphysicae*, disp. XXXI, sect. I, 1.

[6] *Ibid.*, disp. II, sect. IV, 4.

[7] Distinctio et compositio essentiae et existentiae in ente creato.

[8] This vocable was uttered first, as we know, by the goddess of Parmenides' metaphysical revelation. The Stranger in Plato's *Sophist* says to Theaetetus: "it seems to me that Parmenides and all those who ever undertook an attempt to take the beings (τὰ ὄντα) on trial... have talked to us rather carelessly. [...] Every one of them seems to tell us a story as if we were children. [...] Then since we are in perplexity, do you (sc. Parmenides and the other great thinkers of the past) tell us plainly what you wish to designate when you say 'a being' (τὸ ὄν). For it is clear that you have known this all along, whereas we formerly thought we knew, but are now perplexed" (242c–244a). These last words were chosen by Heidegger as epigraph to *Being and Time*.

what the entity is (*extra rerum quidditatem*), irrespective of its definition as a particular existing thing, having its definite what-character. A participle, formally connected with the grammatical category of tense, always *implies temporality* in its meaning (*tempus consignificat*). Therefore *ens* means a "being-now" or *actual exercising or accomplishing of being* (*actuale exercitium essendi seu existendi*). Due to this fact *ens*, taken as a participle, refers to existence of an entity irrespective of its "whatness."

2. Yet *ens* can be used in another — in the *secondary*, as Suarez puts it,— sense, being taken as a *noun* [substantive], *vi nominis*. Then the term *ens* "has the power" to name a thing in its semantic delineation, in its whatness (*quidditas*), thought separately from the act of existence.

It seems that this semantic analysis already constitutes a certain progress towards distinguishing between essence and existence. Yet we meet with serious difficulty here. There is a difference, Suarez says, in the usage of the words *ens* and *res*, even if the former is taken as a noun. *Ens* does not signify an arbitrary, more or less vaguely represented, semantic construction; it refers *only* to something which has a "real essence" (*essentiam realem*). In other words, an essence "apt to exist in reality" (*essentiam veram et aptam ad realiter existendum*), an essence which may happen to exist, a "true" essence, not a "fictional or chimerical one" (*non fictam, nec chymericam*).[9] So, even if the word *ens* is taken as a noun (*sumptum nominaliter*), its connection with the "verbal character" of being is not completely lost: *ens* always implies an ability, a readiness, an aptitude, a fitness to *be* actually (this *being* is expressed by the verb in present tense). The word "essence "(*essentia*) itself points, in a way, to the verb "to be" *(esse)*. That is why, as Thomas Aquinas remarks, the concept of essence does not completely coincide with the concept of "whatness" (*quidditas*). "Whatness" is the content expressed in the definition, and the essence is so called because a thing has its being through it and in it (*per eam et in ea ens habet esse*).[10] Yet if the definition is not an arbitrary invention of the mind immersed in unfounded semantic constructions, but actually shows and brings to light the essence of the entity, as it is "in the nature of things" (*in rerum natura*), then the border line between *ens* taken as a noun and *ens* taken as a participle gets blurred. The noun becomes not just a noun, but also a noun referring to and implicitly containing in itself the verb "is," the aptitude to be.

Thus the "being-ness" (*entitas*) of a being (*ens*), if the latter is taken as a noun (*sumptum nominaliter*), means that the entity is comprehensible in a strong sense of the word: it is intelligible in perfect clarity, non-deceptive, free from illusions of language and imagination. Here the concept of *entitas* ("being-ness") and that of *essentia realis, realitas*, reality are getting as close as possible.

Leibniz carried this point of view to its extreme:[11] he identified the entity (*ens*) and the possible (*possibile*). The entity is characterized as a conjunction (*conjunctio*)

---

[9] Disp. II, sect. IV, 5.

[10] Thomas Aquinas, *De ente et essentia*, ed. 3 (Torino: Marietti, 1957), cap. I, sect. 2.

[11] The fact that Leibniz (as well as Descartes) was strongly influenced by Suarez' *Disputationes* is well known and amply discussed by the historians of philosophy.

of simple and mutually independent qualities or distinctive marks (*notae*). One can speak of a *possibile*, when the combination of marks or constituents (*requisita*) within the same medium (i.e. their predication relating to the same subject) does not lead to a logical contradiction. In this way the law of contradiction becomes the umpire establishing the distinction between *essentia realis* and *essentia ficta seu chymerica*.

Hegel seized on this incipient movement and developed it further in the *Phänomenologie der Geist.*[12] Here he still understands "thing-ness" or "pure essence" as a "medium," in which, as in a kind of a simple unity, various simple qualities "penetrate each other without, however, coming into touch with one another or influencing each other in this mutual penetration." A thing is *das Auch*, a mere conjunction (*conjunctio*) of mutually indifferent (and therefore obedient to the law of contradiction) qualities. Such is the replica of the scholastic concept of *essentia realis* in Hegel's phenomenology.

1.4. *Essence, Whatness, Nature*

The *essentia realis* allows the form of a thing to be present in the "perceiving intellect" (*intellectus percipiens*). This permission-to-be-present is, to the Scholastics, an expression of the thing's *nature*.[13] According to the definition given by Thomas Aquinas, a thing reveals its nature in the way it influences other things: the nature structures actions or "operations" (*operationes*) of the entity and marks them as proper precisely to this thing (*propria*). One of these "proper operations" is, as was mentioned above, the permission to be present to the mind. A thing is revealed to the mind as a definite "what," *quid*, as "whatness," *quidditas*. Whatness is expressed in the definition and presupposes grasping of the proximate genus and the relevant specific difference. That is why the essence means something which is proper to any nature and through which various entities are distributed between various genera and species.[14] So, the terms "form," "whatness" and "nature" correspond to different closely related and yet non-identical meanings of "essence." We see that even this preliminary analysis includes in its scope a certain system of ontological ideas. Essence (*that in which and through which the entity possesses being*, the core of the meaning of being) is thus very closely connected with the ability to be present to the mind as something definite and distinct. Being-ness implies as its counterpart a possibility of adequate, that is, thoroughly clear knowledge where confusion is no longer possible. Through its essence, the entity possesses a *definite* being (*esse quid*), a "being-something"; after Aristotle, essence is referred to as *quod quid erat esse*, τὸ τί ἦν εἶναι.

[12] See section II entitled "Die Wahrnehmung, das Ding und die Täuschung."

[13] This interpretation of the concept of *natura* can be traced back to Boethius. In the treatise entitled *Liber de persona et duabus naturis* Boethius says that in whatever sense one may understand a being (*ens*), it is due to the nature of this being that the intellect is capable to grasp it. Cf. Migne, *Patrologiae cursus completus*, Ser. latina, t. 64, col. 1541 B.

[14] Thomas Aquinas, *De ente et essentia* I 2.

1.5. *Essence is Different from Being (Existence)*

One of the "ontological axioms" formulated by Boethius in his work *Quomodo substantiae in eo quod sint bonae sint cum non sint substantialia bona*, reads as follows: "being (*esse*) and what-is (*id quod est*) are different." It impossible to say: "Being is." On the contrary, the entity, once a form of being (*essendi forma*) has been received, is and remains in being (*consistit*).[15] Further we read: "for all complex entities being is one thing, and the entity itself is another thing."[16] This last axiom unequivocally implies the "ontological difference" for the complex entity. The first axiom can be read in various ways. Thomas Aquinas cites this passage of Boethius to confirm his own thesis stating that it is necessary to posit a "real" distinction (*distinctio realis*) between essence and existence.

Although essence is the ontological source of the entity and of its intelligibility, nonetheless, Aquinas says, the cognition of the essence or of the whatness in itself (*suo facto*) does not imply that we are simultaneously in a position to assert the thing's being as well.[17] That is why being (*esse*) is not a component, an aspect or a formal characteristic of the essence; to be brief, it is not something inseparable from the essence. For it is impossible to contemplate any essence without its components, though "we can understand what a man or a phoenix is, remaining at the same time ignorant concerning their being in the nature of things."[18] Yet everything which is not comprehended within the notion (*non est de intellectu*) of essence or whatness, is accidental and forms a composition with the essence. This mental experiment allows us to say, according to the author, that being, *esse, existentia*, is something different from essence.

Leibniz could have argued that the understanding spoken of could be illusory. Understanding is not in itself (*suo facto*) identical to an adequate knowledge.[19] We can believe that we know what the word "phoenix" signifies, but it is quite possible that in fact we have only an illusion of knowledge. Knowledge must be conceived as *cognitio adaequata*; to know in this way is to have carried out a final analysis of a concept back to its irresolvable constituents and to have made sure there are no contradictions, in other words to distinguish the nominal definition (capturing some of the features distinguishing a phoenix as a phoenix) from the real definition, from which it follows that the thing defined is possible.[20] Still the question whether "such

---

[15] Cf. Boethius, *Quomodo substantiae in eo quod sint bonae sint cum non sint substantialia bona*. Migne, PL 64, 1311B.

[16] "Omni composito aliud est esse, aliud ipsum est." (*ibid.*)

[17] Thomas Aquinas, *op. cit.*, V 3: "Omnis autem essentia vel quidditas intelligi potest sine hoc quod aliquid intelligatur de esse suo facto..."

[18] *De ente et essentia* III 3: "...possum enim intelligere quid est homo vel phoenix, et tamen ignorare an esse habeant in rerum natura."

[19] G. W. Leibniz, "Meditationes de cognitione, veritate et ideis" in *Die philosophische Schriften von G. W. Leibniz*, ed. G. L.Gerhardt, Bd. IV (Hildesheim: Georg Olms Verlag, 1962), pp. 422–426.

[20] "Whenever our knowledge is adequate, we have *a priori* knowledge of a possibility, for if we have carried out the analysis to the end and no contradiction has appeared, the concept is obviously possible." (*ibid.*) Christian Wolff writes in his manual *Philosophia prima sive ontologia* (Frankfurt, 1730): "Quod

an analysis is within human reach, i. e., whether people are able to reduce their thoughts to first possibilities and to the simplest and primitive qualities, or, which is the same thing, to the absolute attributes of God, and consequently to the first causes and the ultimate bases of all things,"[21] remains unsolved and hardly solvable. For Leibniz the distinction made by Thomas Aquinas between essence and existence is simply a reference to the finite nature of the human mind. On the other hand, Leibniz writes elsewhere: "And as an entity (*ens*) is expressed by means of a distinct definition, so what exists (*existens*) is expressed by means of a distinct perception." Distinct (for Leibniz this means internally harmonious, proceeding in accordance with laws, coherent) perception can to a certain extent free us from the burden of carrying out the final analysis of the concept. But clear and distinct perception implies the presence of its object "in flesh and blood" here and now, whereas Thomas' mental experiment deals with the possibility of removing temporal connotations of the participle *ens*, i. e., the meaning of participation in the *actus essendi* taking place at present; it deals, in other words, with the possibility of using *ens* as a noun.

Suarez remarks quite correctly that the distinction of essence and existence, as posited and articulated by Thomas Aquinas, is expressed much more precisely by the classical Aristotelian terms of *potential* and *actual being*. A new problem is encountered when we ask *whether essence and existence are different in the actual entity*; to be more precise, when we ask whether it is possible to distinguish between *existentia*, and *actualis essentia existens*. Suarez believes that the latter distinction can be drawn not in the thing itself but belongs in a sense to the distinguishing (analysing) intellect: the two members distinguished correspond simply to different ways, different strategies of thinking about the entity. The distinction in this case is nothing but a *distinctio sola rationis*, a difference merely in reason, in concept or interpretation, though it has a meaningful motive outside of pure intellect. The existing thing is one, but it is grasped by the mind and compared with itself as if there were two of them.[22] Existence adds nothing to the actual "what."[23] Yet it is not arbitrarily that the mind proceeds in this way. As we already said, there is a certain reason for this within the nature of created things as such. We shall describe this reason later.

---

impossibile est, existere nequit. Impossibile est id, quod contradictionem involvit." (§132) "Quod possibile est, illud existere potest." (§133) "Ens dicitur quod existere potest, consequenter cui existentia non repugnat." (§134) "Quae in ente sibi mutuo non repugnant, nec tamen per se invicem determinantur, essentialia appellantur atque essentiam entis constituunt." (§143)

[21] G. W. Leibniz, *op. cit.*

[22] "quia non sunt duae res sed una tantum, quae per intellectum concipitur, et comparatur ac si essent duae." Suarez, *op. cit.*, disp. XXXI, sect. III, 1.

[23] This thesis as to its form is very similar to Kant's famous sentence (in the section "On the Impossibility of the Ontological Proof of God's Existence" of the *Critique of Pure Reason*) stating that existence is not a *real predicate*. Yet Kant makes one essential addition to that: existence is the *absolute position* (within the horizon of experience). Later on we shall analyse this statement of Kant in more detail.

1.6. *The Simple and the Complex*

The finite or created entity (*ens finitum = creatum*), as opposed to the First (i.e. Divine) entity,[24] which is characterized as being infinite (*ens infinitum*) and absolutely simple, is complex *in more than one sense*. Manifold distinctions can be drawn in it, e.g., the distinction *of matter and form*. It is true that this distinction is not inherent in all created entities, for in pure forms or intelligences, which subsist in themselves (*intelligentiae subsistentes*),[25] it does not make sense to search for a composition of matter and form. It follows that a *universally complex character*, or (which is the same) a *universal principle of internal difference* must be asserted, which declares all created, or finite, entities to be complex. For Thomas Aquinas and his school, this role is played by the principle of distinction and composition of essence and existence. The distinction between essence and existence, according to the Thomists, must be accepted as *real*. Suarez formulates this doctrine (or *sentence*) as follows: "existence is a thing (*res*), distinct from the essence of the creature in an absolutely real way."[26] We have seen already that for the author of the *Metaphysical Disputations* himself the distinction between essence and existence is, on the contrary, a distinction drawn by reason alone (*distinctio sola rationis*), just a distinction of concepts. This divergence of metaphysical positions becomes the matter of his argument against the Thomists. Yet before we analyse some of Suarez' ideas, we must touch on the scholastic classification of distinctions.

1.7. *Distinction as Such*

Distinction as such is a negation of identity. Yet one must add that, on the one hand, distinction may be inherent in the thing itself, irrespective of the intellect and its differentiating activity, whereas, on the other hand, it may be a result of a purely intellectual operation. In the first case the distinction is termed "real" (*distinctio realis*), and in the second, "mental" (*distinctio rationis*). A real distinction may be major or strong, minor or modal (*minor seu modalis*), and minimal (*distinctio minima*). The second distinction is also called formal, following from the nature of the thing (*distinctio formalis ex natura rei*); it was introduced by Duns Scotus within the context we are concerned with here, the problem of essence and existence, and that is why it is sometimes referred to as *distinctio Scotistica*.

A major or strong distinction is based on a difference between two separate and independent parts belonging to the same whole (Peter and Paul among the Apostles), or between entities which are not actually separate, but can be separated (body and soul).

---

[24] "Ens primum ac praecipuum, quod et est totius metaphysicae primarium objectum, et primum significatum et analogatum totius significationis et habitudinis entis." Suarez, *op. cit.*, disp. XXXI, "Introductio."

[25] Cf. Thomas Aquinas, *Summa contra gentiles* II 49. Thomas speaks also about "separate substances" (*substantiae separatae*), which means substantial forms without (or separated of) the matter. To this latter class belong the First Cause (i.e. God), the pure intelligences (angels), and the souls.

[26] Suarez, *op. cit.*, disp. XXXI, sect. I, 3.

A minor or modal distinction is made between the entity and its modus, so that the former can exist without the latter, but not vice versa (body and its posture).

A formal distinction can be established between two formal aspects or capabilities of the same subject: none of these aspects can exist independently from the subject, and that is why they are ontologically inseparable from the thing itself and from each other. A formal distinction, nonetheless, follows from the very nature of the thing (*ex natura rei*), it is inherent in the thing itself according to its natural formal foundation and its whatness,[27] not simply asserted by the perceiving mind (*per actum intellectus percipientis*) or by the comparing and distinguishing will (*per actum voluntatis comparantis*).

In its turn intellectual distinction can be understood in two different ways. First, it can be purely mental (*distinctio rationis pura*, or *distinctio rationis ratiocinantis*). This distinction is drawn by the mind alone according to certain strategies of distinction, which have their foundations or motives only in the mind itself (for instance, two different ways of naming or defining the same thing). Second, a mental distinction may have its foundation or meaningful motive in the thing itself, and then it is termed *distinctio rationis ratiocinatae* or *distinctio rationis cum fundamento in re*. This foundation may, again, be either external, i.e. bear on various activities of the thing (*varietas effectuum*), or internal, i. e., related to its *intrinsic* power or ability (*virtus*) to carry out various actions.

Such is the scholastic classification of distinctions. As a matter of fact it can hardly be considered quite clear. Does not, for instance, the presence of a meaningful motive or foundation, which compels the mind to make a *well-founded* distinction (*distinctio rationis ratiocinatae*), mean that there is, in the thing itself, a naturally inherent formal basis (*ratio formalis et quidditativa*) for such a distinction, i. e., a kind of differentiation of *the thing's* formal aspects? Actually Suarez believes that a formal distinction, following from the thing's nature (*distinctio formalis ex natura rei*) and a mental distinction having its foundation in the thing itself (*distinctio rationis cum fundamento in re*) must be considered very close, if not altogether identical, notions. This provides him with a pretext for making Duns Scotus his ally in his polemics against the Thomists.[28]

1.8. *Suarez on the Distinction and Composition of Essence and Existence in the Finite Entity*

1.8.1. *Distinctio Rationis...*

Thus Suarez' view on the nature of the distinction between essence and existence in created entities is that it can be mental only (*distinctio rationis*). In other words,

[27] "Dico esse formaliter in aliquo, in quo manet secundum suam rationem formalem et quidditativam." Duns Scotus, *Reportata Parisiensia* I, dist. XLV, quest. II.

[28] Duns Scotus states *ex professo* in one passage of his *Reportata Parisiensia*: "Dico igitur, loquendo de *esse* actualis existentiae, vel non est res sine actuali existentia, vel non differt existentia a re extra causam suam, nisi sola ratione." *Reportata Parisiensia* III, dist. VI, quest. I, 7. Yet strictly speaking here the difference between the thing itself and its existence is under consideration, and not the one between essence and existence *in* the thing. The former distinction – between a thing and its being – is exactly what Heidegger calls *ontological difference* in GP.

this distinction is not *present in* the entity, but belongs to the way of understanding the finite entity as such. Here we must be aware of the situation in which the problem of distinction is raised.

What is the starting point of the inquiry, the point of convergence of this perspective? It is the really existing thing itself. The comparison is made between actual existence and the actually existing essence.[29] Between essence and existence *understood in this way,* Suarez says, there is no (real) difference in the actually existing thing.

Yet what happens then to Thomas' mental experiment? Suarez' answer is: *a mental experiment cannot be evidence of a stronger form of difference than a mental distinction.* To be more precise, the fact that we can contemplate a thing's essence or whatness without contemplating its existence means only that the essence *in a certain way* has its being in the intellect as an object of cognition and is therefore not an independent self-sufficient entity, but a determination (or a state) of the intellect itself. Similarly, existence, when separated and abstracted from the existing *quid,* is yet another mental definition. The difference between these two definitions is a *mental* difference, i. e., a distinction drawn by the intellect in its object.

Besides, the essence being contemplated, and the essence of an actually existing, really present thing are not the same, as if the contemplated "what" of the thing were a kind of "copy" of the actual essence of (in) the thing, "captured" by the mind and therefore having received an independent existence as an object of contemplation.

The same topic emerges again in the third section of *Disputation* XXXI, in relation to the concept of *objective potency.* Some authors, Suarez says, interpret the composition of essence and existence literally, as a kind of addition (*additio existentiae*). If the mental experiment of Thomas allows us to subtract, as it were, existence from essence, then it seems possible to interpret an entity = an *actual* essence as a result of the addition of existence to the essence as such. And this is a metaphysical model of creation. Suarez cites the words of Aegidius Romanus, a student of Thomas Aquinas, who used a metaphor of the imprint or impression in order to describe the composition of essence and existence: being is impressed (*imprimitur*) as it were on the essence when the correspondent *compositum* is created, and then the essence becomes existent.[30] But, Suarez says, alluding among other things to Thomas' *Gedankenexperiment,* essence, *as contemplated essence only,* does not have independent being, being in itself, *in ipso;* the being of a thing contemplated, as contemplated, depends on the mind's contemplation, and the essence, as contemplated essence only, is nothing — *omnino nihil.* How then can the actual being be "impressed" on what does not exist on its own? In other words, an actually contemplated essence has being only as an accidental determination of the contemplating mind, and is nothing "outside" and without this mental act. It is necessary for

[29] "Inter actualem existentiam, quam vocant esse in actu exercito, et actualem essentiam existentem." Suarez, *op. cit.*, disp. XXXI, sect. I, 13.

[30] "Aegidius... ait, esse imprimi essentiae dum creatur et fit existens."

an entity, in order to bear an imprint of something else, to be in actuality and to possess a certain receptive potency (*potentia receptiva*).

Yet the argument of Thomas Aquinas is not so straightforward. It rests upon the ability of the essence to become an object of thought when the grounds to assert the existence of the corresponding entity *in rerum natura* are absent, and this ability, being *irrespective of any given act of thought or any particular thinking mind,* can claim to possess a special ontological status.

But Suarez insists that the thesis must be given up, for the ability mentioned is just a *potentia objectiva*, i. e., an ability to become an object of thought, and is to be distinguished from the active, as well as from the passive or receptive, potency (*facultas* or *receptivitas*), which means a "power" (to act or to be acted upon). *Potentia objectiva* is not a sign of being: an independent ontological status can by no means be ascribed to a possible object of thought, because it "remains wholly in its cause" (i.e. in the mind contemplating it). Potency understood as a *power* belongs not to this "quasi-thing" itself, but to something else — namely, to the contemplating subject, and is a *potentia agentis*. That is why it is impossible to speak of an "impression" (*imprimatio*), even if the expression is taken as a metaphor.[31] It is to be understood that "an actual and a potential being are distinguished in a strictly formal way — as what is and what is not, and this distinction does not mean an addition of another entity to another [already existing] entity."[32]

1.8.2. ... *cum Fundamento in Re*

Thus essence and existence cannot be considered really distinctive parts of the entity itself. They are simply different reasons (*rationes*) allowing the mind to contemplate the entity as entity in different ways and "articulating being more precisely in its basic constitution." The combination of essence and existence and their separation are operations of the intellect, or rather the two aspects of the same operation (*articulation*), which allow us to "articulate" (i.e. to contemplate and to express in discourse) the being of a finite entity.

We have already seen that even the mere possibility of speaking of the distinction between essence and existence, the possibility of articulating this distinction, depends on the ambiguity of the word *ens*. Taken as a participle, it refers to the actuality of being, *actus essendi*, whereas taken as a noun (*vi nominis*) it refers to the real essence (*essentia realis*), i. e., to the content which *can* (in the sense of the Latin *aptum* — "apt at" and "apt to") actually exist, but *does not posit* the *actual existence* (or leaves this undefined till further clarification).

Ens taken as a noun does not designate a possible entity (*ens in potentia*) *as opposed to* an actual entity (*ens in actu*), but leaves in suspense the *distinction* between these two meanings.[33]

---

[31] Disp. XXXI, sect. III, 5.

[32] *Ibid.*, sect. III, 8: "Est ergo in universum verum secundum principium supra positum, scilicet, ens in actu et ens in potentia distingui formaliter immediate tanquam ens et non ens, et non tanquam addens unum ens supra aliud ens."

[33] Suarez, *op. cit.*, disp. II, sect. IV, 4–12.

Starting from this point of the original "suspense of distinctions" it is possible to go in two different directions. Entity as such can be contemplated *sub ratione essentiae.* The intellect discloses the being of a thing pursuing this strategy while bringing to light *the real concept*, i. e., discovers the rational foundation of being, which constitutes a thing within the scope of "real entities" as distinguished from arbitrary mental constructs (*ens fictum*).[34]

The essence of a created thing thus understood does not yet, however, include a *sufficient* condition of the thing's actual existence. It follows that we grasp the essence as "indifferent" with respect to its actual being or non-being.[35] When the thing receives actual being-ness (*recipit entitatem suam*), we perceive (the mind perceives) another definition of entity as such — and this precisely is the reason and foundation of grasping the actual = autonomous being of the thing apart from its causes[36] (including the formal one). Considering the entity's being in this way, we contemplate it as an existence (*sub ratione existentiae*). The expression "*recipit existentiam suam*" must not be understood as if the essence were already somehow *present* (as God's intention) before the act of creation. Suarez has already amply explained that the intention can be understood only as a *potentia objectiva,* and no positive ontological status must be ascribed to it.

We must not violate our initial premises and confuse the distinction of essence and existence with the distinction of potentiality and actuality. The starting point is the actually existing finite entity, and we ask, whether *in it* the real distinction of essence and existence can be posited. Suarez' answer is: essence and existence are one and the same thing.[37] Nevertheless an entity in its presence-at-hand (to use Heidegger's term) can be conceived either *sub specie essentiae* or *sub specie existentiae.*

To sum up, although the distinction between essence and existence is mental only, it has a foundation in the things themselves (*distinctio rationis cum fundamento in re*). This foundation is the *being-created* of created things: created entities are not self-sufficient ontologically; they do not receive their being from themselves and through themselves (*aseitas* and *perseitas* are attributes of God's being only). In other words, a created entity may exist and may not exist. As *essentia realis,* the essence is indifferent as regards the actuality of being (it neither negates nor affirms it), but allows a further more precise "existential" definition of it. This more precise definition is semantically expressed by a transition from *ens* as a noun to *ens* as a participle.

---

[34] *Ibid.*, sect. VI, 23: "Est enim essentia id quo primo aliquid constituitur intra latitudunem entis realis, ut distinguitur ad ente ficto..."

[35] "Ex hoc enim fit ut essentiam creaturae nos concipiamus, ut indifferentem ad esse vel non esse actu, quae indifferentia non est per modum abstractionis negativae, sed praecisivae..." (*ibid.*)

[36] "formalis ratio essendi extra causas" (*ibid.*).

[37] "Dicendum ergo est, eamdem rem esse essentiam et existentiam..." (*ibid.*)

2. EXISTENCE AS FINITE BEING

Ursprünglicher als der Mensch ist die Endlichkeit des Daseins in ihm.
M. Heidegger, *Kant und das Problem der Metaphysik*.

2.1. *Phenomenological Interpretation*

Of course, Suarez' teaching serves only as a theme for Heidegger's hermeneutical variations and can by no means be regarded as a final answer to the problem of the nature of ontological difference. It cannot be taken as such in the first place because, for the author of *The Basic Problems of Phenomenology* (who considers himself a participant in the phenomenological movement), the opposition of *distinctio realis* to *distinctio rationis* has no ontological sense. For a phenomenologist, any difference is a *distinctio rationis*, a distinction constituted by the mind in the intentional object. Even Duns Scotus, whose concept of formal distinction was later classified as a *distinctio realis minima*, proposed to interpret this distinction as following from the nature of the thing itself (*ex natura rei*) and not as asserted (constituted) by the perceiving intellect or by the comparing and distinguishing will. An extremely important conceptual change or "reversal of philosophical consciousness" has taken place between *ex natura rei* as propounded by Duns Scotus and *zu den Sachen selbst* of phenomenology. Already in Hegel's phenomenology the object's *in-itself* was understood as its *in-itself-for-us*. The nature of the thing itself, its "selfhood," *Selbigkeit*, cannot be contemplated in a position of phenomenological reduction without having recourse to an *actus alicuius potentiae comparantis*, without thematizing the ways of our access to the entity.

2.1.1. *Ontology as Phenomenology*

In one of his Marburg lectures of 1923 — 24 Heidegger says:

Together with the discovery of intentionality, understood as simultaneous thematizing the entity and the ways (*das Wie*) we encounter it, the foundation for a radical ontological research was laid down for the first time in the history of philosophy.

At the same time Heidegger asserts that the way of encountering beings in the world, which is to be regarded as ontologically primary and basic, is by no means connected with the *intellectus recipiens*, with "consciousness," which is a generic term for the ways an *intentional object* is "given" to us. He tries to discover a hidden tendency, an unconscious foreboding of the "fundamental ontology" behind the speculations of medieval metaphysics. Indeed, the difference between essence and existence is treated in Scholasticism within the theological context of *creation*. This idea links the *potentia objectiva*, i. e., the pure ability to be contemplated, to become an object of thinking, with (God's) *creative power* (*potentia creativa*). Heidegger writes:

The verbal definition of *existentia* already made clear that *actualitas* refers back to an *acting* on the part of some indefinite subject, or, if we start from our own terminology, that the present-at-hand (*das Vorhandene*) somehow referred by its sense to something to which, as it were, it *comes to be before the hand*, at hand, to be handled. The apparently objective interpretation of being as *actualitas* at bottom also refers back to the subject, not, however, as with Kant, to the apprehending subject in the sense of relation of the *res* to the ability of the cognitive faculties, but in the sense of a relation to our *Dasein* as an acting *Dasein* or, to speak more precisely, as a creative and *productive Dasein*. (GP 143)

2.1.2. *The Ontology of Creation*

If we attempt now to interpret phenomenologically this "hidden tendency" of medieval ontology, we shall immediately meet with serious difficulties. Husserl's method does not allow us to discuss this first and basic meaning of the entity (which we are searching for and which exists for us as yet only as a vague allusion or presentiment) *as the first*, for the very first thing happening when we encounter an entity in its being is, according to Husserl, its "self-giving" (*Selbstgegebenheit*) to consciousness, and the primordial way of such an encounter is *perception*. This "event-of-encounter" already implies quite a definite way of accessing the entity, the intentional relation, with a threefold internal structure: *ego-cogito-cogitatum* [*qua cogitatum*]. The "field of pure phenomena" is here a kind of "gap," opened up as a result of phenomenological reduction — the (*reell*) variety of *cogitationes*, in the widest possible sense of the word, placed "between" the transcendental *res cogitans* and the transcendent *res cogitata*. This primordial phenomenal field is, on the one hand, a complex *object* of phenomenological reflection, and on the other, is to be interpreted as the sum total of the manifold ways of givenness in which a certain "external object" presents itself to the mind.[38] That is why "the meaning of the entity as entity (*das Seiende als Seiendes*)" is declared as it were from the very beginning. Even if the *ontic* belief (*Seinsglauben*) of "natural" consciousness is reduced and "put between parentheses" within the framework of the phenomenological attitude, the essence, the being-ness, of the transcendent object within the phenomenological parentheses is searched for in a new "immanent" *objectivity*, accessible to reflection. The entity has already been interpreted as something, which can become a pole of intentional relation, an ob-ject of consciousness, as something *objective* or *present-at-hand (das Vorhandene)* in Heidegger's terms. The entity is something exposed to the "gaze of the intelligent," νόησις, for (as has been noted many times) νοεῖν, in its original etymology, is a verb of vision.[39]

Ontology was always developed around this centre of relation of intelligent contemplation (θεωρία) and articulated its theorems within the theoretical logos. Creation and moral action (ποίησις, πρᾶξις) as ways of revealing the entity in its being were already understood by the Greeks as something ontologically secondary. Still, according to Heidegger, they were closer to the primordial — though too rapidly lost in oblivion — meaning of being.

---

[38] To be more accurate, we should express this in the following way: "an object, which is constituted as being external, alongside of its *ontic* claim."

[39] It is quite possible to νοεῖν with one's eyes. Cf. in Homer: ἐνόησεν...ὀφθαλμοῖσιν (*Il.* XV, 422), ἐν ὀφθαλμοῖσι νοήσας (XXV, 294).

The Greeks had an appropriate term for "things": πράγματα that is to say, that which one has to do with in one's concernful dealings (πρᾶξις). But ontologically, the specific "pragmatic" character of the πράγματα is just what the Greeks left in obscurity; they thought of these "proximally" as mere "things" (*bloße Dinge*). We shall call those entities which we encounter in concern "equipment" (*das Zeug*). (SZ 68)

This oldest hypothesis, or, to be more precise, this hereditary amnesia of philosophy, remains unperceived, according to Heidegger, and eludes the phenomenological requirement of "absence of premises." After it is finally discovered, the main task seems to be to meet the entity (or to *understand* it ontologically) in a manner as unprejudiced as possible. Understanding must not be constructed anew or "purified," since the fullness of presence depends on the fullness and the *wholeness* of understanding. We must rather understand our understanding as it reveals itself in our everyday encounter with the things of the world. The project of creating a new starting point for ontology, which determines the content of *Sein und Zeit*, is an attempt to explain the classical subject-object relation of the thinking subject to the object of thought or cognition as derived from encounter-in-concern (*Begegnung im Besorgen*) with entities in the world. The "intelligent contemplation," which the Greeks believed to be a kind of sacred "carelessness" or serenity,[40] must be explained on the basis of *care* understood as the *existentiale* of *Dasein* — more than that, understood as *Dasein*'s being itself (SZ, § 41).[41]

As we have heard, Heidegger calls the entity coming to light within the scope of concern, *das Zeug* — "tool," "implement" or "equipment,"[42] and the way of being peculiar to equipment — *das Zuhandensein, being-ready-to-hand, being-handy.* The self-revelation of equipment within the horizon of concern is essentially different from the self-giving of the object to the perceiving consciousness. Equipment reveals itself as such only in the "world's" totality, and this means that its character of being (not to say its "essence," returning in this way to the main *terminus technicus* of classical ontology) does not lie in its "what" *(quidditas)*, but in its "for-what." The world's wholeness is not thought of any more as the totality of objects, the totality of Husserl's "objective (= intersubjective) nature," the totality of transcendent objects exposed to intersubjective (let us say, public) viewing (mental contemplation). For Heidegger, it is the manifold of references or assignments (*Verweisungsmannigfaltigkeit*), which bear the character of "in-order-to."[43] The equipment's "within-the-

[40] Aristotle writes in the *Nicomachean Ethics* that the wise men like Anaxagoras and Thales may (or even must) be ignorant of their own interests, because they possess knowledge that is not an everyday cleverness, but rare, marvellous, difficult, and even superhuman. But people declare the wise (σοφοί) to be not prudent (φρόνιμοι), because these sages do not seek to know the things that are good for human beings (ὅτι οὐ τὰ ἀνθρώπινα ἀγαθὰ ζητοῦσιν), and so their knowledge is useless (*Eth. Nic.* VI 7, 1141b1–10). Aristotle insists that the "wondering" (θαυμάζειν), which is declared to be the origin of philosophising, must be in a sense care-less or serene. To think that prudence is in authority over wisdom would amount to believing that political science governs the gods, because it gives orders how to worship them (VI 13, 1145a10ff.).

[41] We shall come back to this topic in chapter 6.

[42] The latter, being a collective noun, has become the traditional English translation for Heidegger's *Zeug*.

[43] We shall discuss this structure in detail in chapter 6, where the ontology of the *being-ready-to-hand* will become the subject of our special interest.

world-ness" (*Innerweltlichkeit*), i. e., its being included and interwoven into the totality of "in-order-to" references is disclosed primarily to concerned dealing (*Umgang*).

As something ready-to-hand, the entity possesses a special "inconspicuousness" (*Unauffälligkeit*) to the extent that it does not resist the habitual course of affairs, is obedient and is fitted to its neighbour (another entity within the world's totality) in the system of worldly concern. The equipment's unobtrusiveness is similar to the inconspicuousness, the transparency of acoustic signifiers obedient to the intention to express the meaning signified: before becoming an object of semiological reflection, the signs are a medium in which the meaning becomes possible, a medium of "concerned expressiveness."

The self-manifestation of equipment is not "intentional" in the Husserlian technical sense of some *noematic* content being objectively present to and constituted by consciousness. Equipment's "in-itself" is rather its "in-the-world," within the variety of its "in-order-to," of its fitness, its usefulness, its handiness. Heidegger writes:

> The kind of being which equipment possesses — in which it manifests itself in its own right — we call "*readiness-to-hand.*" Only because equipment has *this* "being-in-itself" and does not merely occur, is it manipulable in the broadest sense and at our disposal. No matter how sharply we just *look* at the "outward appearance"[44] of a thing in whatever form it takes, we cannot discover anything ready-to-hand. If we look at things just "theoretically," we can get along without understanding readiness-to-hand. But when we deal with them by using them and manipulating them, this activity is not a blind one; it has its own kind of sight, by which our manipulation is guided and from which it acquires its specific thingly character. Dealings with equipment subordinate themselves to the manifold assignments of the "in-order-to." And the sight (*Sicht*) with which they thus accommodate themselves is *circumspection (Umsicht)*. (SZ 69)

This latter "sight" does not belong to θεωρία, it is rather an integral part of the structure of ποίησις, *creatio*, insofar as it can become a *phenomenon* of finite human being (*Existenz*).

The "thesis of medieval ontology" interests Heidegger because he sees in it an intention (not quite conscious or carried out) to get away from the "classical" understanding of being as *a presence for senses and mind.* Heidegger is interested in the Scholastic attempts to construct metaphysics after the pattern of creation and not of contemplation (θεωρία), i. e., to take into account creation-creativity, which means, in the case of finite human being, to include in ontological consideration the primordial phenomenon of care. Such is at least the explanation Heidegger gives *ex professo* for his interest in Suarez' treatise. Yet there is, it seems to me, one more circumstance bringing close together Heidegger and the author of the *Metaphysical Disputations*, which the former does not mention.

---

[44] "... das Nur-noch-hinsehen auf das so und so beschaffene 'Aussehen'..." The word 'Aussehen' is regularly substituted by Heidegger for the Greek εἶδος.

2.2. *Being received and not received*

2.2.1. *Objection to a Purely Mental Character of the Ontological Distinction*

One of the arguments in favour of Thomistic doctrine characterizing the distinction of essence and existence in a created entity as being *real*, the argument which Suarez has to refute, proceeds as follows.[45] If creation is a *communication* of being, it follows that the being of created entities must be *received* in something (*esse receptum in aliquo*). Received by what? or — into what? Received by the essence or into the essence, for nothing else can be thought of as receiving it. It follows that being (existence) is really different from essence, since nothing can receive itself into itself (this contradicts the semantics of the verb *to receive*).

But why is the being of created entities a *received* being (*esse receptum*)? If it were otherwise, the being of created entities, as opposed to the entities themselves, would have to be considered self-sufficient and autonomous (*esse in se subsistens*). This means that this being, as completely separated from any potency or subject, which can receive it, would be pure actuality (*energeia*) or pure act (*actus purus*). Thus we would then deal with the highest and most perfect being, unrestricted in its beingness,[46] which belongs to God only, not to the creature.

This thesis can be confirmed with one more argument. Being (in the sense of *existence*) which is not received in something must be unlimited. For what could set limits to it? It is not limited by the potency "receiving" it (as an act of this potency), it is not limited, on the other hand, by any act, for it cannot be a potency of any act. Indeed, existence is already perfect actuality. For the same reason it cannot be restricted (defined) by any specific difference, for in this case the latter, as a form, would be an actuality of the potential, whereas existence as such does not allow any more perfect actuality, in relation to which it could be regarded as its potency. Yet unlimited being (*esse infinitum*) belongs only to God.

It follows from this reasoning then that, once we refuse to accept the difference between essence and existence as real, we thereby ascribe to the creature an unlimited or infinite being, which contradicts the "ontological status" of the created entity as *ens finitum*.

2.2.2. *Suarez' Answer to this Objection*

First, given that the relation between existence and essence *is not* the one between receiving potency and act, it is hardly possible to draw, from this premise, the conclusion that the being of created entities must be characterized as non-received (*irreceptum*), autonomous (*esse subsistens*), perfect and infinite.

a) Indeed, this conclusion is not correct as regards the accidentals (*accidentia*), for every such entity is received in the substance as in the subject.

[45] Disp. XXXI, sect. I, 5.

[46] "Esse perfectissimum et summum, atque adeo purus actus, et infinitum quid in ratione essendi" (*ibid.*).

b) It is not correct as regards the substantial form in matter (Aristotle's εἶδος ἔνυλον), for it is received in matter and, consequently, limited by it.
c) It is not correct as regards matter, because the existence of matter can be spoken of only insofar as matter has form, i. e., is determined and limited by it.
d) It is not correct as regards a composite whole (compositum), for the latter is a unity of matter and form, the actuality of matter having form or of the form having determined matter. Therefore the being of a *compositum* cannot be unlimited, since the being of its components (matter and form) is mutually limited.

However, there is a difficulty presented by pure created (finite) minds (*intelligentiae*), free from matter. It is not without interest that the question of the difference between essence and existence was first raised by Thomas Aquinas in his *opusculum De ente et essentia* precisely in connection with the necessity of positing an internal difference (a negation of simplicity) in such matter-less entities.[47] It is difficult to understand what could be the reason for an internal complexity in pure intelligences. Of course, one might argue that they are complex because of the difference of their states:[48] for instance, because of the variety of contemplated objects (*cogitata qua cogitata*). Yet these determinations or states of the mind are accidental, whereas a finite intellect is to be interpreted as a *substantial pure form* without matter, and we are looking for the *internal* complexity properly (not accidentally) belonging to it. The argument stating that the being of such a substance is limited by its accidentals is erroneous, for the being of a substance is by definition independent of the being of its accidental determinations.

Thus, there still remains a "region of being" which resists Suarez' assault. Besides, I cannot help mentioning that this way of answering the opponent's objections is not quite satisfactory, for it has recourse to arguments of a different nature and structure in relation to different ontological notions (substance, accidentals, matter, form, etc.), while the objection itself rests on a universal characterization of the created as finite.

2.3. *Received in Something Other and Received from Something Other*

But Suarez also has a more elegant answer. When speaking about received being (*ens receptum*), two meanings have to be distinguished: (i) the being received *in(to)* something other, and (ii) the being received *from* something other. Correspondingly, the being characterized as non-received (*esse irreceptum*) must also be understood in two ways: (i) non-received *into* something other, and (ii) non-received *from* something other. Even if existence is not received *into* something other, it is received *from* something other and limited in the sense that it is not a pure act, but a "form" of participating in a pure act (in God's being), from which it is received.

---

[47] *De ente et essentia*, cap. V.

[48] It is not difficult to recognize here one of Leibniz' basic theses in his *Monadology* concerning the mode of being of the monads.

One might object by asking: what else can participate in pure act except essence? Does not this explanation again introduce, though implicitly, a real difference between essence and existence?

The answer is *no*. One cannot say that a "thing" (essence) participates, while another "thing" (existence) is participated in. The same "thing," a created entity as such, insofar as it is created and insofar as it is entity, *is* in its thingness (reality) due to participation. This participation means that the entity *is* due to (*per vim*) an active agent (*sicut per vim agentis*). An actually existing thing is actual due to the activity of the Divine agent. Thus the finite entity is finite because it is created and therefore limited by God.

2.4. *Existence as the Finite Being (esse) of Finite Entities*

Even though this argument shows that the being of created entities is not an infinite being (*ens infinitum*), for it is received *from* something other, the argument refers, however, to an *external* limitation only. The existence of created entities still remains un-defined internally (*esse indefinitum*). What has been said, introduces only one, negative, definition or delimitation: the being of created entities is *not* God's being understood as *actus purus*. Thus the whole reasoning turns almost into tautology. The main question remains unanswered: is an intrinsic definition of a created existence possible? and how is it possible? Thus the question has for its target not so much an external limit and restriction as the *internal form* of finite (created) existence as such.

Suarez' answer seems, at first sight, little more than a formal juggle:

> To answer this [objection], it is sufficient to remark that [the being of created entities] is limited by itself, it is limited and finite due to its own way of being.[49]

But then a further explanation of this statement follows. There are two ways to understand contraction or limitation (*contractio vel limitatio*), physically and metaphysically. To speak of limitation in the metaphysical sense, there is no need for what is limited and for what limits to be *really* (*realiter*) different "things." It is sufficient (since a "metaphysical limitation" is connected with an internal form of what is limited) for them to be different concepts having the foundation of their difference within the thing itself. Then we can allow the essence to be limited and defined by its destination for being, and vice versa, the being (existence) of created entities itself is limited and finite because it is not the actuality as such (*actus purus*) but the actuality of a definite essence.[50]

Insofar as pure intelligences (minds), pure substantial forms, are concerned, one can say that in the physical sense also there is no need to look for another limitation of existence besides these intelligences themselves. Just as composite substances are

---

[49] "Ad hoc uno verbo sufficienter responderi posset: seipso, et ex vi entitatis suae esse limitatur et finitum..." (*ibid.*, sect. XIII, 18). To be sure, while translating the expression "entitas sua" as "its own way of being," I deliberately shift the quotation a little towards Heidegger's terminological horizon.

[50] "Et hoc modo... admittere possumus, essentiam finiri et limitari in ordine ad esse, et, e converso, ipsum esse finiri ac limitari, quia est actus talis essentiae." (*ibid.*)

internally defined by their constituting causes (form is actual only insofar as it shapes matter, matter is actual only insofar as it is shaped by form), so substantial forms free from matter (= minds or intelligences) are also limited by themselves in the "physical" sense, i. e., by their own being-ness, their proper way of being (*per suammet entitatem intrinsece limitari*). And this limitation applies to potential entities before they are created as well as to actual entities.

The fact that the actuality of being is not "received in," or "impressed on" the self-sufficient entity having power or ability to receive it (*potentia receptiva),* does not make existence simple and indefinite. Since existence is nothing but an actual essence or an actuality of essence (*essentia in actu constituta*), the former gets its internal determinations due to the latter. If the difference between essence and existence is only *mental,* we are fully entitled to assert that existence possesses internal form derived from essence itself and that there are essential differences between various existences.[51]

2.5. *Existentia and Existenz*

Heidegger takes over the term *Existenz, existentia,* the main *terminus technicus* of medieval metaphysics, and inscribes it on the banner of the "newly rekindled γιγαντομαχία περὶ τῆς οὐσίας" (SZ 2), but he gives a new meaning to it: existence now signifies the being of *Dasein,* the being to which this entity (*Dasein* itself) is "delivered over" (SZ 42). Existence is that "which is at issue for" *Dasein.* Yet the "diachronic homonymy" *existentia/Existenz* is not accidental and by no means a mere coincidence of names.

The question of the meaning of being should be approached through the analysis of existence, through the *analytic of Dasein*, which "is there" (*ist da*) through the opening to this same question, through the precomprehension of being. Existence "should be [phenomenologically] *uncovered*" in the undifferentiated character (*Indifferenz*) of everydayness, and because *Dasein*'s characters of being are defined in terms of existentiality, Heidegger calls these characters "existentialia" (cf. SZ 44).

Existentialia, Heidegger says, "are to be sharply distinguished from what we call categories "(*ibid.*). Categories are *essential* definitions of a thing — *essentialia*, definitions of its real content or, in Kant's terms, the most general "real predicates." *Existentialia* are named so because they express (as the main concepts of existential analytic of *Dasein*) the most important structural moments of *existence* in their interrelation.

But if the project of existential analytic is from the outset related to the question of the *internal structure* of existence itself, described in terms of existentialia, our claim to establish a more close connection between *existentia* and *Existenz* than that of homonymy is threatened by the following observation. It seems at first sight that for medieval ontology a discussion of the internal structure of existence is an absurdity, for all structural aspects, all distinctive marks and characteristics defining an entity as such and such a thing (*res*), defining an entity as something real, are related

[51] "[Concludere possumus,] inter existentias dari essentialem diversitatem." (*Ibid.*, sect. XIII, 19)

to its whatness, and ultimately (if we waive the accidental features) to its essence (*essentia*). Existence (*existentia*), which in Thomas Aquinas' school was believed to be distinct *realiter* from essence, is an additional (in relation to essence) "dimension" of being. It is not by chance that Thomas calls existence an accidental determination of an entity, for everything not pertaining to a thing's essence is, by definition, an accidental determination.[52] According to the Scholastic formula, essence (*essentia*) gives the answer to the question "*quid sit res*" — "what a *thing* is," whereas existence answers the question "*an sit*" — "whether something is." "To each being the *what-question* and the *whether-question* apply" (GP 123).[53] And while the first question allows various answers, the last one can be answered only with yes or no, *tertium non datur.* Existence is a simple *yes*, and it would seem that it is impossible to speak of its "internal constitution," of *such* or *such* an existence, i. e., there can be no formal difference at all between various "existences."

The following argument from Thomas Aquinas' work *On Entity and Essence* seems also to suggest that existence is to be thought of as a simple concept.[54] If there is such an entity of which the essence *is* identical with its being, this entity must be numerically unique (and it is God). For how could such an entity be different from another similar entity? — Through the addition of a difference only. Yet being *as such* does not tolerate any internal difference (*non recipiet additionem differentiae*), for otherwise it would no longer be just being or being as such; it would be being *and* form. It follows from this reasoning that being as such has no form besides "to be" itself, *actus purus* = *forma essendi.*[55]

Yet already in his *Summa theologica* (I 3, 4) Aquinas clearly distinguishes God's being and universal being; i. e., the being, which is, according to Thomas, a predicate of any thing when we say that this thing *is.* Universal being *admits of further definitions* (just as a genus admits of the addition of specific differences), whereas God's being does not (see *ibidem, the answer to the first objection*). God's being as an *actus purus* is absolutely simple, and of course all schools of Scholasticism are unanimous in this respect. But is being simple too?

The acceptance of a real distinction between essence and existence certainly does not mean that this *distinctio realis* was taken by the Thomists in the major or strong sense. The existence of a created entity is not a created thing; otherwise the question of the difference between essence and existence ought to be raised again regarding existence itself, and thus *ad infinitum.* Yet it is not an *actus purus* either, for in that case the statement "a thing exists" would mean "a thing is God."[56] Johannes Capreolus, *princeps Thomistarum,* says[57] that existence or being in the sense of exis-

---

[52] "Omne quod praeter essentiam rei dicatur accidens." (*Quodlib.* II, quaest. II, art. III.)

[53] We find both questions as corresponding to different species of knowledge in Aristotle's list of questions in the beginning of the *Anal. post.* II 1.

[54] Cf. *De ente et essentia* V 3.

[55] The latter term (*forma essendi*), as the reader might remember, goes back to Boethius' treatise *Quomodo substantiae in eo quod sint bonae sint...*

[56] Thomas Aquinas, *Summa theologica* I 3, 4.

[57] Johannes Capreolus, *Questiones in quattuor libros Sententiarum*, sent. I, dist. VIII, qu. I, art. II (solutiones IV).

tence cannot be thought of as an *ens* or a *res* in the proper sense. *Existentia* is not an entity; it is something *within* entity. One cannot say either, that existence exists. Still this does not mean that existence is nothing. Of course, properly speaking it *is* not, it does not *come into being*, it is not *created*, for then it would be a created entity. One can only say concerning it that it is created along with other things (*quid concreatum est*). This "con-," "cum" means of course something more than just a simple and uniform (i.e. independent from essence) external addition of the "existential quantifier" (in the sense of modern logic) to the listing of the thing's meaningful characteristics. Suarez, founding his opinion on the commentary on *De ente et essentia*, cap. IV, by Cardinal Tommaso Cajetan, believes that the Thomists acknowledged formal differences within existences, particularly if the entities in question are pure substantial forms, like intelligences. For it is stated that such forms are designed to possess various being, and this *cannot be true unless differences within existence itself are allowed.* If existence is created *along with* such or such essence, it has to be commensurable with this essence. And this means that existences (a plural is justified here) have to be different to the same extent as essences (*necesse est, quod tanta sit distinctio inter existentias, quanta est inter essentias*).[58]

Whatever the opinions of the Thomists may be, the position of the author of the *Metaphysical Disputations* compels the endorsement of an internal form of existence itself.

This [admission] becomes all the more necessary in connection with our statement [that the difference between essence and existence is purely mental — *A. Ch.*]: for if existence is, in the things themselves, nothing but actual essence, existences must be different just as actual essences belong to different kinds.[59]

What is stated here seems at first sight incredible: existence as such is internally structured by the existing essence; there are formal differences between existences. Moreover, existence limits essence to the same extent as essence limits existence, and therefore *essence receives its determination through its destination for a certain mode of being.* Yet it is only at first sight that this statement seems extraordinary. The idea that being is a simple concept, a simple *yes* or the existential quantifier, is a result of the *oblivion of being*, as Heidegger believed; many crucial steps in the history of philosophy as a history of the question of being sank into oblivion as well, or were reduced to oversimplified schemes. As we have seen in chapter 2, Aristotle already maintains that *being* (τὸ ὄν) is not a universal genus, that being has meanings, and there are as many ways *to be* as meanings of *being* (*Metaph.* 1017a23f.; a31; a35). Aristotle's *Metaphysics* is nothing but a development of this central thesis.

For Heidegger, as we already explained, the *existentialia* are characteristics of the internal structure of *existence*, i. e., *being*, not of the whatness of *Dasein*. This circumstance seems to our oblivious age either an unheard-of novelty or an absurdity. But what is novel here is only a special "phenomenological interpretation" of the ever-present metaphysical attempts to think of being as internally distinguishable. The "destruction" of traditional metaphysics as a task of *Being and Time* and the breakthrough into "post-classical" ontology does not amount to a reversal of

[58] F. Suarez, *op. cit.*, disp. XXXI, sect. XIII, 19.

[59] *Ibid.*

the traditional philosophical concepts or their manipulation with an arbitrary light-mindedness. On the contrary, these concepts are interpreted by Heidegger *strictly in accordance* with their formal structure. It is true, though, that Heidegger's interpretation, his rich hermeneutical armoury, embraces a scope of phenomena new to transcendental philosophy and to Husserl's phenomenology as well — the phenomena which characterize *Dasein* as being-in-the-world: those of having-a-mood or being-attuned, of concern and care etc. Some of these matters are on the agenda of subsequent chapters.

It is not by chance either that Heidegger's definition of *Dasein* formally reproduces (all necessary reservations made, of course) the medieval definition of God's being (being of the absolute substance = the absolute subject) as an *actus purus*. Heidegger says about *Dasein*:

The "essence" (*Wesen*) of this entity lies in its "to-be." Its being-what-it-is (*essentia*) must, so far as we can speak of it at all, be conceived in terms of its being (*existentia*)." (SZ 42)

According to the medieval formula *in ente a se sunt essentia et existentia unum idemque,* essence and existence are the same in the entity receiving its being from itself (that is, in God). Yet these two definitions do not coincide completely. Strictly speaking, Heidegger does not affirm here that *Dasein*'s essence is absolutely identical to its existence, for the question of "who" (of "who-ness" and "self-ness")[60] of the entity existing in this way can and must be asked. It is only stated that existence comes first, and essence comes second. *Dasein*'s essence is to be understood according to its existence. In this definition we can rather recognize Suarez' formula:

Essence is defined and limited due to its destination for a certain being (per ordinem ad esse), and, vice versa, being itself is defined and limited (de-termined) because it is the actuality of such an essence.

Existence is being-definite and being-limited, and thus, being-finite; it is the finiteness of a finite entity. In a word, existence is finite being. Thus Suarez. But thus Heidegger also, because *Dasein* is itself a finiteness of being disclosed in itself and for itself. *Dasein* is finiteness, which has become existent. This finite being-in-itself and *for-itself* is precisely what *existentia* sounds in German — Heidegger's *Existenz*. "Existence is in itself, as a mode of being, finiteness..."[61]

More primordial than man himself is the finiteness of *Dasein* in him. (KPM 222)

---

[60] SZ §§ 25, 64.

[61] "Existenz ist als Seinsart in sich Endlichkeit und als diese nur möglich auf dem Grunde des Seinsverständnisses. Dergleichen wie Sein gibt es nur und muss es geben, wo Endlichkeit existierend geworden ist." (KPM 222) In KPM the finiteness of human being becomes the central theme of the "metaphysics of *Dasein*" and the clue to Heidegger's interpretation of Kantian metaphysics (cf. §§ 39–41). The finiteness of *Dasein* underlies the project of fundamental ontology as it is presented in this book (§§ 42–45). Of course, Heidegger recognizes the fact that medieval metaphysics asserted the finiteness of human being as an important thesis, but at the same time, he says, the medieval thinkers did not acknowledge "its rightful due" (KPM 213).

CHAPTER FOUR

ONTOLOGY OF HUMAN ACTION

(*Aristotle's Eth. Nic. VI and Heidegger's commentaries*)

ἐφηῦρέ σ' ἄκονθ' ὁ πάνθ' ὁρῶν χρόνος
Sophocles, *Oedipus rex*, 1213

1.1. *The Topography of the Truth. How the Soul "Discloses the Truth"*

Ontology understood as phenomenology starts with the idea that entity reveals itself, lets itself be encountered, goes out to meet us. In the midday epoch of philosophy the *place* of this meeting is designated by the two most important words with "concentrated" meaning: ἡ ψυχή and *das Bewußtsein.*

'Αληθεύει ἡ ψυχή, "the soul achieves or discloses the truth," says Aristotle.[1] As an "area" where the truth is disclosed the soul has its own topography, described (or rather constructed) by Aristotle in book VI of the *Nicomachean Ethics.* Aristotle mentions here five ways in which the soul *discloses the truth* (ἀληθεύει) through affirmation or negation. These are: art or technical skill (τέχνη), science or scientific knowledge (ἐπιστήμη), prudence (φρόνησις), wisdom (σοφία) and intelligence (νοῦς).[2]

In the winter semester of 1924–25 Heidegger delivered lectures on Plato's *The Sophist*;[3] almost one third of this lecture course is devoted to interpretation, or rather variations on the theme, of Aristotle's reasoning in *Eth. Nic.* VI. Here Heidegger discusses the ontological foundation of the concept of *truth* and interprets the Greek ἀ-λήθεια (α *privativum*) so: *nicht mehr verborgen sein, aufgedeckt sein* ("to be no more concealed," "to be uncovered") (PS 16).

[1] *Eth. Nic.* VI 3, 1139b15.

[2] *Ibid.* 1139b15ff.

[3] M. Heidegger, *Platon: Sophistes*, GA 19 (Frankfurt a. M.: V. Klostermann, 1992). Cited hereafter as PS.

Later on in Heidegger's language there will be a consistently used term to designate the Greek ἀλήθεια: *Unverborgenheit*, "unconcealment." Aristotle's phrase ἀληθεύει ἡ ψυχή is now rendered as "human *Dasein* remains in unhiddenness and unlocks the entity"; and ἀληθεύειν means "to be uncovered," "to free the world from closeness and concealment."[4]

Thus the truth is one of the characteristics of the entity itself insofar as it goes out to meet us, but in the proper sense of the word it is a determination of being of human *Dasein*. (PS 23)

I intend to show that Heidegger's search for *primordial* thinking about being, which underlies the project of "fundamental ontology," depends in many respects on Aristotle's "onto-psycho-logical" approach developed in the treatise *On the Soul* and, first of all, in the *Nicomachean Ethics* (and which to a certain extent remains faithful to the "cartography" of the corresponding topics). The lengthy commentary by Heidegger creates a common textual space in which the discourses of both philosophers mix together, or rather in which Aristotle's discourse is strangely rendered in Heidegger's "German."

The most important result of the analysis of Aristotle's text is the discovery of different and equally primordial (having fundamental ontological significance) ways of arriving at the truth and abiding by the truth in its uncoveredness, i. e., of encountering entity in its being, of drawing entity in its being from hiddenness into unconcealment.

1.2. *Quarrel of Wisdom and Prudence*

The main collision depicted in book VI of the *Nicomachean Ethics* is a quarrel about primogeniture, about superiority between wisdom (σοφία) and prudence (φρόνησις).[5] Yet as Heidegger has rightly remarked, behind the choice between *phronesis* and *sophia* there is hidden the fundamental ontological decision concerning the primordial, "first" meaning of being, along with another decision, not less fundamental, concerning the ultimate Good, at which all things aim (*Eth. Nic.* I, 1 1094a3). The latter is modelled in Aristotle's text after the divine *summum bonum* and is achieved when the human mind becomes in a certain way similar to the divine mind steadily contemplating the loftiest thing, i. e., itself (*Metaph.* XII 9). Thus along with the ontological and ethical (or "soteriological") decision we also encounter here the most important *theological* decision. These three firmly interrelated decisions have in many respects determined the whole of the European onto-theological tradition.

---

[4] Heidegger refuses to translate in his commentary the verb ἀληθεύειν in order to avoid habitual connotations. He writes: "Wir wollen dies nicht übersetzen. ἀληθεύειν meint: aufgedecktsein, die Welt aus der Verschlossenheit und Verdecktheit herausnehmen. Und das ist eine Seinsweise des menschlichen Daseins" (PS 17).

[5] *Prudentia* is the commonly accepted Latin translation of Aristotle's φρόνησις. I keep the term "prudence" as its English substitution. Ross renders φρόνησις as "practical wisdom" Rackham (*Loeb Classical Library*, vol. 73) prefers "prudence."

1.3. *"Parts" of the Soul and Their Virtues*

At first Aristotle defines two "parts" (μέρη)[6] of the soul: the one "possessing the logos" (the *rational*, as the Latin speaking world used to designate it) and the one "lacking logos" (*irrational*). The first is in its turn divided into the "scientific" or "knowing-epistemic" (τὸ ἐπιστημονικὸν μόριον) and the "calculating," "projecting," "scheming" (τὸ λογιστικόν) parts. Wisdom (σοφία) and prudence (φρόνησις) are "virtues" (ἀρεταί) of these two "parts of the soul"; *sophia* belongs to the scientific, and phronesis to the calculative[7] part. *Virtue* is maturity, completeness, fulfilment (τελείωσις) of a certain ability of the soul, "the best disposition of the soul" (*Eth. Nic.* 1139a16), which manifests itself in an action proper to it, i.e. in the way of arriving at the truth.

1.3.1. *The Soul as the First Entelecheia*

We must not forget that whatever Aristotle deals with, his reasoning always retains a connection with "first philosophy," its system of concepts and its rules of thinking. The metaphysical definition of the soul affirms that it is the form, or the first *entelecheia*, of the living body.[8] Aristotle calls *the first entelecheia* a fully developed, mature ability (δύναμις) so far as it remains *ability*. In some particular cases the formula: "I can... I am able... I perfectly understand how to...., but I do not act at the moment," may serve as an illustration of the meaning of the term.

In *De anima* II, 5 Aristotle explains the difference between the first and the second *entelecheia* as follows: A newborn child is able, as a human being, to read and write, because he can be instructed in grammar. That is to say: he possesses the potency to know grammar = he knows grammar in potentiality = he is "a grammarian in potentiality." And when he actually has learned to read and write but is neither reading nor writing at the present moment but, say, is exercising in a gymnasium, he is also (at this moment) a grammarian in potentiality, though in a different sense: he knows grammar and can at any time, without learning any more, actually begin to read or write ("provided that external causes do not prevent him"). Mastery of grammar is the first *entelecheia* with regard to the capacity to learn which the child possesses. The reading and writing activity is the second *entelecheia* with respect to the "I can" of a person actually knowing grammar. Accordingly two meanings of

---

[6] To be sure, in Aristotle's terminology "parts" (μέρη) rather signify "faculties." To the irrational part of the soul belong, for example, nutrition and growth, and because these functions are common to all living things they constitute the so called *vegetative soul*. The other division of the irrational part is the seat of appetites and of desire in general. In a sense, it is amenable and obedient to the rational principle (cf. *Eth. Nic.* I 13, *De anima* II 2).

[7] We need to be cautious here, because the adjective "calculative" can provoke misleading associations. Aristotle explains that calculation is the same as *deliberation* (τὸ γὰρ βουλεύεσθαι καὶ λογίζεσθαι ταὐτόν, *Eth. Nic.* 1139a12f.). I preserve in my translations the formal distinctions between λογίζεσθαι ("to calculate") and βουλεύεσθαι ("to deliberate").

[8] "So the soul must be οὐσία in the sense of being the form of a natural body, which potentially has life." (*De anima* II 1, 412a20f.) "The soul may therefore be defined as the first *entelecheia* of a natural body potentially possessing life." (27f.)

potentiality can be distinguished (termed *potentia prima* and *potentia secunda* by the Schoolmen). The ability to learn is the first potentiality; with respect to this potentiality the *knowledge* (though not actually exercised) of a person having learned to read and write is the first actuality (*entelecheia*). In its turn this knowledge (or skill) is the second potentiality (*potentia secunda*) as regards the activity of reading and writing. At a certain period of his or her life a person manifests different abilities; he or she manifests these abilities in activity, though even before, and irrespective of, any activity he or she already possesses them in the element of "I can." The soul as the first *entelecheia* is, according to Aristotle, a coherent system of such interconnected abilities (capacities, faculties, functions), which can immediately manifest themselves in an appropriate action — *in* the body possessing organs (412b5f.), *by means of* it or *in relation to* it.

That is why we can dismiss as unnecessary the question whether the soul and the body are one; it is as though we were to ask whether the wax and its shape are one, or generally the matter of a thing and that of which it is the matter. (6 ff.)

1.3.2. *Corporeality and Responsible Act*

The care that Aristotle takes in making distinctions and providing definitions, in order to single out the actions (πράξεις) which can be submitted to moral judgment,[9] merits our close attention. Moral actions are not and cannot be exceptions to the "rule of corporeality," i. e., being manifestations of the soul's dispositions — which means in the language of metaphysics, being definite aspects of form — they cannot be separated from matter, which "incarnates" form. For every *moral action* is an *action*, and from the formal ontological point of view belongs to the category of ποιεῖν. As such it is in one way or another connected with the possibility of manifesting human being *ad extra*, i. e., with his/her corporeality. For Aristotle *the ethics of incorporeal beings* is an impossibility. A moral *action* is a *responsible* action which can be an object of praise or blame. To be responsible, the action must be voluntary. An initial distinction pairing actions performed despite oneself or involuntary (ἀκών, ἀκούσιος) with those performed freely or voluntary (ἑκών, ἑκούσιος) serves in the *Nicomachean Ethics* as the starting point for the subsequent investigation of moral action and virtue. Aristotle asks whether acts performed under the unbearable pressure of external circumstances are to be considered voluntary. His answer is *yes*, though with certain reservations. This is why in such a case "the origin of the movement of the parts of the body instrumental to the act lies in the agent (ἐν αὐτῷ); and when the moving principle is in oneself, it is in one's own power, depends on oneself (ἐπ' αὐτῷ), whether to act or not" (*Eth. Nic.* III 1, 1110a15–18). We see that the fact in itself that the source of the body's movement lies in the human being (and not, say, in the wind which makes its body move along with the ship

[9] In order to stress this peculiar character of human actions (to be subject to moral evaluation whenever it is a matter of submitting the action or the agent to the moral judgment), I call them sometimes in what follows "moral actions." As just a specification of this kind of actions the adjective "moral" does not designate a "positive" quality of the action in contradistinction to the "immoral."

to a certain place against the person's will) does not make the body's movement a moral *action*. The ἐν αὐτῷ is not so important as the ἐπ' αὐτῷ. The difference between the prepositions[10] allows us to designate in a *preliminary* way the difference between a human action bearing an ethical determination, and a deliberate bodily movement, which cannot be an object of ethics. This difference becomes more precise in the subsequent development of the argument. And yet the context of the body, its instrumental parts and their movement determines the direction in which the search for a solution proceeds.

Going back to the discussion of the soul's topography we must remark that Aristotle, of course, cannot and does not mean, to use a Scholastic designation, the "real" parts of it, i. e., existing separately and independently. The soul is *the form* of the living body. The form is the principle of unity and the foundation of the entity's being [as such and such a *definite* thing].

"Unity" has many senses (as many as "is" has), but the proper one is that of *entelecheia*. (*De anima*, 412b9f.)

The soul as *entelecheia* or form is *simple*, and therefore does not have *real* parts. The "parts" Aristotle speaks of (in *the Nicomachean Ethics* too) are structural moments (faculties or functions), which cannot exist by themselves separated from the whole[11] just as a side of a triangle cannot exist as a side of a triangle separately from the triangle.

1.3.3. *Eternal and Temporal Truth* (αἰών *and* καιρός)

Yet since the abilities of the rational soul being discussed are connected with the logos, i. e., with understanding and calculating taken in the broadest sense, Aristotle distinguishes the "parts" (functions) of the soul according to the ontic character of the entity *manifested and brought to light* by their means, i. e., by means of the *energeiai* of manifestation (= perfect actions) inherent in them. "In a sense the soul is all beings," says Aristotle (*De anima* III 8, 431b21). "The soul is all beings" means that the forms of the things without matter are (potentially) located in the soul as in a "place of forms." The "parts" or faculties of the soul have a certain likeness or affinity to their "objects," "for it is in virtue of a certain likeness and kinship with their objects that they have the understanding (γνῶσις) they have."[12] The parts of the soul correspond to the genera of the entities they are "receptive of" and, as a result, also to the ways of arriving at the truth. Now we could also say that they correspond to the ἐνέργειαι *of the truth*, i. e., to the actuality of the disclosed = the activity of disclosing = the specific character of disclosure.

---

[10] Cf. the discussion of this passage in connection with the problem of the "ascription" of action to the agent in P. Ricœur, *Oneself as Another* (Chicago and London: The University of Chicago Press, 1994), pp. 88–112.

[11] The only exception is perhaps the so-called νοῦς ποιητικός discussed in *De anima* III 5. However, the meaning of this exception requires thorough analysis and clarification.

[12] *Eth. Nic.* VI 1, 1139a10f. For the reasons, which will become clear later, I substitute "understanding" for "knowledge" in Ross' translation.

And let it be assumed that there are two parts which possess reason (λόγος) — one by which we contemplate the kind of things whose principles cannot be otherwise, and one by one by which we contemplate variable things. [...] Let one of these parts be called the scientific and the other the calculative, for to deliberate and to calculate are the same thing, but no one deliberates about what cannot be otherwise.[13]

As regards the entity disclosed, the distinction between the scientific and calculative parts of the soul is drawn on the basis of the following differentia specifica: the "scientific part" makes manifest the entity, "the first principles of which cannot be different" (αἱ ἀρχαὶ μὴ ἐνδέχονται ἄλλως ἔχειν). Such principles belong to the invariable and the eternal (τὸ ἀΐδιον). The other part, the circumspective and calculating one, relates to *the first principles which admit of variation*, i. e., it discloses the changeable, the eventful, the situational being, that which gets unlocked in a single "moment (καιρός) of vision."[14] The καιρός of the revelation and recognition of the truth (ἀναγνώρισις) in Greek tragedy is the most striking example of such a disclosure. Oedipus, approaching the culminating point of the plot and the cutting edge (ἀκμή) of his fate, utters: ὡς ὁ καιρὸς ηὑρῆσθαι τάδε (OT 1050).

Wisdom (*sophia*) is the clearest insight into the first principles of what-is, the principles that cannot be different.

It follows that the wise man must not only know what follows from the first principles, but must also disclose truth about the first principles. (1141a18f.)

To see what follows from the first principles means to be able to deduce conclusions from premises, i. e., to master the art of demonstration (ἀπόδειξις). It is intelligence (νοῦς), which is capable of disclosing the first principles. Both these faculties — the art of deduction and intellectual insight — belong to the scientific part of the soul, and *sophia* is their fulfilment and completion.

Therefore wisdom must be intellectual comprehension (νοῦς) combined with knowledge — knowledge of the highest objects, [the intellectual contemplation,] which has received as it were its proper completion. (a19f.)[15]

Aristotle's decision in favour of *sophia* is well known. This verdict determines also the *logos* in which *sophia* discloses the truth (ἀληθεύει ὁ λόγος): this is the *logos* of first philosophy. Its matrix is Aristotle's *Metaphysics*, its text is European metaphysics. Heidegger's fundamental ontology is an attempt to think differently yet, as I intend to show, within the same configuration of topics. Aristotle has already chosen the parties to the suit, and Heidegger wants to appeal against the verdict. The court where the appeal is made is the *crisis* of European metaphysics. Heidegger understands "the first object" of first philosophy in a way contrary to the Aristotelian project; his fundamental ontology is the ontology of *action* (πρᾶξις) and creativity (ποίησις). Prudence and circumspective understanding become the first and the most fundamental way of disclosing beings in its being.

---

[13] 1139a6–14. Trans. W. D. Ross.

[14] I refer here also to Heidegger's concept of "Augenblick," "the twinkling of an eye" which will be of crucial importance for us in chap. 7. The "Augenblick" is "the present that is held in resoluteness and springs from it" (GP 407).

[15] Trans. W. D. Ross.

1.4. *Prudence*

Prudence is, according to Aristotle, a "disposition of the soul" (ἕξις), the state (the formal moment of the "first *entelecheia*") and possession of the soul (*habitus, Habe*). Yet as distinguished from, say, the creative ability-skill or art-mastery, which can be *more or less* developed, prudence by itself is *already* a virtue.

Any virtue is the best disposition of the soul, completeness, maturity, fulfilment, *entelecheia* of its definite constituents, those dispositions "which will best qualify them to disclose the truth" (cf. *Eth. Nic.* 1139b13). And yet a virtue as such is an ability or potentiality (*potentia secunda*) with respect to the actuality of the action proper to it (τὸ ἔργον τὸ οἰκεῖον).[16] In this action the "second *entelecheia*," the fullness of the truth *proper to this part* of the soul, is reached.

With regard to prudence (*phronesis*), ἀληθεύειν means to arrive at truth-unconcealment so that human good acquired in an appropriate action becomes visible (and can be interpreted and articulated in a peculiar way):

ὥστ' ἀνάγκη τὴν φρόνησιν ἕξιν εἶναι μετὰ λόγου ἀληθῆ περὶ τὰ ἀνθρωπίνα ἀγαθὰ πρακτικήν.— It is, therefore, necessary to consider prudence to be a disposition or state of the soul which partakes of logos and discloses the truth, being concerned with action in relation to the things that are good for human beings. (1140b20 f.)[17]

It is true that the "articulation" mentioned above is of a special kind: prudence discloses the truth in the element of its proper logos, of its proper understanding-interpretation. The text itself of the *Nicomachean Ethics* is a theoretical reflection *on* the "practical logos," i. e., a metaphysics of human action, an attempt to grasp somehow the logos of *phronesis* in the theoretical logos of *sophia*, to construct a science (ἐπιστήμη) which has the structure of *moral action* as its object. Yet science is the knowledge of the invariable: τὸ ἐπιστητόν = τὸ ἀεὶ ὄν. A "living" occurrence, a "fleeting instant," "the twinkling of an eye" of the ethical situation cannot be "known" in the sense of ἐπιστήμη, for the preferential choice (προαίρεσις) as the driving force of moral action (1139a31f.) implies precisely an absence of a necessary foundation. Τὸ προαιρετικὸν τῆς ψυχῆς, the choosing power of the soul, the ability to decide or freedom of choice, discloses the truth when there is no advice nor compulsion of the "eternal principles," and belongs to the καιρός, not to the αἰών. Καιρός is proper or critical time, the instant of opportunity, the time which "waits for no man."

It is impossible to deliberate (βουλεύεσθαι) concerning eternal entities outside of the human sphere, "for there are other things much more divine in their nature even than man, e.g., most conspicuously, the bodies (sc. the stars) of which the heavens are framed" (1141b1ff.). That is why the text of ethics does not articulate understanding itself (γνῶσις), which is proper to prudence and circumspectively "calculated" in the element of the λόγος πρακτικός; it is an external reflection *on* ethical "reasoning-projecting" (τὸ λογίζεσθαι), a plastic image (πλάσμα) which strives to stop and to represent the living gesture of the moral action.[18] Aristotelian

---

[16] *Eth. Nic.* VI 1, 1139a17.

[17] My translation.

[18] The Greek word πλάσμα signifies "image," "figure," formed or moulded, but also "figment," "fiction" and "forgery."

ethics as a metaphysics of human action searches for its ontological foundations within the scope of categories and principles of the "first philosophy." The results may be true in their way. Yet this truth is something quite different from the truth, which the soul discloses in action. Philosophical ethics cannot be the "proper logos" (λόγος οἰκεῖος) of action. According to another great text of Aristotle, his *Poetics*, the discourse expressing the truth of moral action must be understood as *poetical* discourse. That is why Heidegger's fundamental ontology, ontology of understanding (P. Ricoeur) or ontology of *human action*, had to become onto-poetics.

The principles (ἀρχαί) of a particular moral action, choice, or decision escape "proper" philosophical speech, proper λόγος τῆς οὐσίας, because they are transient, sudden, unpredictable, and *nulla est scientia fluxorum*. Nevertheless, in Greek tragedy abruptness, the inimitable character of a decision, is elevated to the order of language and thoroughly preserved, because the event of choice is inscribed into the internal temporality of a tragic tale, into speech which has rhythm and tune, into the melodic line of verse. Paradoxically, the tragic event endlessly repeats itself as the *same and still unexpected*, as always self-identical suddenness. Through the peculiar structure of its plot, through internal poetical time, tragedy exposes how an action comes to its fulfilment, its ripeness.

According to Aristotle's definition in *Poetics*, "tragedy is an imitation of action" (μίμησις πράξεως).[19] And we should not forget that poetic "imitation" comes about mainly (though not exclusively) in speech and as speech: within tragedy a character exists only as speech, which discloses and conceals, and postpones the real meaning of events. However *mimesis* here amounts by no means to a clear and faithful *description* in the ordinary sense of the word. The language of tragedy is powerful enough not only to describe but to *excite* pity and fear (as Aristotle says in his *Poetics*), and that is why it is able to express (or authentically "imitate" — καλῶς μιμεῖσθαι) the whole structure of the human action. Not only in its directedness to some end, which is somehow always understood and calculated within the situation of choice, but also in desire, which Aristotle considers to be the *efficient cause* of each human action. The language of tragedy articulates *Dasein's* being-attuned (*Gestimmtsein*), which along with "understanding" (*Verstehen*) discloses the being of entities.

Poetic speech or diction (λέξις) as the most important part of tragedy and the most powerful instrument of *mimesis* has its own way of disclosing the peculiar "meaning of the unique and the transient." While a philosopher, a σοφός, strives to express the essence of what is present-at-hand by means of the most adequate, the only proper term, which codifies the appropriate definition, the poet tries to elevate the entities to the new poetical order out of the indifference of its commonplace existence:

> The perfection of diction (of λέξις) is for it to be at once clear and not mean. The clearest indeed is that made up of ordinary words for things, but it is mean... On the other hand, the diction becomes distinguished and non-prosaic by the use of unfamiliar words [i. e. dignified and outside the common usage]... (*Poet.* 1458a18)

Thus, in poetic speech the clarity of θεωρία is sacrificed for the sake of "dignity" and "height." The poets give to the things strange, unusual names, which are drasti-

[19] 1450b3.

cally different from the proper ones, they use metaphors and metonymy.[20] To grasp strange resemblances, which are far beyond and far above commonplace similarities, is the highest point of *mimesis.* Through the articulation of this resemblance human works, and human moral actions, are inscribed into the whole of the world and history.

Paradoxically enough, in this strange mimetic mirror, through the mediation of narration, through the catharsis of passions, the Self which is the source, the principle of action, which is responsible for it and, thus, is the subject "for praise and blame," discloses itself in the most appropriate manner.

1.5. *Principles of Action*

As we have said already, Aristotle's reasoning always remains connected with the "first philosophy" and its main question: *What is being as being?* Ethical writings make no exception to this rule: Aristotle is interested in the ontology of human action subject to ethical determination. We shall see further on that such an entity as an action or a doer (insofar as he performs moral actions) do not, in a certain sense, "fit into" the system of basic principles of "first philosophy." For Aristotle himself this circumstance is an occasion of reservations and added detail; for Heidegger interpreting the *Nicomachean Ethics* it is one of the motives for a radical revision of the foundations of metaphysics connected with a re-posing of the question of "the meaning of being" in *Being and Time.*

How can the ontology of human action be fitted into the conceptual system of the *Metaphysics*? Aristotle says (1139a31–33):

πράξεως μὲν οὖν ἀρχὴ προαίρεσις, ὅθεν ἡ κίνησις ἀλλ᾽ οὐχ οὗ ἕνεκα. προαιρέσεως δὲ ὄρεξις καὶ λόγος ὁ ἕνεκά τινος.

The principle of action, as the source of movement, not as the "for-the-sake-of-which" [guiding the action], is choice; the principle of choice is desire and the logos manifesting the "for-the-sake-of-which."[21]

---

[20] "And it is a great thing to make proper use of each of the elements [of ornamentation] mentioned, but by far the greatest thing is to use metaphor. That alone cannot be learned: it is the token of genius. For the right use of metaphor means an eye for resemblance."

[21] My translation. To disclose the primordial intentions of Aristotle, those that are the subject of Heidegger's interest, we need to refrain from following the generally accepted ways (*via* the Latin mediation) of translating some terms of Aristotle into English (sacrificing sometimes the smoothness of discourse in order to stress some shades of meaning). The cited passage in H. Rackham's translation (*Loeb Classical Library*, vol. 73, p. 329) reads as follows: "Now the cause of action (the efficient, not the final cause) is choice, and the cause of choice is desire and the reasoning directed to some end." Ross: "The origin of action – its efficient, not its final cause – is choice, and that of choice is desire and reasoning with a view to an end." The connection between this thesis of Aristotle, on the one hand, and Heidegger's *das Worumwillen* as a primordial determination of existence, on the other hand, becomes completely lost. Heidegger himself wrote in the "Introduction" to *Being and Time*: "With regard to the awkwardness and 'inelegance' of expression in the analysis to come, we may remark that it is one thing to give a report in which we tell about entities, but another to grasp entities in their being. For the latter task we lack not only most of the words but, above all, the 'grammar.' [...] And where our powers are essentially weaker (sc. in comparison with the Greeks) [...] the hardness of our expression will be enhanced, and so will the minuteness of detail with which our concepts are formed" (SZ 38 f.).

The action *is* insofar as it is actually performed. The ἀρχή, i. e., the foundation or cause of the action, is designated by the word προαίρεσις, *deliberate* (βουλευτική, i. e., *conscious, responsible, not blind*) *choice*. The choice is based on deliberation (consideration of reasons, assessment of possible consequences of the action), but most of all on consideration of the means of achieving the desired goal. "We deliberate not about ends, but above all about means." (*Eth. Nic.* III 5, 1112b12f.) And the decision is made concerning things, which are in our power (ἐφ' ὑμῖν). Yet all this does not suffice to perform the action. The actual choice adds resolution to decision. The choice is connected with ὄρεξις βουλευτική, a desire (which is in a certain way determined through deliberation) to carry out the decision made, resoluteness manifesting itself in a proper action (ἔργον οἰκεῖον). This carrying out of the decision, which attains and fulfils (in the sense of Aristotle's term ἐντελέχεια) the goal, is the κίνησις of the action; this word is used strictly as a special term and designates actuality (*energeia*) or completeness (*entelecheia*) of a potentiality, qua potentiality.[22] Yet now a slight transformation of meanings of the key metaphysical terms must not escape our attention. "Potentiality" is to be understood as something disclosed and projected in the element of *I can*, insofar as it is revealed to me as *my possibility*, as something "in my power," as something dependent on the agent himself: ἐπ' αὐτῷ.

As we have seen, every movement takes place "from something into something else" and has *that-towards-which* as one of its formal determinations. And this "towards-which" is nothing but the *fullness of presence* at the end, as the result of the movement, of what had been predelineated at its beginning. This is the goal (τέλος), the *for-the-sake-of-which* (τὸ οὗ ἕνεκα) of the movement.

Aristotle distinguishes movement from coming into being and perishing because the former supposes an invariable substrate, an entity of the first category (*substantia*), that underlies the changes in quantity, quality or location, whereas coming into being or perishing, understood absolutely, is a change in the category of substance, i. e., appearance or destruction of a certain (always self-identical) form *in matter*.

Can we really say that moral action is movement (in Aristotle's sense)? Are we not dealing here with coming into being of a certain state of affairs, of certain circumstances, which are considered good or bad? No, Aristotle answers, since when an action aims at creating or destroying something which lies outside of this activity itself, we are dealing with making or production (ποίησις), not with action (πρᾶξις) as subject to a possible moral evaluation.

"Among things that can be otherwise," Aristotle says, "are included both things made and things done; making and acting are different" (1140a1f.). What is done is done by art, mastery. The ontological meaning of art is that of ἀληθεύειν; τέχνη is a way of disclosing the truth, the faculty of the soul to which the *first principles* of the entity, which is brought to light in (creative) making, are revealed. And these first principles may be different each time, because such an entity appears neither necessarily nor naturally, but originates with the maker (*Eth. Nic.* VI 4), in an arbitrary

[22] Cf. our discussion of the Aristotelian notion of *movement* in chapter 2.

way. This holds true for moral action too, only the acts of the master, his work (ἔργον), in contradistinction to the doer's deed, are carried out towards a result lying outside of the activity itself. Each making is directed to an external goal. He who makes something always has some further end in view. The act of making is not an end in itself; it is only a means and only proceeds for the sake of something else (1139b1ff.).[23] Aristotle says that a work of art, when it is completed, exists παρά — *apart* or *alongside* of the act, not *in* it.[24]

τῆς μὲν γὰρ ποιήσεως ἕτερον τὸ τέλος, τῆς δὲ πράξεως οὐκ ἂν εἴη. ἔστιν γὰρ αὐτὴ ἡ εὐπραξία τέλος.

For while making has an end other than itself, action cannot; for goodness of the action itself is its end. (1140b6ff.)

The thing created (ποιητόν), alienated from the act of creation, acquires its own being independent of the master. The work when it is made, the ἔργον, exists separately from activity (ἐνέργεια) of making as its result, but with moral action the situation must be different; human action considered from the ethical point of view is itself a certain *energeia*. In what follows I shall make this statement more precise.

6. Εὐπραξία, Εὐδαιμονία *and Eigentlichkeit*

Let us consider what the term εὐπραξία could signify. This is neither just "well-being" or "well-doing" in the everyday sense nor "benefaction" as alms or a service rendered kindly; it is a characteristic of *the way of doing itself.* We translate it "the goodness of the action." But what does the goodness of action consist in? Let us come back to the "specific difference" which underlies the Aristotelian distinction of making and acting. We find the ground for this distinction in the first lines of the *Nicomachean Ethics*. The specific difference is manifested in relation to the ends to which human activities aim. In some cases the activity is itself the end, whereas in other cases the end is some product (the "work" — ἔργον) over and above the mere exercise of the art. The *ends* which are themselves activities must be called *energeiai* in complete conformity with the metaphysical scheme discussed in the chapter 2, section 1.2.[25]

Εὐπραξία is the goal of the action (i.e. one of its ontological foundations) and at the same time *the doer's way of being* manifested in every action, a special *energeia* of human action and a special *energeia* of the agent's self-manifestation. The "goodness of the actions" defines the character of my being, i. e., of my life (since for a living creature, as Aristotle remarks, *to be* means *to live*). Εὐπραξία belongs to the *ends*

[23] Let me quote Heidegger's commentary on this passage: "Das ἔργον hat in sich die Verweisung auf etwas anderes; als τέλος ist es *von sich wegweisend*. Es ist πρός τι καί τινος, 'zu etwas für jemanden.' Der Schuh ist hergestellt *zum* Tragen, *für* einen Anderen" (PS 41). We recognize immediately here the theme of the "ready-to-hand" (*des Zuhandenseins*), developed in *Being and Time*.

[24] Cf. *Eth. Nic.* I 1, 1094a4ff. M. Heidegger, PS 42.

[25] Let me cite the Greek text here because of its importance: διαφορὰ δέ τις φαίνεται τῶν τελῶν τὰ μὲν γάρ εἰσιν ἐνέργειαι, τὰ δὲ παρ' αὐτὰς ἔργα τινά. ὧν δ' εἰσὶ τέλη τινὰ παρὰ τὰς πράξεις, ἐν τούτοις βελτίω πέφυκε τῶν ἐνεργειῶν τὰ ἔργα (*Eth. Nic* I 1, 1094a3–7).

which cannot be separated from (do not exist παρά) activity itself, and which according to this peculiar inseparability determine the action as being *perfect* or *complete.* Thus *eupraxia* is a certain *energeia* of human action. As activity-actuality it presupposes a definite potentiality. The potentiality in question must be understood as "disposition" (ἕξις) or a state of the soul. This state is, of course, a virtue or excellence (ἀρετή). And this is what the term *prudence* refers to.

Now it is thought to be a mark of a prudent man to be able to deliberate well about what is good and expedient for himself, not in some particular respect, e. g., about what sorts of things conduce to health or to strength, but about what sort of things conduce to the good life in general (πρὸς τὸ εὖ ζῆν ὅλως).[26]

The prudent person *is able to deliberate well* because he clearly sees the grounds and sources of his decisions in his whole life. In Heidegger's translation the words of Aristotle πρὸς τὸ εὖ ζῆν ὅλως are transformed as follows: (... *was zuträglich ist*) *für die rechte Weise des Seins des Daseins als solchen im Ganzen* ("what is advantageous for the correct way of *Dasein*'s being"). Later on in Heidegger's commentary and finally in *Being and Time* the "correct way of being" will become *Eigentlichkeit*, meaning *authenticity.* In the following chapters we shall discuss this concept in more detail. Yet some remarks are already necessary at this stage of our analysis. The formal characteristic of *Dasein*'s being is its peculiar character of *Jemeinigkeit* of being "in each case mine" (SZ 42); the "essence" of *Dasein* lies in the fact that in each case it has its being to be, and has it as its *own* (SZ 12f.). "And because Dasein is in each case essentially its own possibility it can, in its very being, choose itself and win itself" (SZ 42). To choose itself means to choose *was zuträglich ist für die rechte Weise des Seins des Daseins als solchen im Ganzen.* But this is nothing else as a translation from Aristotle's Greek. Only in so far as *Dasein* has this possibility of a radical choice, radical προαίρεσις, can it be *authentic.*

As modes of being, *authenticity* and *inauthenticity*... are both grounded in the fact that any *Dasein* whatsoever is characterized by "mineness" (*Jemeinigkeit*). (SZ 42f.)

Here *authenticity* amounts to a certain transparency of existence, which makes manifest "my own ἀρχαί" and allows *my own* being to be fulfilled in the world into which it has been thrown. In his "Greek," etymologically dissected and transformed into German, Heidegger renders thus Aristotle's definition of prudence (slightly paraphrased): ἕξις ἀληθὴς μετὰ λόγου πρακτικὴ περὶ τὰ ἀνθρώπῳ ἀγαθά: *ein solches Gestelltsein des menschlichen Daseins, daß es über die Durchsichtigkeit seiner selbst verfügt*[27] (such an attitude of the human *Dasein* in which it possesses the transparency for itself). Εὐπραξία is thereby identified with transparency, complete visibility of *Dasein* for itself — to be more precise, with such a character of actions which proceeds from this transparency and manifests it. In SZ such a character of existence is called "anticipatory resoluteness" (*vorlaufende Entschlossenheit*).

---

[26] 1140a25ff. Trans. W.D. Ross (slightly modified).

[27] PS 50.

Aristotle actually does say (*Eth. Nic.* VI 5, 1140b11–20) that temperance preserves (literally "saves") prudence (σωφροσύνη σῴζει τὴν φρόνησιν),[28] that is, preserves the notion (ὑπόληψις) of human good. It follows that intemperance (say, excessive pleasure or suffering) destroys this notion. And this destruction is of a special kind: for the notion that the sum of the angles of a triangle equals two right angles is in no way subject to destruction as a result of excessive pleasure or suffering. Only notions connected with action are subject to such "destruction." What happens when they are "destroyed?" What gets lost in the process? – The understanding of the foundations and the first principles of the action.[29] What are these foundations? – That *for the sake of which* the action is done. So temperance preserves the unconcealment (ἀλήθεια) of the first principles of the action. This unconcealment, this manifest character of the *for-the-sake-of-which* is the activity-actuality of *prudence*, not of intellectual contemplation ("*phronetic*," not "*noetic*" *energeia*).

Thus does prudence disclose the truth. Yet the first principles and foundations spoken of here "can be such and can be different," and besides, in contradistinction to creative activity, the action always has the being of the agent himself as its goal, the being of every doer.[30] As was already said, the first principles of an action cannot be known in the sense of ἐπιστήμη, i. e., retained in supratemporal self-identity befitting knowledge. It would seem that this precludes for us any possibility of further analysis. If there actually are no *universal* foundations of good in any sense whatsoever, then only one universal formula (the last refuge of theoretical ethics) still remains true: the only good of man consists in the *transparency* of his own being, his being as a doer, his being for himself.

That is why, Heidegger concludes, prudence as a way of disclosing the truth is connected not with the vision of this or that *eidos* which must be posited as the foundation of the action according to some universal law, but with the *possibility of seeing* itself, with *Dasein*'s transparency for itself. Strictly speaking the metaphor of vision (intelligent vision) must give way here to the metaphor of hearing: the understanding of the authenticity of existence is connected in *Being and Time* with *the call of the conscience* or *the call of care* which one must be able to hear.[31] The conscience *bears witness to* the authentic, proper ability-to-be (*Seinkönnen*). Thus the τὸ εὖ ζῆν ὅλως of Aristotle becomes, in Heidegger, a special internal acoustics of *Dasein*.[32]

---

[28] *Eth. Nic.* 1140b11. Aristotle here follows Plato in the etymological "analysis" of the word σωφροσύνη: σωφροσύνη δὲ σωτηρία οὗ νυνδὴ ἐσκέμμεθα, φρονήσεως (*Crat.* 411e4f.).

[29] "A man who has been ruined by pleasure or pain, entirely fails to discern any first principle, and cannot see that he ought to choose and to do everything as a means to this end, and for its sake; for vice destroys the sense of principles." (*ibid.*)

[30] "Ein Resultat ist nicht konstitutiv für das Sein des Handelns, sondern lediglich das εὖ, das Wie. Das τέλος in der φρόνησις ist der ἄνθρωπος selbst. Bei der ποίησις ist das τέλος ein anderes, ein weltlich Seiendes gegenüber dem Dasein, bei der πρᾶξις nicht." (PS 51)

[31] See SZ, §§ 54–60. "Das Rufen fassen wir als Modus der Rede. Sie gliedert die Verständlichkeit. [...] Die 'Stimme' [des Gewissens] ist aufgefasst als das Zu-verstehen-geben. In der Erschließungstendenz des Rufes liegt das Moment des Stoßes, des abgesetzten Aufrütteln. Gerufen wird aus Ferne in die Ferne. Vom Ruf getroffen wird, wer zurückgeholt sein will." (p. 271)

[32] "Die 'Täuschungen' entstehen im Gewissen nicht durch ein Sichversehen (Sichver-rufen) des Rufes, sondern erst aus der Art, wie der Ruf gehört wird – dadurch, daß er, statt eigentlich verstanden zu werden, vom Man-selbst in ein verhandelndes Selbstgespräch gezogen und in seiner Erschließungstendenz verkehrt wird." (*Ibid.*, p. 274)

The soteriology of "temperance" spoken of by Aristotle consists, according to this interpretation, in acquiring this transparency, "sonority" of the "whole life." The end of the action, its ultimate *for-the-sake-of-which*, is *Dasein itself* insofar as it is disclosed (*erschlossen*) for itself in its resoluteness (*Entschlossenheit*) to be itself. And on this primordial disclosure Heidegger hinges all the other kinds of truth, including "scientific" truth traditionally understood. Actually this is what the project of fundamental ontology consists in. Fundamental ontology, building on the basis of the Aristotelian ontology of human action,[33] has as its object "the how" (*das Wie*) of the disclosure of *Dasein*'s being for *Dasein* itself; the structures of this disclosure are called *existentialia* in *Being and Time*; their prototypes are (among other things) the soul's "ways of disclosing the truth" listed by Aristotle in *Eth. Nic.* VI.

Yet, the subject of ethics proper as understood by Aristotle is in danger of getting lost when things take this turn.

The *Nicomachean Ethics* begins with the words:

Every art and every inquiry, and similarly every action and pursuit, is thought to aim at some good; and for this reason the good has rightly been declared to be that at which all things aim.

Prudence (the ability to determine what is *good for man*) forms the subject matter of the book VI. Finally, the last book (*Eth. Nic.* X 6–7) deals with "happiness" (εὐδαιμονία), which is declared to be the goal of all human activity. According to Aristotle, happiness is not a state or disposition of the soul, which serves as a possibility of activity and determines its specific internal character with respect to the good, but the activity-*energeia* itself. Happiness is listed among the activities (*Metaph.* 1048b18–35) that are defined as *energeiai in contradistinction to* movements. The inner form of such activity is complete at any moment and does not require temporal duration to be fulfilled: εὐδαιμονεῖ (ἅμα) καὶ εὐδαιμόνηκεν — "someone is experiencing happiness and is already happy" (b26). The *praesens* and the *perfectum* are declared to be identical. Thus happiness is activity (*energeia*) in accordance with virtue (*Eth. Nic.* 1177 a 12). This activity is not directed towards something other, but is self-sufficient; and it deserves to be chosen not for the sake of something else, but for itself. Consequently it is this *energeia* that constitutes the altimate εὖ ζῆν ὅλως, *die rechte Weise des Seins des Daseins.*

The chain of identifications — τὸ εὖ ζῆν ὅλως = εὐδαιμονία; τὸ εὖ ζῆν ὅλως = *Eigentlichkeit* — inevitably leads to Heidegger's interpretation of Aristotle's "happiness": "εὐδαιμονία" is to be translated into the language of the existential analytic of *Dasein* as "authenticity" (*Eigentlichkeit*) of the being of *Dasein*. This interpretation is by no means a pure arbitrariness or exaggeration. After having identified εὐδαιμονία with the contemplation of the first ontological ἀρχαί Aristotle proceeds with the statement[34] that the ability to abide (in θεωρία) by the first principles of being is the highest thing in man; and one ought to do all that man may in order to live in accordance with the highest thing in him.

---

[33] Among other things, to be sure.

[34] Cf. *Eth. Nic.* X 7, 1177b29ff.

It may even be held that this is true *self* of each, inasmuch as it is the dominant and better part; and therefore it would be a strange thing of a man to live not *his own* life but the life of some other than himself.[35]

Now, if βίος is "the being of *Dasein*," then ὁ αὐτοῦ βίος is *Dasein's own, authentic* being; and εὐδαιμονία constitutes the authenticity of this authentic being:

Sie (sc. εὐδαιμονία) macht die Eigentlichkeit des Seins des Daseins aus. (PS 172)

For Heidegger, it is *conscience* understood ontologically that bears witness to the *authenticity*, which amounts to the disclosedness of the ownmost ability-to-be. Yet in the conscience *Dasein itself* bears witness to itself; it is at the same time voice and hearing; it is, according to Heidegger, an entity projecting itself, stretching itself towards the future, and thus always possessing irreducible temporal "length," a lengthy call from afar unto afar (SZ 271). Heidegger does not admit openly (at least in *Being and Time*) a *universal* source of the calling voice. *Virtue* and the *good* of Aristotle become, as was already said, not that which must be heard by all together, and not even that which must be heard by every one, but a special acoustics of *Dasein*, a permeability for giving-to-understand which originates with *Dasein* itself. Virtue as mastery in disclosing the truth becomes the absence of noisy talk (*das Gerede*) in which the articulation of one's own ability-to-be is constantly lost in everyday idle chatter of the "they." Parmenides' *mortals* who listen with an ear deafened by noise and Heraklitus' *the many* or *sleeping ones* unable to hear the Logos, become Heidegger's *das Man*.

On the contrary, for Aristotle himself happiness as *energeia* gets its status of a higher goal from the ontological hierarchy of the principles, the hierarchy which has already been established in the *Metaphysics*. Happiness must be connected with the most divine part of the soul, which can perceive the most divine. This part of the soul is the intellect, which contemplates the first principles of entity as entity. It is contemplation by *sophia* as "first philosophy" fulfilled, i.e. actualised in the "intellect of the soul," that constitutes happiness according to Aristotle.[36]

On the other hand, an attempt to understand ontology as a special Heideggerian version of phenomenology is connected with the necessity *to reject* (or rather to refrain from accepting) any previously established ontological hierarchy and to bear the burden of solitude, the burden of responsible thinking which cannot be shared with anybody. To be more precise, the hierarchy of foundations is constructed anew, and in this hierarchy the being of *Dasein*, non-indifferent-for-itself, becomes the first principle. Within the framework of Heidegger's existential analytic, it is *Dasein* itself that gives witnesses on behalf of itself and bears witness to itself.

Coming back to the ontology of moral action we have to admit that all that Heidegger says concerning *Dasein* itself can be applied to the way of being characteristic of action also: action's essential determination is not exhausted by providing a certain external "what" (a description of the corresponding circumstances, the per-

[35] δόξειε δ᾽ ἂν καὶ εἶναι ἕκαστος τοῦτο, εἴπερ τὸ κύριον καὶ ἄμεινον. ἄτοπον οὖν γίνοιτ᾽ ἄν, εἰ μὴ τὸν αὐτοῦ βίον αἱροῖτο ἀλλά τινος ἄλλου. (1178a2f.). H. Racham's translation, italics mine.

[36] Cf. *Metaph*. XII 7, 1072b13–30. In particular Aristotle says: ἡ θεωρία τὸ ἥδιστον καὶ ἄριστον (l. 24).

sons involved and their roles, a *record*). Its essence consists rather in the fact that it is a moment and a foundation of the *Jemeinigkeit* belonging to existence. The *whatness* (*essentia*) of this entity, insofar as it can be spoken of, must be understood from its existence (*existentia*) (cf. SZ 42). The essence of action, insofar as we can speak of it, is uncovered only as a meaning, which outlines itself, projects itself into the future, postpones itself ("temporizes" and "temporalizes," *zeitigt*) and starts to run in pursuit of itself. The action is one of the moments of the "projection" (*Entwurf*) of the future, of the opening-up of the temporal horizon into which *Dasein* as *desiring intelligence or thinking desire* (*Eth. Nic.* 1139a4f.) projects its being upon possibilities (SZ 148). This projection upon the possibilities is called βούλευσις (deliberation) by Aristotle. We deliberate about things that are in our power (ἐφ' ἡμῖν) and are attainable by action (1112a30f.), and in addition not about ends, but about means (1112b11f.) according to our possibilities. The future is the only dimension in which desire (ὄρεξις) can exist.

He is prudent who can fore-see the development of the meaning of action in the open scope of the future, who prudently includes his choice in the temporal whole as an event of convergence, meeting or union of the three dimensions of time, the past, the present and the future. He is prudent who can inscribe the extremely concrete (τὸ ἔσχατον) instant (the "twinkling of an eye") of choice into *the whole of all life*. The meaning of action is performed (as a tragedy of Antiquity is performed) in the field of tension between the suddenness of choice and the whole of fate.

Still, the visibility, the transparency of this field with which the soteriology of temperance is concerned, belongs, according to Aristotle, only to the sphere of human things (τὰ ἀνθρώπινα, 1141b8f.). Outside of it there are objects "rare, marvellous, difficult and divine" (b6f.) with which the "higher" soteriology (that of *sophia*) is connected. In *Metaphysics* the principles of being as being are proclaimed as the highest of the high. In Aristotle's philosophy the object itself, being itself, which reveals itself in and through the activity of the soul (in *the soul's disclosing of truth*), but keeps *in itself*, independently of the "human," its own eternal principles and foundations, becomes the criterion of the "lofty character" of *contemplation*. For, indeed, the faculties of the soul have "a certain likeness or kinship with their objects." That is why σοφία is said to be the highest virtue of the soul, and at the same time the moments (χρόνος μικρός)[37] of perfect contemplation are the glimpses of the highest pleasure and the best life, which God possess perpetually (ἀεί) and we, the mortals, do only from time to time (cf. *Metaph.* XII 7, 1072b13–30).

Still, what is "divine" (τὰ δαιμόνια) is not human (τὰ ἀνθρώπινα). People may declare these marvellous and difficult things to be useless (ἄχρηστα), they say that the sages like Anaxagoras and Thales "display ignorance of their own interests" (1171b4f.), they may be wise but are not prudent (b5).

---

[37] 1072b15.

1.7. *Noema and Phronema*

Despite the unusual and "unclassical" character of the logos in which prudence calculates and articulates the stages of its concern, phronesis as a special "ethical noesis" must have, as it were, a "noematic counterpart" of its own, which I shall call *phronema.*

The primordial foundation of an action, as Aristotle says, is the for-the-sake-of-which (οὗ ἕνεκα). To the extent that it is already pre-determined, the "for-the-sake-of-which" must be *already* manifest, visible. Yet it is only with essential reservations that we can call this manifest, visible thing an *eidos*. An *eidos* is that towards which, to fulfil which and for the sake of which, genesis takes place. What-is-done, τὸ πρακτόν as πράξει ὄν, comes into being through a preferential choice (the choice is said to be the *causa efficiens*), and the principle of the choice, in its turn, is *that for the sake of which* it is made. But the "eidos" of the "for-the-sake-of-which," the "eidos" of care, is manifested in a special way. The external "what" of the action, the context of events (of which the doer is one of the parts), its record, the distribution of the participants and even the motives and circumstances which are given to the agent in "intelligent" reflection (νόησις) do not determine the essence of the πρακτόν as πρακτόν. To be more precise, the metaphysical concept of essence is nothing more than an imperfect substitution, a theoretical imitation of the being of the moral action and of what-is-done, something external with regard to it — to the same extent that philosophical ethics is an imprint, an *imitation* of the *practical* logos. The latter is concerned with circumspection and "striving" by means of which I find or lose my way in the confusion of conflicting reasons which affect and wound me in my being at the moment of making a choice. The manifest character of the "for-the-sake-of-which" for the sight (*Sicht*) of circumspection (*Umsicht*), concerned with "what is good and advantageous for the human being"; the sight we refer to saying that the preferential choice is not blind, does not amount to "awareness" (*das Bewußtsein, Bewußthaben*) in the classical sense. The "for-the-sake-of-which" as the principle of action, that which it cares for, is not a *noema* (νόημα), but a *phronema* (φρόνημα).

Of course, this usage of quasi-phenomenological terminology is conventional and even paradoxical, for it is introduced to discuss certain things that elude the phenomenological reduction and to understand the character of this elusion. Besides, in Aristotle the word *noema* has a perfectly precise meaning and signifies form without matter as it is in the *nous*; the actuality of this pure form as we have seen earlier is at the same time activity (*energeia*) of the *nous*, *noesis*. Yet *phronesis* is by no means a *kind* of *noesis*, it is defined as *energeia* of a different "part" of the soul. However,

εἶδος μὲν οὖν τι ἂν εἴη γνώσεως τὸ αὑτῷ εἰδέναι, ἀλλ᾽ ἔχει διαφορὰν πολλήν. καὶ δοκεῖ ὁ τὰ περὶ αὑτὸν εἰδὼν καὶ διατρίβων φρόνιμος εἶναι... (1141b33–1142a2)

Now seeing [what is good] for oneself will be one kind of understanding, though it is very different [from the other kinds], and the man who sees what is good for himself and is occupied with these [matters] is thought to be prudent.[38]

Being *analogous* to *noema*, *phronema* must have a certain "universal" character, but its universality, its ἓν ἐπὶ πολλῶν consists in the possibility of subordinating every time the *Augenblick* of the choice to the ultimate οὗ ἕνεκα, in the possibility of hearing the call of care, of appropriating the ownmost way of being *(das eigentlichste Seinkönnen)*, which possesses its own special internal unity. This unity as we have seen determines the internal *energeia* of my proper actions, and can be manifested in the multiplicity of my actions as ἓν ἐπὶ πολλῶν. It can become a common (καθόλου) property (a written rule, norm, precept) only as an imprint (πλάσμα) or inscription, and this inevitably results in a kind of forgery.

The paradox pursuing the special universality of "phronematic" understanding has two sides to it.

First, in contradistinction to the universal as the *ideal*, *phronema* is not, as already mentioned, "elevated" into the supratemporal order but is interwoven with the temporal flow, as a projection towards the forthcoming. He is prudent who sees his own good and occupies himself with that which belongs to this good: ὁ τὰ περὶ αὑτὸν διατρίβων — "he is prudent who spends his time in these occupations." As a result of the enigmatic and far from accidental coincidences, which appear in the concentrated (*gedichtet*) multilingual space that has acquired a poetical (*dichterisch*) dimension, the Greek διατρίβειν may have the same meaning as the German *zeitigen*. Penelope postpones ("temporizes"), puts off the wedding festivities (γάμον διατρίβει) by secretly unweaving the tissue and thereby letting fate attain its maturity in this temporization (*Zeitigung*) — the fate of which Moira weaves the tissue.

The temporizing of such a kind is connected, according to Heidegger, to the primordial "temporalizing" of time (*das Zeitigen der Zeit*). *Primordial temporality* is the main topic of our investigation. We postpone its detailed discussion till chapter 7. Yet what has been said up to now already enables us to hear and understand a remark of Heidegger: the "for-the-sake-of-which" (*das Worumwillen*) is the source from which the world "temporalizes" itself, which opens up the horizon of time.[39] He who is prudent strives to attain a certain *energeia* of his actions called εὐπραξία. He is forced to linger and wait, to spend his time in waiting, to bear time's wilfulness, its torturing delays and absurdly swift fleeting, for the sake of εὐπραξία, the good-doing itself, the authenticity and the transparency of *Dasein*. *Dasein* temporalizes time by temporizing its own fulfilment and by projecting itself into the fu-

---

[38] My translation.

[39] "Temporality temporalizes itself primarily out of the future. This means that the ecstatic whole of temporality, and hence the unity of horizon, is determined primarily out of the future. That is the metaphysical way of saying that the world, which is grounded in nothing but the ecstatic totality of the time horizon, temporalizes itself primarily out of the *for-the-sake-of*. This for-the-sake-of is, in each case, the for-the-sake-of willing, of freedom, i.e., of the transcendental being-towards-oneself." (M. Heidegger, *Metaphysische Anfangsgründe der Logik im Ausgang von Leibniz*, GA 26, V. Klostermann, Frankfurt a. M., 1978, § 13, p. 273.)

ture. The meaning of the action and the "essence" of *Dasein* arrive to themselves from the opened-up horizon of the future.

Second, the universality of *phronema* must be in a way the universality of desire (ὄρεξις). That for the sake of which the action is outlined in deliberation; that towards which it projects itself, must have "power" to initiate and to attract my action. I must not simply *know* (have the knowledge, ἐπιστήμη, the idea of) *what* is good for me, *what* is due etc., and *how* I must or must not act according to it. For the way of being of such *knowledge*, its "showing" (*apophansis*) in the logos, is (to use the logicians' phrase) a *deontic* modal judgment, i. e., ultimately the classical τὶ κατά τινος (something *about* something) modified. *Phronema* must be charged not only with meaning (*Bedeutung*) but also with significance (*Bedeutsamkeit*). The latter is not something external and additional in relation to the meaning, not a secondary noematic layer enveloping the central *noematic* core. Significance as a correlate of the "deliberating desire" belongs to the "phronematic core"; it attracts my choice and generates my action.

Heidegger defines the primordial meaning of meaning as follows:

> Meaning is the "upon-which" of a projection in terms of which something becomes intelligible as something; it gets its structure from a fore-having, a fore-sight, and a fore-conception. (SZ 151)[40]

The term "*phronema*" intends to express exactly this side of *meaningfulness*,[41] it is brought to use in this intertextual place in order to exhibit some tendencies of Aristotelian "ontology of human action" operative in Heidegger's fundamental ontology-as-phenomenology.

The true τὶ ἕνεκά τινος is disclosed originally for *care and in care*, it is not a *noema* of the good and of the due to which a special "way of givenness" (*Gegebenheitsweise*) is "added" externally, generating *deontic* modality in the "layer of expression" (to use Husserl's terms). Such a *noema* would be only a plastic imitation of the "for-the-sake-of-which" of care, a result of *theoretical* reflection. Yet the "unconcerned contemplation" (θεωρία) of "intelligent" reflection is in its turn a certain "conscious choice" which identifies the truth-rightness-righteousness (ὀρθότης) with the truth of theoretical thinking (θεωρητικῆς διάνοιας καὶ μὴ πρακτικῆς μηδὲ ποιητικῆς);[42] making this choice the thinker strives to go beyond his "human" origin, the *eros* of desiring thought = thinking desire. That for the sake of which the choice is made are divine objects themselves, far more divine in their nature than man, complex and worthy of admiration (1141b1). For example, the lofty stars which make up the heavens.

---

[40] Later on we shall discuss this definition in detail.

[41] "Meaning is the *existentiale* of *Dasein*... *Dasein* only 'has' meaning, in so far as the disclosedness of being-in-the-world can be 'filled in' by the entities discoverable in that disclosedness. Hence only *Dasein* can be meaningful or meaningless." (*ibid.*)

[42] 1139a27f.

CHAPTER FIVE

GOD WITHOUT BEING AND THOUGHT WITHOUT THINKER

1. THE SOURCE OF BEING WHICH "IS NOT"
(*ON THE DIVINE NAMES* V 5)

We have already said that for "late philosophy," the history of philosophy is not a mere illustration, a sort of complementary addition to the development of discourse: a method of enriching and specifying a meaning otherwise completely expressed in itself. History is the diachronic dimension of philosophical process, an essential aspect of positioning new meaning. "The history of philosophy," G. Deleuze wrote in his dissertation, "in our opinion, has to play a role similar in many respects to the role of collage in painting."[1] The collage allows us to make a new meaning out of meanings already expressed and historically connected with a certain context; the new meaning then appears as a figure of meanings, an arrangement or rather a consonance of meanings. I think we are dealing here not so much with *collage* as with a polyphonic development of the perennial philosophical theme — *what is a being as being?* We shall begin with a graphical illustration, at first taking the term "textual collage" almost literally.

1.1. *What Never Was Nor Will Be; What Is Not*

Parmenides, fr. 8DK, 5–6:

οὐδέ ποτ᾽ ἦν οὐδ᾽ ἔσται, ἐπεὶ νῦν ἐστιν ὁμοῦ πᾶν ἓν, συνεχές. τίνα γένναν διζήσεαι αὐτοῦ;

"It never was nor will be, since it is now, all together, one, indivisible. For what parentage of it will you look for?"

Plato, *Parmenides* 141e:

οὔτε ποτὲ γέγονεν οὔτ᾽ ἐγίγνετο οὔτ᾽ ἦν ποτέ, οὔτε νῦν γέγονεν οὔτε γίγνεται οὔτε ἔστιν, ἔπειτα γενήσεται οὔτε γενηθήσεται οὔτε ἔσται [...] Οὐδαμῶς ἄρα τὸ ἕν οὐσίας μετέχει.

Dionysius the Areopagite, *On the Divine Names* V 4:

οὔτε ἦν οὔτε ἔσται οὔτε ἐγένετο οὔτε γίνεται οὔτε γενήσεται, μᾶλλον δὲ οὔτε ἔστιν.

"[He] has not been, will not be and never was, He did not come [into being], does not and will not; more than that — He is not."

Apocalypse 4, 8:

ἅγιος ἅγιος ἅγιος κύριος ὁ Θεός ὁ παντοκράτωρ, ὁ ἦν καὶ ὁ ὢν καὶ ἐρχόμενος.

---

[1] *Différence et répétition* (Paris: PUF, 1968), "Intodiction."

"... it never came into being and never was coming into being and never was, it does not come into being now, and is not, and will not, and it will not be... So that the One in no way participates in being."

"Holy, holy, holy, the Lord God the Almighty, who was and is and is to come."

Parmenides' fragment speaks of τὸ ἐόν. We are quick to pronounce the word "being," we are used to the fact that Western metaphysics is centred on this term. The tradition has taught us to define and position by means of this name the main (and essentially the only one) subject of thinking. Yet, as we have seen, the name itself, τὸ ἐόν, was "established" by Parmenides in his "ontological revelation," it was revealed to him, was encountered by him on his wanderings "beyond the day and the night." For a philosopher, to wander means to question. In his wandering along the *way of being* (i. e., while asking what an entity is in its being, what it means for the beings to be), Parmenides finds many signs — σήματα πολλά — pointing to the "meaning of being."

One of these signs is τὸ ἕν, *the One*. This is the word taken up by Plato's Parmenides and it is the subject matter of the fragment cited above (*Parm.* 141e). This text, however, is only one of the subsequent steps of a dialectical argument, and can hardly be considered Plato's final judgment concerning the One. In the *Timaeus* it is said that the verb "to be" in the present tense (as opposed to "was" and "will be") expresses the eternal essence in an appropriate way (*Tim.* 37d — 38b).

The One (ἕν) belongs to the series of *Divine names* discussed by Dionysius in his treatise *On the Divine Names* (XIII 2).[2] In chapter V, from which the fragment we are dealing with ("...[He] has not been...") comes, another name is spoken of, *[He] Who-is* (ὁ ὤν). The author calls this name, following his (and his time's) predilection for pleonasms, poetic repetitions and alliterations, "the truly existing God-naming essential name of Him Who truly is" (V 1). This name is the first in order of superiority (πρεσβύτερον)[3] in the series of theological terms for God, because it is contained in the Scripture as God's own answer to the question "Who art thou?"

[2] "He is called *One* since He is singly all in one pre-eminence of unity; for he is (without becoming multiple Himself) the cause of the unity of all [multiple beings]. There is nothing among beings that is without participation in this One. [...] Whatever is a being is so by being one." The Greek text of the treatise *On the Divine Names* and the classical *Scholia* of John of Skythopolis and Maximus the Confessor (included in Migne's *Patrologia Graeca*) are cited after the edition by G. M. Prochorov: Dionysius the Areopagite, *The Divine Names and Mystical Theology* (St. Petersburg: Glagol, 1994). The text reproduces the critical edition: *Corpus Dionysiacum* I. Pseudo-Dionysius Areopagita: *De divinis nominibus*, B. R. Suchla, *Patristische Texte und Studien* 33 (Berlin–New York, 1990). The fact that throughout this chapter I write "Dionysius" and not "*Pseudo*-Dionysius" does not mean that I believe (against all contemporary philological evidence) in the authenticity of the authorship of Dionysius the Areopagite mentioned in *Acts* 17, 34. I am not at all occupied here with the problem of authorship. It is enough for me that the *Corpus Areopagiticum* has become one of the most influential theological treatises in the Christian East (and in Russia in particular), as well as in the West. Already this "external" circumstance, to say nothing about the content of the Treatise, justifies one more attempt to understand it. The treatise of Dionysius is cited hereafter as DN.

[3] DN V 5.

(*Exodus* 3, 14).[4] In it the essence itself, the *quidditas*, "what-ness" or, rather, the *quissitas*, "who-ness," of God, is referred to (though, as we shall see further, not *shown*, nor *expressed*). It is not accidental that this name is connected with the *essential naming* or *naming of the essence*.

Yet besides that, the "supremacy" of this name is proved by the author with arguments of quite a metaphysical turn, which partly remind the reader of Aristotle's reasons for establishing being *qua* being as the subject of "first philosophy," and are borrowed in part from the Platonic and neo-Platonic philosophical armoury. Any other name, be it Good, Life or Wisdom, by creating a certain perspective for contemplation and by revealing God as being this or that, automatically reveals Him as an entity, for only an entity can be revealed or, rather, un-hidden. Even life itself and wisdom itself become entities through participation in being as such.[5]

Yet, in a wonderful way, the name *Who-is* (ὁ ὤν) impels the text in a direction where it "transgresses" itself — more precisely, since we are speaking of names, not of "concepts" in the usual sense, it would be more correct to say that it loses its naming ability. For if Parmenides says about being, "it has not been at any time and will not be, because it *is now*, whole, all at once," and Plato states (*Tim.* 37c — 38a) that the expressions "was" and "will be" are legitimate in relation to *becoming* only, whereas "is" befits the *eternal essence*, Dionysius finds an insurmountable obstacle to a correct theology in the grammatical category of tense, in the fact that verbs *per definitionem* imply time.[6] And then he says "He-Who-is is not." Designating being by means of "is" already defines, i. e., limits being in a way.[7] The grammatical terms for the forms of the verb themselves affirm it: "is" is a *verbum finitum;* the way to convey a meaning, proper to this verb, the grammarians call *modus finitus* (finite, i. e., limited, de-termined mode), whereas "to be" represents the *modus infinitivus*, infinitive, or rather undetermined, mode.

> God is not somehow be-ing (ὤν), but simply and unlimitedly be-ing, comprehending and anticipating the whole being in itself... [He] is the Being for beings. Not only beings but even the Being itself for beings is from the be-ing before eternity. For God is the eternity of what is eternal, the One whose being precedes eternity. (V 4)

We see that in the text the distinction between being and beings is asserted (the ontological difference). More than that: as it seems, the author of the treatise distinguishes the being of an *entity* from a primordial and undetermined source of being. Any finite (created) entity *is* "somehow" (πώς ἐστιν ὄν); for it, "to be" means "to

---

[4] It goes without saying that Dionysius in his speculations leans upon the text of the *Septuagint*. The answer of God to the second question of Moses ("[If] they ask me, 'What is his *name*?' what shall I say to them?") reads in Greek: ἐγώ εἰμι ὁ ὤν. ὁ ὤν is the participle of the masculine gender derived from the verb "to be." In order to render the structure of this word in English I use sometimes the expression "[He] Who is," and sometimes an artificial construction "be-ing."

[5] DN V 5.

[6] This is the definition of the verb in general, given by Aristotle in *De interpretatione* 3.

[7] This is a sort of explanation that can be traced back to Plato's *Parmenides* and can be found in an explicit form in connection with the problem under consideration in the *Sentences* of Peter Lombard. Cf. Francis Ruello, *Les "noms divins" et leur "raisons" selon St. Albert le Grand, commentateur du "De divinis nominibus"* (Paris: Librairie philosophique J. Vrin, 1963), p. 51.

be present," whereas the *pre-eternal* Being (is) simple and unlimited (un-determined). Yet whatever the mode (sense) of being spoken of, we cannot say "being is," for *what is* is an entity. In particular this means that although present being is correctly *expressed* by the verb "is," "is" cannot *be predicated of* being (*esse*) as such. Yet here a kind of hermeneutical ἐποχή is to be accomplished and the temptation overcome to interpret Dionysius in the sense of Hegel's logic, which distinguishes between pure being (*Sein*) as "the simple immediacy" and being-extant (*Dasein*) as a result of mediation. We have to let the text speak for itself, and to do this we have to understand *how* it speaks and *how* the Divine names name.

1.2. *On the Poetic Way of Naming*

The expression "divine names" (τὰ θεία ὀνόματα) as a designation of a philosophical topic originated in *Cratylus* (401b and further). It was the wisdom, Plato's Socrates says, of the first name-giver,[8] i. e., of the first poet, of the proto-poet, to give names to the gods which in one way or another articulate, show, express their essential features. Dionysius' assertion is quite the contrary:

> The aim of our discourse (including naming God — *A. Ch.*) is not to manifest (ἐκφαίνειν) the Being beyond being as beyond being[9] for this is ineffable, unknown, and completely non-manifest... [but] to celebrate (ὑμνῆσαι) the being-producing procession of the thearchic source of the essence in all beings. (V 1)

For Dionysius naming is not the same as predication, i. e., logical clarification or *apophansis*. Certainly naming is a special method of showing, and consequently of manifesting, discovering, laying bare the thing named, making it accessible to the mind's eye. It seems that the Divine names are designed, by naming God, to display (ἀποφαίνεσθαι) Him as being the Good, Life, Wisdom, the Word... A name refers to something else as itself, it names *the thing named* (for otherwise, Plato says, it would be the name of a name);[10] and so it seems that the name somehow retains the formal structure of predication: *something of or about something* (τὶ κατά τινος). Yet the paradox of using the Divine names is that the "thing" named is never present as

---

[8] ὁ θέμενος πρῶτος τὰ ὀνόματα (436b).

[9] ὑπερούσιος οὐσία. I avoid rendering the prefix ὑπερ- traditionally as "super-," because by doing so we lose the negative, *apophatic* character of many of Dionysius' terms. ὑπερ- as "over" and "above" refer not only to excess, surplus, preeminence, superiority etc., but alludes also to "beyond," "outside the limits." In this case ὑπερ- has the same meaning as ἐπέκεινα (*praep. cum. gen.*). The relationship between the ὑπερούσιος οὐσία of Dionysius and Plato's ἐπέκεινα τῆς οὐσίας (*Resp.* VI, 509b) is rather obvious and perfectly explicable historically. Another morphological part of the adjective ὑπερούσιος is derived from οὐσία. The latter is not to be understood in a too technical Aristotelian sense as *essence* or *substance*. It signifies just being-ness of beings. That is why I refrain from translating ὑπερούσιος as "superessential" or "supersubstantial." Calling the divinity "superessental essence" would mean that the divinity transcends every finite being and is preeminently essential or substantial. In my opinion, this is not what Dionysius intends to say. The ὑπερούσιος οὐσία is rather "beingness beyond being," "otherwise than being or beyond essence" (to reproduce the title of one of Levinas' books). Of course, all these attempts to find English counterparts of the main terms in Dionysius can only provide a space where interpretation *begins* its work.

[10] *The Sophist*, 244d.

something, it is beyond essence and *a fortiori* beyond *whatness* (it has no "what" of its own, it is without a "what," ἄνευ τοῦ τί). The naming *apophansis* is only a reference to, or a designation of, the *locus* of what forever remains hidden in its essence.

We must, with all possible earnestness, consider the word ὑμνῆσαι, "to celebrate," "to glorify in words," as the designation of a *method*. Dionysius speaks of a poetic clarification, a *poetic apophansis*. Is not *glorifying* and *glory*, in the sense of the Greek δόξα θεοῦ, actually the primary way of showing, manifesting, bringing to light? In Plato's *Cratylus* the name is a tool in the hands of the person who seeks and attains knowledge, which enables him "to catch up with" (440a) the object of his investigation. The name enters as it were into conversation with beings and charms them by the very fact of being enunciated: an entity stands still then, "without transcending the boundaries of its idea," its meaningful aspect, *eidos*, is moulded into a firm and immutable shape, and now the entity is accessible to the mind's eye. "How can a thing," Plato's Socrates asks, "which never remains in the same condition, be some*thing* (τί)?" Such an entity "leaks like a cracked pot[11] and is just like a person with a cold in his head" (440c). Only *a something* (τὶ) always-remaining-in-itself can be named, and, conversely, to be named means for an entity to keep in itself an *eidos* accessible to the mind, and this means to attract and to keep directed to itself mental vision, the *noesis*. *He Who is*, however, has no "what" of His own (He is ἄνευ τοῦ τί). That is why for Him to be named is to remain in glory; this is a latent method of revealing and being revealed for a "content," which eludes the direct vision.

Establishing names is the eternal task of poets. "What remains, is established by the poets," Hölderlin says.[12] The art of giving poetic names has a nature of its own. The poet frees the entity out of the indifference of common sense, by calling it *not by its own name*, but by using a *metaphor*, remarking and establishing ὁμοίωσις, similarity. The metaphor manifests *this* through *something other*, and *the one* through its connection with *the other;* it weaves the thing into the world's whole. The metaphor, according to Aristotle, the author of the first *Poetics* in history, is a *transfer*, "giving the thing a name that belongs to something else" (*Poet.* 1457b6f.). The metaphor allows the poet to name by analogy, in a special way, even things which have no name.[13] Yet the not proper, the alien (ἀλλότριον), exists only in opposition to the proper (τὸ ἴδιον). Of course, Aristotle is interested not in the habit or rules of *usage*, but in an ontological understanding of an entity's proper logos (λόγος οἰκεῖος). This *logos* must express the *ontic* foundation of the entity, that is τὸ τί ἦν εἶναι, the true "what," the internal form of the thing. To express in discourse the being of an entity in its proper logos is to express τὸ τί ἦν εἶναι, to express *what* it means (and has always meant, and always will mean) *to be this thing*. A person skilled in logic expresses the essence in words, brings it to light, not merely by giving *a name* to the en-

---

[11] Plato imitates here the famous πάντα ῥεῖ ascribed to Heraclitus and says: πάντα — ὥσπερ κεραμία — ῥεῖ, "everything leaks like cracked pottery."

[12] "Was bleibet aber, stiften die Dichter." *Andenken*. See F. Hölderlin, *Gedichte*, hrsg. v. J. Schmidt (Frankfurt a. M.: Deutscher Klassiker Verlag, 1992), pp. 360–362.

[13] Aristotle refers to the verse: "the sun sows the god-given light." The action of spreading the light and warmth has no proper name, but it bears the same relation to the sun as sowing to the sower, and thus acquires its poetical name (*Poet.* 1457b25–30).

tity, but giving *a definition* (ὁρισμός). The definition is precisely λόγος οἰκεῖος, the entity's proper logos, the logos of the essence (λόγος τῆς οὐσίας). Aristotle says that if we know the definition, we can easily cope with such an unpleasant feature of language as equivocalness, and then *the proper name as a definition code* establishes a close and inseparable connection with the thing itself.

It is true, that the poet has to give up using "domestic" words which name the present state of things, for such naming means getting away from the essence of poetic *apophansis*. Here clarity as a merit of discourse (λέξεως ἀρετή, lexical virtue corresponding to wisdom as the virtue of the soul's epistemic part), comes into collision with the lowness (ταπεινότης) of commonplace language (*Poet.* 1458a18).[14]

Yet where there is no "logos of the essence" and can be none, since the thing named is beyond essence, poetics, as it seems at first sight, becomes the only possible logic. That is why Dionysius says: not to clarify, not to manifest, but to celebrate and to glorify. Where there is no definition, the philosophers say, opinion (δόξα) predominates. But δόξα means also glory as conspicuousness, splendour — the only way to resist sinking into oblivion (Λήθη). Un-forgettable-ness is ἀ-λήθεια in Greek; the poetic logos has its own way of abiding by the truth and in the truth.

1.3. *Naming by Analogy*

Thus the difference between proper (τὸ ἴδιον) and not proper (τὸ ἀλλότριον) names draws, in Aristotle's works, the boundary between poetic and properly philosophical discourse. This circumstance became a very serious problem for the medieval Latin commentators on Dionysius' work, for, as it has been said, it is impossible to make this distinction in the case of the Divine names. Albert the Great's commentary on this topic says:[15] the name refers to the same meaning (*ratio*), which is expressed in the definition. The definition only explicates (*explicat*) what is implicitly contained in the name. That is why the absence of definition means the absence of a proper name. The Divine entity cannot be defined by any means since it is not limited.[16] It follows that it has no proper name.

Yet it is impossible to accept that the Divine names are just metaphors, that they name God only figuratively (*figurative*). Such an admission would destroy the very basis of the scholastic, "reasoning," theology. Albert calls such an opinion "the most absurd heresy" (*absurdissima haeresis*). The solution of this problem is seen by him (following Aristotle), as well as by his disciple Thomas Aquinas, in a detailed concept of *analogy*.

Already the Philosopher himself, when he discovered the irreducible equivocalness of the name τὸ ὄν, the principal term of metaphysics established by Parmenides, had to search for the means of resisting the arbitrariness of poetic discourse,

---

[14] Cf. chap. 4, sect. 4.

[15] In DN XIII, q. 12 (*Utrum Deus possit nominari...*), *solutio*. Cited after: Francis Ruello, *Les "noms divins"...*, p. 89, n. 48.

[16] "...quod autem nullo modo finitum est, definitionem non habet, quare nec nomen. Deus autem nullo modo finitus est." (*ibid.*)

i. e., to look for a possibility of prohibiting the metaphorical arbitrariness where there can be no univocal discourse. "The word 'being' is used in many senses" (*Metaph.* VI 1, 1028a10).[17] It is important, however, that these multiple meanings of being do not indefinitely grow in number; according to Aristotle; as we have seen in chapter 2, section 3, they remain within the hold of an easily countable listing. The attempt, made in the *Metaphysics*, to limit the discourse on "being" within the limits (scheme) of a controlled polysemy,[18] does not allow it to lapse into a confusion of undifferentiated meanings. Yet a difference can be posed only in relation to a certain, although *non-generic*, semantic unity. For Aristotle this unity, since there is no common genus, is the unity of *analogy*, a new domain for philosophical discourse based on the existence of non-accidental homonyms referring πρὸς ἕν, *ad unum* (*Metaph.* IV 2), pointing to a non-metaphoric properly transcendental resemblance among the primary significations of being. *Analogy* lies midway between pure univocity (a complete uniqueness of meaning) and simple "accidental" dispersion of equivocal meanings, caused by the facticity of language.

Yet, while for Aristotle the goal was to find the middle way between a) accepting the generic unity for the meanings of being (the main Aristotelian thesis: "being" is not the designation of the highest genus) and b) recognizing these meanings as merely "accidental homonyms," the schoolmen were facing another dilemma. To accept the univocity of the terms naming *Him Who is* and designating created entities, would be to destroy the incommensurability of God and creatures, i. e., to reject the *superessentiality*[19] (or, to use a later Latin term, the transcendence) of God. On the other hand, assuming total incommunicability of meanings from the "celestial" to the "terrestrial" level would "condemn" the discourse about God to the arbitrariness of poetic celebration and reject any possibility of a proper onto-theology. We cannot here investigate in detail the complicated medieval teaching on *analogia entis* in all its diversity, and shall limit ourselves to what seems relevant now, outlining very briefly some theses of Thomas Aquinas.[20]

---

[17] See our discussion of this topic in chap. 2, sect. 3.1.

[18] I borrow this term from Paul Ricœur, *The Rule of Metaphor*, trans. R. Cherny (London: Routledge, 1994), Study 8, "Metaphor and philosophical discourse," sections 1–3.

[19] We have already mentioned that the adjective "superessential" or "supersubstantial" is somehow misleading as the translation of the Dionysius' term ὑπερούσιος. This circumstance had not escaped the attention of Scholastic thinkers. The problem here was actually of a more general kind, *viz.* it was necessary to clarify to what extent the concept of substance is applicable to God. As we have seen in chap. 2, the Aristotelian term οὐσία is a cross-section of different ontological intentions. On the one hand, substance is defined as the ontological principle, as something independent in its being of anything else, *ens a se et per se*. On the other hand, substance is the "primal subject," i. e., the name of a substance cannot be predicated of anything else, that is to say, substance is the ultimate subject of all possible determinations. The medieval writers insisted that in characterizing God as "substance" only the first meaning of the term must be retained. It is necessary to reject any complexity in God's being. And that means that the divine substance cannot be the subject of manifold accidental determination, the difference of *substantia* and *accidentia* is unthinkable in God, for it would imply His complexity. In other words, God cannot be "subordinated" to the *category* of substance.

[20] Cf. P. Ricœur, *op. cit.*, Study 8, sect. 2, "Metaphor and *analogia entis*: onto-theology," pp. 171–280.

First, the ontological basis of analogy lies in the fact that created entities participate in Divine being. Such is the meaning of the term *analogia per participationem*. In the earlier works (*Commentary on Book I of the Sentences of Peter Lombard*) participation is understood as *imitation*. God *communicates* His image to finite beings, and the creature *imitates* the First being (*ens primum imitatur*) *as far as it can*. We must not think that analogy implies the existence of a common term that precedes God and creatures or a common form that things possess with a radically different measure of perfection — God fully and perfectly, and creatures with an insurmountable deficiency. "The diminished image ensures an imperfect and inadequate representation of the divine exemplar, halfway between fusion in a single form and radical heterogeneity."[21] God Himself communicates His image to created things, and this leads to a paradoxical asymmetric relation of resemblance: *the creature resembles God, but not the other way around.*[22]

Still it remains unclear what the dictum, "God communicates His image," could mean if no common formal aspect, no image at all common to Him and to His creature follows. Does not any analogical relation, any resemblance or communication suppress the infinite distance separating finite beings from God? In its later version (especially as we find it in *De potentia* and in the two *Summae*), the Thomist doctrine of analogy interprets the *analogia entis* not as an incomplete isomorphism of a kind, a deficient formal resemblance between the pre-eternal pattern and the created copy, but as a participation in *being*, and being is conceived now less as form than as act, in the sense of *actus essendi*. As we have seen in the preceding chapter, the scholastic tradition holds that an absolute identity of essence and existence is to be found in God. God is pure actuality-activity (*actus purus*). Translated back from Latin into Greek, this means that He is ἐνέργεια ἁπλῶς; we must not search for any other form or essence in Him except pure *energeia* (a formula completely unacceptable for Eastern theology, as we shall see later). It follows that created entities can become similar to God in their *being* only, in their *way of being*. And here *to imitate* or *to resemble* means to accept or to receive what is given and communicated by God Himself, i. e., to receive being. It is creative causality that makes the relation by analogy ontologically possible.

Second, the efforts made by Thomas Aquinas to render the concept of analogy more and more precise testify to the rejection of a compromise with the poetical form of discourse. Although poetics allows metaphors based on analogy, an onto-theological text, which inevitably speaks about God by analogy, has to be drastically different from a metaphorical text. Ricœur writes: "Such is the magnificent exercise of thought, which preserved the difference between speculative discourse and poetic discourse at the very point of their greatest proximity."[23] It seems that he con-

[21] *Ibid.*, p. 274.

[22] In this solution Thomas was to a considerable degree influenced by Albert the Great. The Master wrote, interpreting the same text of Peter Lombard: "non est aliquid univocum creaturae et creatori, sicut unius rationis existens in utroque, sed convenire potest aliquid secundum prius et posterius." See Albertus Magnus *in I Sent.*, d. 8, a. 7, ad 3; d. 46, a. 12, sol. Cited after F. Ruello, *Les "noms divins"...*, p. 82.

[23] P. Ricœur, *op. cit.*, p. 280.

siders this effort to be perfectly victorious. And yet the analogy always remains dangerously close and suspiciously akin to the metaphor, and theology, therefore, cannot separate completely from celebration.[24]

1.4. *To See the Invisible*

If theological name-giving is possible by analogy only, and if the latter is ontologically based on creative causality, on the communication of and participation in being, then the knowledge of God, guided by theology (or, as Dionysius terms it, "God-naming"),[25] is possible to the extent to which the finite created mind becomes similar to the Divine mind. Dionysius says: "What is divine uncovers itself and is inspected according to the analogy of each intellect" (I 1). The "uncovering" of what is concealed is of a special kind. Calling it *knowledge*, Dionysius adds that this knowledge is possible only in the form of ignorance;[26] speaking about vision he refers to *Divine darkness*; theology and celebration are both connected with silence. This profusion of *oxymora* is characteristic of the strategy of poetical (rhetorical) *apophansis* chosen by the author of the *Corpus Areopagiticum*.

Albert the Great writes:

> God is not a term (limit), nor something de-termined, since He is simple, and His being is not comprehended by anything else; He is a pure act, free from any potency; He cannot be conceived in anything else according to His being; and therefore the mind cannot grasp Him according to what He is (*intellectus non potest comprehendere ipsum secundum id quod est*). In quite the same way no name can designate (*significare*) Him and express completely what He is. Thus His whatness (*quidditas*) is not commensurable to anything. However, the mind, after coming into touch with His substance (*attingendo ad substantiam ipsius*), cognises Him, either in similitude, as it happens in this life (*in via*), as in a mirror and in conjecture, or immediately, as it happens after death (*in patria*).[27]

What follows is a system of oppositions concerning the ways the mind participates in *Him Who is*, as it appears in Albert's commentary.

Impossible:	Possible:
1. *comprehendere ipsum secundum quid est* — "to comprehend Him as what He is."	1. *attingere ad substantiam ipsius* — "to come into touch with (or to reach) His substance."

---

[24] As we have seen, Dionysius insists on the identity of theology and celebration. I would like to mention here one more "function" of the Divine names which has hitherto escaped our attention. The name allows us to address God in prayer. This side of Divine naming has become a most important subject in Orthodox theology. And not only theology. Many Russian philosophers at the beginning of the XX century (A. Losev and V. Ern among them) were influenced by the theological discussions of a mystical ontological bond between the name and the named, so called "onomatodoxia" (Greek) or "imyaslavie"(Russian).

[25] θεολογία = θεωνυμία.

[26] "For concerning the beyond-beingness (which remains beyond logos, intellect and essence) only un-knowing (or ignorance) is possible, and to this un-knowing the knowledge-beyond-being is to be relegated." (DN I 1)

[27] In DN XIII, q. 12 (*Utrum Deus possit nominari*...), *solutio*; Francis Ruello, *Les "noms divins"*..., p. 89, n. 48. Reference in the passage is to 1 *Cor.* 13, 12.

2. *significare exprimendo totum id quid est* — "to designate, to express completely what He is."	2. *significare nomine attingente ad ipsum* — "to designate by a name which comes into touch with Him, or reaches Him."
3. *videre Deum quid est* — "to see *what* God is.")	3. *videre Deum ut (quia) est* — "to see that God *is*"; *videre Deum per substantiam immediate* — "to see God immediately according to His substance."

The scholastic formula, *videre Deum per substantiam suam immediate*, intends to express a kind of immediate (not merely referential) relation between a perfect intellect and God who manifests Himself in silence, not in words and symbols. "Not in words" does not mean "without words altogether," since the word (the name) directs our vision, by tracing the perspective for it in advance.

As we have seen, the deep similarity, which we have graphically collocated and compared at the beginning of this chapter, between the texts of Dionysius and that of Parmenides lies in the fact that neither Parmenides' "signs" along the way of being nor Dionysius' Divine names are, strictly speaking, predicates relating to a subject. The σήματα πολλά of Parmenides are what one sees when looking attentively at being and discovering it according to being itself. At the same time the "signs" give a perspective to vision; they allow it to reach being as such and to come into touch with it. Of course, the *apophantic* proposition allows us to see something as something; this "as" expresses vision's perspective, directs vision and, if you like, generates a certain effort, a tension, an intention of looking attentively at the object of vision. Parmenides' "signs" and the Divine names require in addition a special effort of a different kind, since what has been said (a sign or a name) must be immediately un-said, crossed out, must be interpreted as a perspective mark only, as a topological characteristic of a kind, as the direction of the line of vision, not as a designation of its object. The word must be left behind in silence. As Heidegger rightly remarks commenting on fr. 8DK of Parmenides' Poem, we must, when looking attentively at being, separate in our vision all becoming and all transition, any "was" and any "will be", from being, eliminate them actively, not letting them into the field of vision.[28]

This applies to a great extent to the Divine names also. They are not simply a pursuit of the impossible, not mere imperfect attempts to express the inexpressible, not just signs of piety, "an offering of celebration," but every one of them defines an effort to look attentively at the thing named, which makes the mind become, as far as possible, similar ("analogous" as Dionysius says) to its Divine "object."

Albert the Great, citing the First Epistle of John (3, 2) (*videbimus eum sicuti est*),[29] interprets the Apostle's words as follows: *Est* [this refers to *sicuti est* of the passage just cited] *autem per suam substantiam. Ergo videbimus eum secundum suam substantiam.* This means that John's words "we shall see Him as He is" are taken by Albert to mean "we shall see Him according to His substance." We shall not see *what*

[28] Cf. EM 104.

[29] In English translation this verse reads as follows: "...when He is revealed, we will be like Him, for we will see Him as He is."

His substance is (*quid est*), but we shall see God, having reached Him, without semiotic or symbolic mediation, for to see is to come into touch with what is seen, *videre est attingere ad rem visam*. Albert explains in detail this metaphor of vision in the following remarkable passage: [30]

Thus, to take a view of a thing means to rake (*pertransire*) it (*sc.* to look at it from every quarter — *A. Ch.*). Yet all that is visible, as Euclid already demonstrated, is seen at a certain angle of a triangle, which has its apex in the eye, while its base forms the object seen. [...] Consequently the thing is looked at along the line which divides the triangle, and that is why we do not see the whole thing at the same time, but [this happens] only as a result of a movement (*sub discursu*) [of the eye] from one point to another, and once our look has thus wandered over the whole thing, we say that we have seen it all. God, however, cannot be seen in this way, i. e., by wandering with one's vision over all that He is. [...] What is not created cannot be seen by the created [intellect] so that the uncreated thing be comprehended (*pertranseatur*) by the created [intellect], but [the created intellect] can only as it were come into touch with the essence [of the uncreated] without mediation (*attingat ipsam substantiam sine medio*).[31]

The formula *videre secundum suam substantiam* rejects mediation. Yet what kind of mediation can be meant here? Aristotle discusses the following situation: when we see something white, we *accidentally* (through the mediation of an accidental feature) see a human being, since this something white is (*per accidens*) the human being, and the name "white" can be truly predicated of it. To see (mentally) a human being *secundum substantiam* means to perceive immediately in the thing present at hand the *eidos* of a human being: the mind "comes into touch with" the internal form of the object and "says" *what* this object is (*Metaph.* VIII 10, 1051b24). But nothing accidental, as we have seen, can be opposed to the Divine substance. There are no accidental properties in God. This means that the formula "*secundum substantiam*" rejects semiotic, symbolic substitutions, deductive mediations, etc. Yet it is impossible either to "come mentally into touch with" the Divine essence in the sense of Aristotle's θιγεῖν καὶ φάναι. Albert explains that the difference here is the same as between to "look at" (to come into touch by means of vision) and "to rake," "to look from all angles" (to comprehend by means of vision). Vision "reaches" (*attingit*) the Divine essence, and the name defines the perspective of vision and names the visible aspect of what cannot be completely comprehended by vision. This analogy has a serious deficiency, however: the Divine essence itself *cannot have various aspects, since it is absolutely simple.* Neither can we consider Albert's opposition between *videre ut* (*quia*) *est*, the one hand, and *videre quid est*, on the other hand, as contrasting the possibility to grasp immediately God's *existence* with the impossibility to comprehend His *essence*. According to a fundamental thesis of medieval metaphysics, as we have already seen, in God essence and existence are one and the same thing (*unum idemque*). Thus, Albert's analogy of "coming in touch" (*attingere*) remains a little too vague.

[30] Which is all the more important for us in connection with our attempts in the next section to propose phenomenological variations in connection with our theme. The passage resembles very much a quotation from a modern treatise dedicated to the phenomenology of visual perception.

[31] In *DN* I, q. 10: *Utrum divina substantia possit ab aliquo intellectu creato videri*. F. Ruello, *op. cit.*, n. 64, p. 99.

1.5. *Energeia and Essence*

The "coming into touch" just mentioned is, however, possible only for a mind which has become similar to the Divine mind, for a *blessed* mind: *Substantiam Dei quia est, omnes beati videbunt...* My aim in this chapter is to try to construct a "phenomenological image," a phenomenological interpretation of this similitude or "analogy" of minds. Thus I am interested in the shift, the transformation of *analogia entis* into *analogia mentis*. The fragments of texts analysed earlier (by composing a "textual collage") are designed to be "signposts" and "landmarks" organizing the movement of my own text and directing the effort of understanding; their choice is determined by my personal interests and by certain goals of this work in particular. That is why I shall not undertake here a systematic discussion of the differences in theological opinions concerning the Divine names and of the theme, so important to the whole tradition in the West and in the East, of the "vision of God."[32] I cannot, however, help mentioning here an important fact: in the East, the tradition of the Church relates the Divine names not to the "perspective of intelligent vision" which reaches immediately the essence of God (*attingit ipsam substantiam [Dei] sine medio*), but to the *Divine energeiai*. Gregory Palamas writes:

> Neither uncreated grace, nor eternal glory, nor life, nor anything like this is the superessential essence of God,[33] for God as the Cause is superior to all these; yet we say that He is Life, Grace, and the like... Since God, however, is totally present in every one of the *energeiai* befitting Him, we name Him after every one of them.[34]

God is inaccessible to participation in His essence, yet He "becomes manifest because He acts" (ἐκ τοῦ ἐνεργεῖν). The *energeiai* spoken of here are identified by the tradition with Dionysius' "being-producing procession (πρόοδος) of the thearchic source of beings in all beings" (DN V 1). These *energeiai* cannot be understood to be created (the opposition created/uncreated within the context of Byzantine controversies of the XIV century needs special interpretation too), yet they are varied manifestations of God's being (literally, they are "forces" manifesting God's being) or the "actions" in which God discloses Himself in His being, but by no means are they created signs or symbols of God (as Palamas' opponent, Akyndinos, believed). They are not beings among beings, but the ways of being of *Him Who is*: "these *energeiai* have no *hypostasis* of their own." This means, among other things, that the *energeiai* can be thought only as *energeiai* of the essence yet different from the essence. And, indeed, Gregory Palamas writes: "An unintelligible Divine essence having no *energeia* different from itself will be absolutely un-

---

[32] See the lectures of Vladimir Lossky delivered at Sorbonne in 1945–46 and published under the title *The Vision of God* (St. Vladimir Seminary Press, 1983).

[33] What is rendered here as "the essence of God" is οὐσία τοῦ Θεοῦ, and the Latin translation of this Greek expression is exactly *substantia Dei*.

[34] The writings of Gregory Palamas (1296–1359), one of the most influential Orthodox theologians, are cited after J. Meyendorff, *Introduction à l'étude de Grégoire Palamas* (Paris, Du Seuil, 1953), part. II, chap. V. Translation from Greek is mine. Concerning the concept of the Divine *energeiai* and its role in the so-called *hesychast controversies* of the XIV century in Byzantium cf. J. Meyendorff, *Byzantine Theology. Historical Trends and Doctrinal Themes* (N.Y.: Fordham Univ. Press, 1979).

hypostatical [here this means 'having no autonomous being' — *A. Ch.*], i. e., a pure representation of the mind." This, of course, would be quite an absurdity. It follows that the essence differs from its *energeiai*, and the *energeiai* are different from the essence. The aim of the passage cited is obviously to provide an argument in favour of the existence of uncreated *energeiai*, which manifest or un-cover God's being. Yet in an indirect way the difference, διαφορά, between the *energeia* and the essence of God is posited here. More than that, the plurality (a "spectrum") of Divine *energeiai* is allowed. And this, as it seems, is the really crucial point of divergence between Byzantine and Western theologians, for the latter consider God to be (in His onto-theological appellation) the *actus purus* (ἐνέργεια ἁπλῶς in Aristote's Greek).

It is true that these two ways of speaking of God cannot be related to each other directly as if there existed already a common ground of thought where they could meet. The texts of Albert the Great and Thomas Aquinas exist on the whole (I am not taking into account now the neo-Platonic influences) within the sphere of concepts and constructions of Aristotelian metaphysics and psychology, including their subtlest details and shades of meaning. Of course, Gregory Palamas writes in Greek and naturally uses Aristotelian metaphysical terminology (precisely the terms of Aristotle, not their Latin counterparts: ἐνέργεια, and not *actus* or *actualitas*; οὐσία, and not *substantia* or *essentia*). Yet in spite of all that his position is much more distant from Aristotelianism than that of Albert or Thomas. To illustrate this, I shall cite two important quotations. Gregory Palamas argues as follows:

> God, when speaking to Moses, did not say: "I am the essence." He said: "I am who I am" (in *Septuagint*: ἐγώ εἰμι ὁ ὤν, "I am the be-ing"). This means that the *I-am* (the be-ing) is not derived from the essence, but the essence is derived from the *I-am*, for the *I-am* comprises all being (τὸ εἶναι). [...] An essence is necessarily an entity, but an entity is not necessarily an essence.

It is evident that within the framework of Aristotle's metaphysics these statements either have no sense at all (the thesis *essence is derived from entity*, i. e., *essence is originated and grounded in entity* is a conscious or unconscious reversal of Aristotelianism), or leads to logical consequences such as St. Gregory could not have meant. As we know, the term οὐσία (*essence* as we have translated it here) has several meanings in Aristotle's works, but whichever of them we take here, an entity can differ from its essence only in that it has matter, which is unthinkable in God. This means that the same Greek words, used by Aristotle and Gregory Palamas, are *homonyms*. It is an extremely difficult philosophical task to understand their usage in St. Gregory's writings, and this task is as yet very far from being accomplished. In any case before contraposing texts containing homonyms, one must be aware that we are dealing here with different types and strategies of meaning-constitution, meaning-generation and meaning-expression. It is well known that to create a semblance of contradiction on the ground of the use of homonyms was a favourite method of the sophists, fought against by Plato and Aristotle. This is a matter of sophistry and ideology, not of thinking. We have to take seriously the "self-definition of discourse" which we find in the writings of Gregory Palamas: "We believe that the opinion is true which is not acquired by means of words and reasoning, but is proved to be so by the deeds of life... Every word, they say, refutes another word; yet what word can refute life?" It is true, though, that to take these words *too seriously and too literally* would be, for philosophy, tantamount to committing suicide, which, in my opinion, is not a very pressing matter.

1.6. *Analogia Mentis*

In chapter V, section 6 Dionysius goes back to the Platonic theme of the good beyond essence as the source of being. The good (τὸ ἀγαθόν) is one of the Divine names (IV 1). In the text under consideration the author speaks about the autonomous good beyond good.[35] The first gift given to us by this hiding giver is the gift of being itself. "The Being itself, and the first principles of beings, and all beings, and all that which is constantly retained by being this way or that, comes from it and is in it."[36] An extremely important *Scholium*[37] interprets the *good beyond good* as the intellect (νοῦς) and constructs a *noetic* counterpart to the ontological text of Dionysius. Here is what the *Scholium* says:

> The autonomous good beyond good, being an intelligence and an *energeia* directed (lit.— *turning round*) towards itself, abides in the state of actuality, not in the state of potentiality. Primarily it is *thoughtlessness* (ἀφροσύνη), and then becomes intelligence in actuality. That is why it is pure intelligence, which has no understanding brought from without, but thinks entirely proceeding from itself. For if its essence is one thing, and what it thinks is something else, it will itself always remain unthinkable, or such (sc.— unthinkable) will be its essence. If it possesses something, it has received it from itself, not from the outside. And if it thinks by itself and [proceeding] from itself, then it is what it thinks. [...] And so it thinks itself in itself. And since God is the Creator of all beings, they are in His not-yet-existent (not-yet-) thinking. [...] So that neither He is in a place, as it were, nor they are in Him as in a place, but He possesses them possessing Himself and is inherent in them. He is the totality of beings, and beings are in Him inseparably. [As the "form" of thought is present inseparably and indivisibly in the varied subjects of thought– *A. Ch.*] [38] For beings are in His thinking.

In this difficult passage many motives are interwoven: the Aristotelian theology of the Divine intelligence (*Metaph.* XII 9), which is a pure noetic *energeia directed towards itself*, and the Platonic *good beyond essence*, and, in particular, the neo-Platonic version of the traditional discussion dedicated to the question, how the thinking of thinking (νόησις νοήσεως) is possible. In this connection the text of *Enneads* V 3, under the general title of "On the knowing hypostases and that which is beyond," is of special interest. Here is what Plotinus writes (V 3, 7, 14–22):

---

[35] The expression, which I translate as "autonomous good beyond good," reads in Greek αὐτοϋπεραγαθότης.

[36] Cf. Plato, *Republic* VI, 509b: "So then you must also say that the known not only receives its being-known from the good, but also it has thence its being and its essence, in such a way indeed that the good is not itself an essence, but lies beyond the essence and outstrips being in dignity and power."

[37] The classical *Scholia* had been traditionally attributed to Maximus the Confessor (ca. 580–662), though in fact, as H. Urs von Balthasar has proved, they were, at least partly, the work of John of Scythopolis (†ca. 540).

[38] Cf. chap. 2, sect. 1.

But silence (ἡσυχία) for the intellect is not the same thing as the cessation of intellect; on the contrary, silence is an *energeia* which accomplishes the abstinence (liberation) of the intelligence from the other. [...] So the being of the intellect is an *energeia*, and there is nothing [external] towards which this *energeia* is directed. It follows that it is directed towards itself. Thinking itself, the intellect is thus in itself and retains its *energeia* in itself. For if anything comes from it, the intellect itself must first of all remain in itself and directed to itself.

What matters for us, however, is that the text of the *Scholium* connects the ontological event of communicating or giving being with these noetic structures. The Divine being beyond being is linked to "thoughtlessness" (ἀφροσύνη), the being of beings in their indivisible totality to the noetic (intelligent) *energeia* directed towards itself, the varied (determined and delimited) entities to the thinking of the objects of thought.

	Ontic dimension	
The Divine primal being beyond the essence.	Being as such	The whole of beings (entities)
προεῖναι καὶ ὑπερεῖναι	τὸ εἶναι πᾶν (αὐτὸ καθ' αὐτὸ τὸ εἶναι)	πᾶν τὸ ὁπωσοῦν ὄν
	Noetic dimension	
The primary thoughtlessness as the source of thought	Self-thinking or self-consciousness as pure *energeia* (activity) directed towards itself	Thinking of thinkable objects
ἀφροσύνη	νοῦς ἐνεργείᾳ, ἐνέργεια εἰς ἑαυτὴν ἐστραμμένη	νόησις τοῦ νοητοῦ

The term ἀφροσύνη, which we translate as *thoughtlessness*, meant in everyday usage *unreasonableness*, even *folly*. Usually ἀφροσύνη is opposed to σωφροσύνη and σοφία, but not in the *Scholium*. This thoughtless intelligence seems to be an unimaginable construct, a rhetorical figure, one more *oxymoron* intended to express an "unconscious awareness of itself" or a motionless self-identity remaining in itself. Nonetheless this identity is quite different from the opaque unity of a lifeless thing-in-itself. The primal thoughtlessness is interpreted as a source of intelligence lying beyond intelligence. The paradoxical character of this definition corresponds to the paradoxical being of God, who "is not," but communicates being to beings. The God without being of Dionysius the Areopagite corresponds to the intelligence without thinking or the intelligent thoughtlessness spoken of by the Scholiast.

This method of exegesis (and its content also) will become somewhat more understandable if we take into account other texts of Plotinus belonging, as has been said, to the same horizon of concepts. In the *Enneads* the *ontic* melts into one with the *noetic*. Plotinus' *One*, which cannot be said either *to be* or *to be the good* is completely inexpressible in itself and manifests (reveals) itself only in something "posterior" in relation to the One — first of all, in the intellect. Without the mediation of the intelligence the One cannot "be," the intelligence affirms (posits) the being of the One, and it is this *energeia* of affirmation that can be interpreted as the *being* of the One. This reasoning is based on the fundamental onto-theological axiom: *to be means to be an object* (at the very least an object of intelligent contemplation), *being* means *being-what.* The One, on the contrary, (is) beyond "whatness," ἄνευ τοῦ τί.[39] That is why Plotinus borrows and reinterprets Plato's ontological "beyond the essence," turning it into a noetic formula "beyond the intelligence." To reach up to the One the intelligence must abandon itself, reject its nature, since according to nature it is never one, but always a plurality. The plurality is for Plotinus the necessary form of the thinking intellect as such.

The thinking intellect must not remain simple, especially insofar as it thinks itself. For it will duplicate itself by so doing... That is why [the One] does not need to query itself about itself: for what can it learn by thinking itself? For what it is will belong to itself before the intellect thinks. Also, knowledge is a kind of longing [for the absent] (πόθος), and like the discovery made by the seeker. But what is completely other remains itself by itself, and seeks nothing about itself; but that which explicates (ἐξελίττει)[40] itself, must be many. (*Enn.* V 3, 10, 44–53)

That is why the autarchy of the One, its being-by-itself-beyond-the-intelligence (for according to Plotinus the intellect proceeds from the One) is not-yet-thinking. Yet strictly speaking one cannot say *the One is by itself* or *the One is.*

The One can never become an object of intellect. But can the intellect itself become its own object or, in other words (these are two *different* questions), can the intelligence think itself? Aristotle asks in his treatise *On the Soul* (III 4, 429b23–28):

If the intellect is simple, and not liable to be acted upon, and has nothing in common with anything else... how will it think, for to think means in a certain sense to be acted upon? For one thing acts and another is acted upon insofar as both have something in common. And our second problem is whether the intellect itself can be an object of thought?

Plotinus' question is somewhat different:

Does that which thinks itself have to be complex, in order that it may with one of its constituents contemplate the rest, and so be said to think itself, on the supposition that the absolutely simple would not be able to revert to itself and so an intellectual grasp of itself is impossible for it? (*Enn.* V 3, 1, 1–5)

What contemplates itself needs, as it seems, to be split, to be divided so that one part becomes the seer and the other the object seen. But if the one contemplates the other, this act cannot be called *self*-contemplation, unless the contemplating-contemplated thing spoken of consists of identical parts, so that the seer does not differ in any way from the seen (5, 5, 1–5). But then the next question arises: who or what divides the whole into the contemplating and the contemplated parts, and *how* does this division take place? Finally, even if the seen is essentially indistinguishable from the seer, how does the seer recognize itself in the seen? Does it not follow hence that there must appear, besides the seer and the seen, something dividing and something identifying? Further (5, 3, 13–16), even if the seer recognizes itself in the seen, this recognition cannot be complete. For the seer cannot grasp itself as a seer, i. e., as the one who accomplishes the act of seeing. This means that the seer recognizes not *itself*, but something else. Or perhaps he will add (προσθήσει) from itself the one who

[39] *Enn.* V 3 (49), 12, 50–52. The One is without the "what," for it were something (some "what") it would cease to be the One itself, because it is "itself" before the "what."

[40] This verb means also "to unfold," "to unroll."

has already completed the contemplation (τεθεωρηκότα) in order to possess a perfect knowledge of itself? One might want to ask at this point whether this will not be yet another splitting of the seer, as a result of which the original situation of incompleteness will repeat itself. Yet the argument in the text discussed, as far as I can understand it, proceeds in another direction. For Plotinus there arises a difficulty of another sort (ll. 18–19). "To add," to posit as an object the seer, which has already seen, is impossible unless at the same time what has been seen (τὰ ἑωραμένα) is retained too, for otherwise the seer that has been deprived of his object will not appear as such (as the seer). What is the "object" that has been seen and how can it be retained as such in the "addition" (πρόσθεσις) discussed? Plotinus constructs the following dilemma: there is in contemplation either the contemplated thing itself or its imprints (τύποι). If there are only imprints in contemplation, then by adding the one who has seen we shall add only the stamps of the original object of contemplation; in other words, by adding the seer we shall lose what it has really contemplated, and so the completeness of self-contemplation will not be reached. If the second possibility takes place and the contemplated thing itself is somehow included into contemplation, then the contemplator "sees" itself *not as a result* of splitting into a part which is contemplating and a part which is contemplated, but can, before any division, be said to possess himself and contemplate himself (ἑαυτὸν καὶ θεωρῶν καὶ ἔχων). If this is so, then contemplation is identical to the object contemplated, and the thinking subject is identical to the object of thought. Yet, Plotinus continues to ask, can this identity, in which the intelligence, as it were, comprises itself, be called thinking? The identity itself as such does not amount to the perfection of *intelligent contemplation*. For things without life, a stone, for instance, are identical to themselves, though they do not think themselves. Plotinus' answer is worth paying attention to. Not only is the thinking subject identical to the object of thought, but both are identical to the act of thinking (νόησις). In our case this means that what is thought is a kind of activity-actuality, i.e., in Greek, *energeia* (ἐνέργεια γάρ τις τὸ νοητόν), and therefore the thing thinking=thought is not alien to life; more than that, it *is* life and *energeia*, and, as it were, the first and the best *energeia*.

Such an intelligence is to be called the first intelligence, for in it there is nothing potential (a repetition of Aristotle again). It is not true that it is one thing and its thinking is another thing, for in this case it would be different from the pure activity of thinking and there would be present in it a certain possibility of thinking not yet turned actual. Thus such an intelligence is *energeia*, its *essence is its energeia* (ἡ οὐσία αὐτοῦ ἐνέργεια). In it the thinking subject, the object thought and the *energeia* (the act) of thinking are identical. This *energeia*, to speak a somewhat different language, being-in-itself and being-for-itself is altogether essence (form), being and thinking. For such an intellect *to be* means *to think* and *to be thought*. So as we see now, it is not for nothing that Plotinus interpreted fr. 3DK of Parmenides in this way: "to think and to be is the same."[41] We must not forget, however, that for Plotinus the situation described where the object of thought, the subject thinking

[41] In chap. 1, sect. 2 we discussed another possible interpretation of this fragment.

and the thinking itself are identical, is an extreme case of thinking (or perhaps its self-negation), because here it overcomes its own plurality. According to Aristotle and Plotinus, such thinking belongs to the Divine intelligence. For "that which explicates itself, must be many."

Let us now go back to the text of the *Scholium*. The noetic analogy constructed in it must serve for the finite intelligence, which always thinks the thinkable and therefore always duplicates itself, as a guideline in fulfilling its own task of imitating the Divine archetype to the extent of its ability — to the extent of its "analogy" or similarity. In this mimetic effort the intelligence has to go, in reverse order, through all the steps of communicating the gift of being and "reach up to" the giver. This means that the finite intelligence must ascend from the thinking of objects to the thinking of thinking (noetic *energeia* directed towards itself) and then further to the primordial thoughtlessness, from intelligible forms (present εἴδη) to their pure presence itself (intelligent *energeia*) and then further to the absent condition of any presence.

This ascension reveals itself impossible in a certain sense. While the intelligence remains what it is, it moves endlessly around an inaccessible point, from which, however, it is not separated. This topological metaphor can be found in the *Mystagogia*[42] of Maximus the Confessor (chap. V, PG 91, col. 676 D): Θεὸς γὰρ ἡ ἀλήθεια, περὶ ὃν ἀκαταλήκτως τὲ καὶ ἀλήστως κινούμενος ὁ νοῦς, λήγειν οὐκ ἔχει ποτὲ τῆς κινήσεως μὴ εὑρίσκων πέρας ἔνθα μὴ ἐστι διάστημα — "God is the truth, and the intellect moving around it incessantly and without lapsing into oblivion, cannot stop (complete) its movement because it finds no limit where there is no distance (interval)."

1.7. *Maximus the Confessor on the two parts of the human soul*

In connection with this text of St. Maximus I would like to touch upon a theme, which still awaits a thorough elaboration.[43] The Eastern Christian world expresses its inmost hope in St. Athanasius the Great's maxim: "God became man in order that man might become God."[44] Christian theology therefore required a new anthropology, embodying the fundamental role of the integral human personality, its being in the world and within the flow of time. Simultaneously it required a new ontology capable of recognizing the ontological significance of moral action. The gulf, the ontological rupture between σοφία and φρόνησις, which we find in Aristotle, became a sign of the "outer philosophy." For the tradition of Eastern Patristics, σοφία and φρόνησις grow together, in permanent interaction.

In chapter V of his *Mystagogia*, Maximus the Confessor discusses the structure of the human soul. The "topography" of the soul, as described here, is strikingly simi-

---

[42] This extremely authoritative and influential text is closely connected with the *Corpus Areopagiticum*.

[43] Cf. L. Tunberg, *Microcosm and Mediator. The Theological Anthropology of Maximus the Confessor*, second edition (Chicago and La Salle, Illinois: Open Court, 1995); A. G. Chernyakov, "The consolation of philosophy today." SYMPOSION. *A Journal of Russian Thought*, 1 (1996), pp. 19–34.

[44] The expression "might become God" refers to the notion of θέωσις ("deification"), one of the most important concepts of Orthodox theology, which I cannot discuss here.

lar to that of book VI of the *Nicomachean Ethics*. Moreover, it is due to this similarity that the fundamental differences between them can be clearly seen.

The comparison of these two texts is both exciting and productive. In my view, it is the details and subtleties that constitute the most important object of hermeneutic work with the text. It is the details, rather than a general overview, that resist the arbitrariness of the interpreter and direct hermeneutic work.

St. Maximus, like Aristotle, discusses theoretical and practical acts of the human soul, calling the faculty of the former σοφία, and that of the latter φρόνησις. *Sophia* and *phronesis* are determined by St. Maximus as the two primary *energeiai* of the soul. Of course, this way of characterizing them is not foreign to Aristotle either: as a virtue, or excellence, each of them constitutes the end, the goal to be achieved by the soul, its *telos*. The "being-already-at-the-goal" = ἐντελέχεια refers to the soul's completed and perfect form. But, as we have seen in chapter 2, Aristotle declares this *entelecheia* to be only *potentia secunda* in relation to the *energeiai* of human action. While for Aristotle the theoretical or scientific part of the soul, on the one hand, and its practical or moral counterpart, on the other, constitute separate structural (formal) moments, for St. Maximus they are rather forces in permanent and complicated *interaction*. The formation, the growth of contemplative capability, its movement towards its highest object, which is God Himself, is impossible without the perfection of the practical part, without moral insight, without, as St. Maximus says, the "habit of good deeds." "The two combine," writes St. Maximus, "to form the true science of things both divine and human, the truly infallible kind of knowledge (γνῶσις), the goal of all sacred philosophy among Christians."

To *sophia*, God discloses Himself as the Truth, to *phronesis* — as the Good. These two events of disclosure are unthinkable in separation. This thesis implies that the dead ends, the paradoxes of reflection, which we shall discuss in the next section, cannot be overcome by reflection itself. We need another more complicated concept of "reflection" as interaction between the two parts of the soul. The *noetic* act must somehow be reflected in *phronetic* action and vice versa.

But the way out of the circle of reflection is only outlined here. Moreover it is said (in the passage cited above) that intellect (νοῦς) as such cannot complete its movement in an "infinitesimal circle," so to say, around the truth, "because it finds no limit where there is no distance." In what follows I attempt a "phenomenological" interpretation of some *aporiai* touched upon.

2. THE PARADOXES OF REFLECTION IN HUSSERL'S PHENOMENOLOGY (THE TRANSCENDENTAL SUBJECT LOST)

> Cum in reflexione sit quaedam similitudo motus circularis,
> in quo est ultimum motus quod primo erat principium,
> oportet sic dicere in reflexione, ut illud quod primo erat prius,
> secundo fiat posterius.
>
> Thomas Aquinas, *De veritate* 22, 12, 1.

After Descartes the question *Sed quid igitur sum?* becomes the most important question of *first philosophy*. Husserl defines his phenomenology as transcendental *egology*. Heidegger in *Being and Time* formulates as a fundamental problem of the *existential analytic* the "existential question" about the *Who* of *Dasein*. Throughout this book I refer to different historico-philosophical contexts and apply different hermeneutical strategies in order to emphasize that the concept of the human Self, the concept of being, and the concept of time are inseparably linked with one another. In the previous section I attempted to add to this triad (Being — Time — Self) one more topic of a theological nature, *viz.* the relationship between God and man as it has been developed, on the one hand, in Orthodox *apophatic* theology and, on the other hand, during the Scholastic controversies concerning the problem of *visio Dei*. We are now passing on to the phenomenological interpretation of some of the problems discussed above. I must say now that I do not propose to consider phenomenology the "only correct medium of interpretation, adequate to the subject matter," in which the "true meaning" of the texts we are interested in will be finally made clear. We have already stressed that such a position seems an impossibility to "late philosophy."

On the other hand I do believe that phenomenological essays are not alien in nature to the problems being discussed, for we are interested in "analogy," in the process of making the finite intelligence similar to the infinite intelligence. Already Aristotle was constructing his theology of the Divine intellect *(Metaph.* XII 7; 9*)* on the basis of the phenomenology of the finite intellect. The meaning of this similarity of the finite intelligence to the Divine intelligence is, as we have seen, twofold. On the one hand, the perfect intellect is an entity which realizes in its being the Divine archetype to the greatest extent, the perfect intellect imitates God, *ens primum imitatur*, as a Scholastic formula says. On the other hand, this mimetic effort and its result are described as a special kind of contemplation (*contemplatio*, *visio*), in which the intelligence in a sense "comes into touch" with the Divine substance without mediation or, in other words, according to the topological metaphor of Maximus the Confessor, moves endlessly "around the truth" without reaching the truth but coming so close to it that there is no interval or distance in between. On the other hand Husserl's transcendental phenomenology is precisely an archeo-logy of consciousness, one of the ways to move towards the primordial source of all sense-constitution (including the positing of "*ontic* sense"), of all activity (or, as Husserl likes to put it, of all *life*) of consciousness.

2.1. *Da-sein and Bewußt-sein*

To what extent can the "ascension to the primordial source of being" be considered a phenomenological topic? Fully and entirely, if we take phenomenology in its Heideggerian version. Heidegger articulates his understanding of the phenomenological method in the famous formula (which seems simply to retell the Greek term): "to let that which shows itself be seen from itself in the very way in which it shows itself from itself" (SZ 34). And if we ask now what phenomenology is to let us see, we will learn that it is nothing but the being of beings, which "at first and for the most part does not show itself at all," though it does not hide itself either, and which belongs in such a way to all other phenomena that it serves as the reason and the foundation for the latter (SZ 35). In marginal notes Heidegger adds: "the truth of being," i. e., the original becoming present in, or pro-cession into, unconcealment. So that πρόοδος, the term of Plotinus and Dionysius, which in the treatise *On the Divine Names* refers to the procession of the super-essential into all entities, seems to be quite fitting here. Ontology, Heidegger says, is possible as phenomenology only (SZ 35), and vice versa, "phenomenology, from the viewpoint of its content, is a science dealing with being, ontology" (SZ 37). Yet in phenomenology-as-ontology or ontology-as-phenomenology, i. e., in the fundamental ontology of Heidegger, an "ontological reduction" to an entity takes place, "which in each case we ourselves are" and which is called *Dasein*. The analysis of the *ontic* structures of this entity must be the starting point for revealing being itself. The main characteristic of Heidegger's method is that he opposes its analytic of *Dasein* to the classical reflective analysis of the mind, including Husserl's *archaeology of consciousness*. The actuality of the entity called *Dasein* is not thinking (being conscious of), but is rather existence (having to be). The method to work with it is not the thinking of thinking, but the unveiling of the structures of existence (unveiling *in thinking* as well). The *ontic* economy of *Dasein* is defined by means of the "originally unified and unifying phenomenon of care." "I-care" has to be able to accompany all the events of my understanding of being. "I-think" is thereby understood as a kind of secondary structure of *Dasein*'s being. This being, directed towards itself and carrying in itself the primordial phenomenon of non-indifference, is internally structured such that it allows the phenomenologist to *exhibit* the being of beings, the being which neither conceals itself nor discloses, being as such. According to Heidegger's ontological project it is *Dasein* (primarily the "I-care" and not the "I-think") that *esse ipsum imitatur.*

Yet the intention of the texts, which have been the object of our interpretation, is to represent the perfect *intellect* as the special finite being which imitates *Him Who is* with the highest possible degree of resemblance. The intellect ascending to the source (i.e. the beginning and the foundation) of thought retraces the steps of the "procession of the super-essential essence into all entities."

The *intelligence*, the consciousness in Husserl's phenomenology is a region of *absolute being*. Yet this thesis implies a pre-determined meaning of being which is expressed by the fundamental axiom of onto-theology: *to be* signifies to *be an object* revealed to the mental vision according to its "what," in its appearance of meaning.

We shall try to understand how this fundamental premise generally works within the context of Husserl's phenomenology.

Husserl has never seriously discussed the problem what "to be" signifies in general. An outline of "phenomenological ontology" that we find in the first volume of *Ideas*, the contrast between the absolute being of consciousness and being as reality (§§ 41–44), the project of *regional ontologies* — all proceed from an already determined meaning of *being*.

Husserl writes in *Ideas*, § 142:

As a matter of principle, *to every "truly existent" object* (*wahrhaft seiendem Gegenstand*) corresponds (in an *a priori* of unconditioned essential universality) the *idea of a possible consciousness*, for which this object itself is originally graspable (*originär erfassbar*) in a *perfectly adequate* way. And vice versa, if this possibility is granted, the thing itself is *eo ipso* truly existent.[45]

Further, Husserl explains that the term *idea* in this passage is to be understood in the Kantian sense, as a regulative principle, for what is meant here is, generally speaking, an *infinite* synthesis of the object's aspects and profiles, a continuum of *phenomena* internally linked and structured by the *essential type* of the thing itself (§ 143). In the perception of a transcendental thing an approximation[46] of the existence of this particular thing takes place, the law of approximation is determined by the *essence*, the *whatness* of the thing which *in concreto* is given in the synthesis of perception. For the so called "internal objects" (the immanent content of the stream of consciousness) such an approximation is not necessary, since, as Husserl supposes, the moments of the internal life of consciousness (absolutely transparent for itself) are adequately given in the internal (simple) perception, and for them the formula *esse=percipi* is true.[47]

As a result of phenomenological reduction and "non-adherence" (ἐποχή) to *ontic* positing the phenomenologist discovers the intentional object as *intentional*. In other words he discovers not a thing-existing-in-itself, but a thing, which is present for the consciousness positing it as existent. In this presence-for the *duality of the presence of the present* is disclosed, and thus the presence itself, the being-present. The intentional object is not only present as a certain content, but is present *in a certain way*, so that one may speak of a determined character of presence as such or, as we could formulate it after Aristotle,[48] of the internal form (*eidos*) of the *energeia* of being-present. Phenomenology treatises the ways of being present, the modes of givenness (*Gegebenheitsweisen*) of the intentional object and the correspondence, the correlation, of the content present (*noematic* sense) and the coherent manifold of the varied modes of its givenness (*noetic* dimension).

---

[45] E. Husserl, *Ideen zu einer reinen Phänomenologie und phänomenologischen Philosophie*, 5. Aufl. (Tübingen: M. Niemeyer Verlag, 1993), p. 296.

[46] The concept of the *approximation* of a transcendent subject in perception is developed by Husserl in the lectures of 1925–26 entitled "Grundprobleme der Logik." See E. Husserl, *Phänomenologie der Lebenswelt*. Ausgewählte Texte II, hrsg. v. K. Held (Stuttgart, 1986), p. 75.

[47] *Ibid.*, p. 73. Cf. *Ideen I*, § 42.

[48] Cf. the detailed discussion of Aristotle's concept of the internal form of *energeia* in chap. 2, sect. 2.

To conclude, Husserl's idea of the *truly existing* thing includes two aspects:

First, conformity to the laws, absence of mutual contradictions, internal arrangement, coherence of the two-dimensional (*noetic-noematic*) manifold of given contents (which form the "internal horizon" of the thing) and of the corresponding modes of givenness.

Second, a primordial character of the presence itself or, as Husserl says, "self-givenness" (*Selbstgegebenheit*),— the way in which an object in its givenness can be characterized relative to consciousness as "itself-there," "there in the flesh," in contrast to its mere re-presentification (*Vergegenwärtigung*), the empty, merely indicative idea of it.

The primordial and the original are opposed here to the modified, the mediated, and the derivative. The bottommost layer of the objective synthesis, its ultimate basis (*Grund*) is the synthesis of temporality, and *modification* means, first of all, *temporal* modification — the passing away of the immediate present experience, of the "now" — or, to be more precise, of the content of this "now" called primal impression — into "the past." According to Husserl, this passing away is still to be considered an "accomplishment of the *ego*," albeit of a profoundly enigmatic nature. It has a double character: to let slip away and to hold back, this retaining-alienation is called "retention," and the change of the mode of presence is to be termed "retentional modification" of the primal impression.

So, the object X *is* if its internal horizon is structured according to the essential type of what we think in X and, second, if the utmost proximity of the now-presence is included in the variety of the modes of givenness corresponding to various aspects of *this* X. There is a passage where Husserl says that the primal impression leaves on the object "the stamp of existence." Such is, in general terms, the phenomenological theory of distinction and composition of essence and existence (*distinctio et compositio essentiae et existentiae* [*in ente* "*extra mentem*"]).

Thus the being of the entity is interpreted in phenomenology by means of schematisation of the ways in which the intentional object is given, i. e., the meaning of *est* is definitely and resolutely reduced to the various meanings of *cogito*; the ways of being present, the ways of being given are themselves forms of pure consciousness, the *energeiai* of its "universal life." In this perspective *the field of pure phenomena* is a kind of interval appearing as a result of phenomenological reduction, the "real" (*reell*) variety between the transcendental *res cogitans* and the transcendent *res cogitata*; it is, on the one hand, a complex object of reflection, and on the other, a totality of the modes of givenness of some other object put between phenomenological parentheses. That is why the meaning of entity as entity is declared as it were from the very beginning. Even if the *ontic* faith of the natural consciousness is reduced in the phenomenological attitude, the essence = "being-ness," of the transcendent object is sought within the scope of a new "immanent" objectivity, accessible to the gaze of reflection. The entity has already been interpreted as that which can become an objective pole of intentional relation, an object of consciousness, as something *objective* or *extant.*

During the preparation of the article for the *Encyclopaedia Britannica* (1927), as a result of a violent polemic with Heidegger (which as we know, ended in a breach),[49] the task of a comprehensive phenomenological analysis of the meaning of being as such became more urgent than ever before, and Husserl had to speak his mind in more definite terms on the subject of phenomenological ontology.

On the one hand he defines phenomenology as universal ontology in which the problem of the *being of transcendental subjectivity itself* becomes central.[50] On the other hand, it has already been decided that *being* means, to use Husserl's pleonastic formula, being "positively posed" (*Sein = positiv Gesetztsein*). Thus the problem of being is the problem of being-positive (= posited) and subordinate to the problem of *distinction* between the positing and the positive (posited). The answer is to be found in the concept of pure *ego* as the last instance of the transcendental constitution, the absolute transcendental functional centre. The being of transcendental subjectivity signifies the "apodictic essence [of the pure *ego*] to be transcendentally constituted in itself and for itself."[51]

Yet this solution, this "in itself and for itself," attempting to unite the being of the *ego for itself* (the object, the posited) and the being of the *ego in itself* (the positing), is menaced by an extremely difficult *aporia*: if every knowledge is the knowledge of a constituted object, how can the constituting *ego* possess phenomenologically reliable knowledge of its own constituting activity as activity? Can the *ego* — to use Fichte's phrase — catch itself red-handed?

The question is then: how is the phenomenologically sound cognition of the fundamental functioning *ego*, the ultimate source of the positivity of the positive, possible? Any phenomenologically certain fact must in this or that way bear the characteristics of intuitability, of self-evidence. The constituting subjectivity in order to be something more than merely a result of empty meaning-formation must become an object of reflection, and the reflection must be understood not in the Kantian sense — as the *necessary thinking of the transcendental unity of apperception*, to which no intuition does or can correspond — but just the other way around, as a "fulfilled consciousness," an immediate self-awareness of the transcendental *ego*. Thus the problem of the possibility of ontology as phenomenology (as far as we are remaining within the framework of *Husserl's* phenomenological project) must be reformulated in this way: How is the fulfilled consciousness of the pure *ego* as the ultimate constituting instance possible?

This "fulfilled consciousness of..." presupposes a process of constitution, and the *ego* here appears simultaneously as constituting and constituted. In this situation the ambiguity of Husserl's concept of constitution becomes apparent. Of course, the term "constitution" signifies here passive constitution and passive synthesis, which is prior to all active object-positing and sense-formation, and, in fact, is its

[49] W. Biemel collects important documents relevant to this polemic in his edition of *Husserliana* IX: *Phänomenologische Psychologie*. Vorlesungen Sommersemester 1925 (Den Haag: M. Nijhoff, 1962).

[50] *Ibid.*, p. 296f.

[51] "...das erweisbare Wesen, transzendental in sich und für sich konstituierte zu sein" (*ibid.*, p. 297).

presupposition. Still the question remains: how can the *ego* of reflection "overtake" itself in its — to use Husserl's term — original or primal passivity (*Urpassivität*), which is characterized also as its standing-streaming life.

At the deepest level of original passivity, according to Husserl, lies the synthesis of the time-consciousness in which the *ego* constitutes itself as temporal and becomes aware of itself as a stream of consciousness. It is precisely the self-constitution of the *ego* in its temporality that became the core problem in Husserl's later reflections in the thirties.[52] To be sure, we are dealing here with the "primal phenomenon in which everything else that can be called phenomenon has its source. It is the standing-streaming (*stehend-stroemend*) self-presence of the *absolute ego* present to itself as itself streaming in its standing streaming life."[53] But how are these *oxymora* — "standing-streaming life" and "passive constitution" to be understood? No wonder that Husserl's own statements on this subject sound contradictory.[54] The streaming life is passive, "the stream proceeds without the activity (*Tun*) of the *ego*."[55] On the other hand, "temporality is in every respect *ego*-accomplishment."[56] The relationship between the streaming temporal flux, on the one hand, and transcendental *ego* which endures through the intervals of internal time and is supposed to be constituted as such in it, is characterised by no less a paradoxical formula: "The stream is always ahead (*im Voraus*), but the *ego* is also ahead."[57] The sober conclusion to be drawn has been amply established by K. Held and L. Landgrebe: "the depth-dimensions of the process of constitution cannot be attained by phenomenological reflection."[58] In what follows I shall try to comment on this thesis and to connect it with the problems discussed in the previous section.

2.2. *Empty and Filled*

As a result of phenomenological reduction every object or theme of consciousness is understood as an intentional object, and this means that it is constituted or posited in acts of consciousness itself, as such and such a meaning-for-consciousness, and provided with this or that ontological status (i.e. interpreted as an entity belonging to a definite *ontic* "region": an "external thing," an "ideal object," an "internal object" = "the content of consciousness," etc.). Therefore the intentional relation of consciousness to its object cannot be understood as a relation in the Aristotelian sense. For Aristotle a relation is an accidental property of its terms. The terms belong in their turn to the ontologically prior category of substance. In the

[52] A detailed analysis of the decisive C manuscripts from 1931 to 1933 devoted to this subject can be found in: K. Held, *Lebendige Gegenwart*, Phenomenologica 23 (The Hague: M. Nijhoff, 1966). Cf. also an extremely elucidating article by Ludwig Landgrebe "The Problem of Passive Constitution," in *Phenomenology of E. Husserl. Six Essays*, ed. D. Welton (Cornell Univ. Press, 1981), pp. 51–56.

[53] MS C 7 II, p. 12 (1932) after Held, *op. cit.*, p. 99.

[54] Cf. Landgrebe, *op. cit.*, pp. 53f.

[55] MS C 17 IV, p. 1 (1930) after Held, p. 99.

[56] MS C 17 IV, p. 5 (1932) after Held, p.101.

[57] MS C 17 IV, p. 6 (1932) after Held, p.101.

[58] L. Landgrebe, *op. cit.*, pp. 51f.

classical Aristotelian scheme of categories the accidental determinations are *ontically* dependent on substance; it is only the latter that possesses being in itself independently of anything else. Yet in phenomenology the members of an intentional relation cannot be naively *ontified* beforehand; the being-in-itself of the intentional object as an "object-outside-of-me" is nothing but the *ontic* belief characteristic of the natural attitude, and in this sense the intentional relation ontologically precedes at least *one* of its members, the intentional object, since the being of the latter is posited (constituted) in this relation itself. And the intention itself, the "directedness of consciousness towards the object," can be "empty" (*leer*), i. e., referring to the object by means of a sign or an image, or, to the contrary, "fulfilled" (*erfüllt*), i. e., connected with the intuition "of the object itself." It is the meaning of this peculiar primordial self-givenness (*Selbstgegebenheit, Selbsthabe*), in which the object is experienced in the modus "itself-there,"[59] that needs to be clarified.

The important distinction between the empty and fulfilled intention (*erfüllte Intention, Erfüllungsbewußtsein*) is introduced by Husserl in the second volume of *Logical Investigations* (Investigation VI). This distinction succeeds to the difference between the mediated and the immediate. "Fulfilment" (*Erfüllung*) signifies in this text the fundamental characteristic of "self-evidence" (*Evidenz*) understood phenomenologically. To interpret self-evidence as an instant of fulfilled consciousness means to see the play of mutual determinations of the *empty* and the *filled*: the term "fulfilment" itself implies "movement" (in the Aristotelian sense) or a *passage* from an "empty" implication, a preliminary outline of meaning, to direct contemplation or intuition.[60] To be more precise, we are dealing here with a connection or a process of connecting two unequal terms, which are linked together by a relation of

---

[59] "In Erfüllung erleben wir gleichsam ein 'das ist es selbst'..." E. Husserl, *Logische Untersuchungen*, Bd. 2, 2. Teil, hrsg. v. U. Panzer, Hua XIX/2 (Den Haag/Boston/London, 1984), p. 597.

[60] According to a testimony of E. Tugendhat (cf. *Der Wahrheitsbegriff bei Husserl und Heidegger*. 2. Aufl., Berlin, 1970), Husserl's distinction of the empty and the filled consciousness can be traced back to Leibniz' article *Meditationes de cognitione, veritate et ideis* (1684) already cited in chap. 3. Franz Brentano highly estimated this work and recommended it to his students. M. Heidegger analysed and interpreted it in his Marburg lectures of 1928 published under the title *Metaphysische Anfangsgründe der Logik im Ausgang von Leibniz* (§ 4). In this small article Leibniz distinguishes between *cognitio caeca vel symbolica*, on the on hand, and *cognitio intuitiva*. Let us consider an important example. When we think of a *chiliagon*, a regular polygon with thousand (equal) sides, we are incapable not only of appropriately intuiting such a thing, but also we do not always simultaneously conceive along with the meaning of the term the exact definition or "nature" of side, of equality, of the number one thousand, etc. We use the words or some other "symbols" in order to refer to such an ideal object as *chiliagon* and to operate with it mentally. Nevertheless, we have the accompanying awareness that we know the essence of this object and can at any moment put it before us, that we can at any moment analyse the notions involved, up to their simplest constituents. This state of referring to an object by means of symbols plus the accompanying awareness – "I know that I know" – Leibniz calls *cognitio caeca*, and Husserl would call empty or "signitive" intention. But when it is possible for me to think simultaneously the meaning of all simple constituents (and for the simple, thinking amounts to immediate intuition) I have intuitive knowledge. Leibniz writes: "Et certe cum notio valde composita est, non possumus omnes ingredientes eam notiones simul cogitare: ubi tamen hoc licet, vel saltem in quantum licet, cognitionem voco intuitivam. Notionis distinctae primitivae non alia datur cognitio, quam intuitiva, ut compositarum plerumque cogitatio non nisi symbolica est." See *Die philosophische Schriften von G. W. F. Leibniz*, hrsg. G. L. Gerhardt, Bd. 4 (Hildesheim: Georg Olms, 1960), pp. 422–426.

identity and difference: exactly what was referred to by means of an empty, merely indicative idea of it, is now being intuitively filled. In connection with this Husserl speaks of a filling *synthesis* (*Erfüllungssynthesis*).

The process of fulfilment establishes a relation of the name, commensurable with its meaning, with what is given in contemplation as the thing named. During the fulfilment process we experience [an object] as "itself-there." [...] The synthesis of fulfilment demonstrates the disparity of the two members being related, and this in such a way that the fulfilled act has a certain superiority, which the bare intention lacks; namely, such an act gives to it the fullness [of the presence] of the [intended] "thing itself."[61]

The disparity, the inequality of the members linked in the synthesis of fulfilment determines a certain field of tension, opens a certain dimension where gradations of the fulfilment are possible (the relation of what is more fulfilled to what is less fulfilled, the distinction between fulfilled in this or that way). It is in this dimension that the movement of fulfilment/emptying takes place. Husserl warned against understanding fulfilment as a still-standing, static coincidence of the intending meaning and the fulfilling intuition. The connection between what is empty and what is fulfilled must be understood as a synthesis of identification, which proceeds in time (it is exactly *the same thing* that is now given in a fulfilled consciousness), as a continuous growth of the intuitive fullness. Here "the same thing" must not refer to a complete formal identity, for in the process of identification synthesis some of the layers or aspects of the originally implied meaning can undergo certain modifications (not destroying the meaning of the whole). So instead of a relation of identity and instead of a *still-standing static coincidence*, we are dealing here with a relation of similarity or even of analogy. These relations denote the progress of a content (which may at first be defined only vaguely or incompletely) from an empty implication towards the intuitive proximity of the thing itself (*aus der Sachferne in die Sachnahe*). This incomplete and imperfect coincidence or mere similarity,[62] which may include definite moments of "frustration," does not however cancel the fact that it is the intended state of affairs that appears now in the modus of *es selbst.*

It follows from what has just been said that for Husserl intentionality means much more than a *directedness towards an object*, whatever interpretations may be given to these words. Of course, in any occurrence of consciousness we can grasp in reflection a duality of the appearance of what has appeared. Now, if the modes of appearance are ascribed to the *ego* (which remains one and the same) as its determinations then the duality mentioned above can be called "directedness towards the object or being conscious of something" (cf. *das Bewußtsein von etwas*), yet the variety, or rather active or habitual variation, of the modes of givenness in the process of objective synthesis implies also a certain act of the will.[63] An empty intention (the

---

[61] LU, p. 597.

[62] The extreme case of such an "incompleteness" is represented by the situation of disappointment (*Enttäuschung*): "Eine Intention enttäuscht sich in der Weise der Widerstreits nur dadurch, daß sie ein Teil einer umfassender Intention ist, deren ergänzender Teil sich erfüllt" (*ibid.*, p. 576). The disappointment in this sense implies a fulfilled consciousness of the same (!) object that has been intended but proved to be a different thing, something unexpected or having betrayed my expectations.

[63] Cf. K. Held, "Husserls Rückgang auf das φαινόμενον und die geschichtliche Stellung der Phänomenologie." *Phänomenologische Forschungen*, **10** (1980), pp. 100–102.

intention of a name) is just an outline of a strategy of fulfilment and implies a will to the intuitive proximity, to the fulfilled consciousness of the object intended. This "will to fullness" exists either as a *habitus* (in the natural attitude of consciousness), or as τέχνη, a special reflective skill of the mind, "converted" into phenomenology. The natural consciousness guided by *ontic* belief, strives to comprehend its object *as it is in itself*; it tries again and again to move from the absent, the far, the mediated to the present, the near, the immediate, in a word, from the empty to the fulfilled. The *habitus* spoken of here is to be understood in the Aristotelian sense as ἕξις μετὰ λόγου, *habitus secundum rationem*, i. e., we are dealing here not only with a bare desire, but also with a certain skilfulness, with a project (*Entwfurf*), an outlined strategy of the will. So if the natural attitude and the *ontic* belief are connected with a special *habitus* of the will, then the going beyond the natural attitude as a result of phenomenological reduction and "suspending" of the *ontic* belief imply a certain resoluteness, also understood as an act of the will. And in this case we must speak of a choice of strategy (the strategy of a special experience of reflection). It is only due to this act of the will, to this act of rejection and choice (προαίρεσις) — in a word, due to the transition from the natural attitude to the phenomenological — that the fundamental concepts of phenomenology can be lived through in an experience of the fulfilled consciousness.[64]

The dynamic structure of the empty intention and its fulfilment just described serves as a measure allowing comparison of the content of the intuitive cognition with the preceding expectations, with an empty outline of meaning. More than that, this outline of meaning becomes an outline of will, it leads on and guides the synthesis of fulfilment, i. e., makes possible a corresponding "evidentially" fulfilled mind either in the positive modus of confirmation and attestation, or in the negative modus of frustration or partial destruction of the original outline of meaning. In the fulfilled mind we meet the entity as it is, *secundum id quod ipsum est.* Being given in the fulfilled consciousness is precisely the phenomenological meaning of "the thing itself," of "the selfness" (= *ipseitas*) of an object. The empty intention also is one of the modes of givenness of the intended object, although it may be termed deficient or preliminary with regard to the fulfilled cognition. Nonetheless the empty intention (the deficient, the preliminary) is a necessary condition of the fulfilled (the perfected, the final). When we enter a new area of experience, closed to us before, be it the phenomenological experience of pure subjectivity or what the tradition called *visio Dei*, this very stepping into the amorphous and indefinite is impossible

[64] Through the phenomenological reduction and the discovery of the "sphere of pure subjectivity" all possible excuses are taken away from the *ego* and it now stands placed before the universality of its responsibility for everything that is certain for it. In the lectures of 1923/24 on the *First philosophy* (*Erste Philosophie*, Zweiter Teil: *Theorie der phänomenologischen Reduktion*, hrsg. v. R. Boehm, Hua VIII, Den Haag, 1959) Husserl discusses time and again the shift to phenomenological attitude in terms of a moral action or even religious conversion. He speaks of "escaping from a kinship with the world," a "world denial," etc. Husserl states later in the *Crisis* that the persistent phenomenological reflection can lead to a "complete personal transformation" of man. He is fully entitled to quote repeatedly St. Augustine's "in te redi, in interiore homine habitat veritas." Cf. L. Landgrebe, "Husserl's Departure from Cartesianism," in *Phenomenology of E. Husserl. Six Essays.*

without a preliminary "empty" outline (let it be extremely vague) of the meanings we intend to fulfil. The way phenomena are preliminarily given to us within this new horizon is nothing but an empty intention of the *name* or an empty intention of the *text*, a textual intention. This admission seems to me unavoidable:[65] realization in experience implies a preliminary empty construction or outline of meaning. It is on purpose that I say "the intention of the name," "the intention of the text," and not "the intention of the concept." In this way I want to avoid referring to this or that pre-determined procedure of concept formation. The poetic *apophansis* is yet another way to bring to light what is un-concealed, and this in the modus of *here is the thing itself*. I shall make so bold as to say that the *Corpus phaenomenologicum* as a whole and the individual texts forming it are a kind of textual precept, a method of inciting the reader's will to follow a certain strategy of fulfilment, to carry out a certain work of reflection. Adhering to the intention of the text (which at first is empty for him), the reader, in his experience of reflection, "enters" the field of pure subjectivity and progresses in it, taking part in a complex interplay of interwoven empty constructing and corresponding fulfilment. In the process, new facts of the fulfilled mind and the way of fulfilling (confirmation, similarity, refutation, disappointment) become a starting point for another instance of empty constructing, outlining the meaning, and so on. This is how phenomenological archaeology proceeds through the layers of mind. It seems to me that this wandering of reflection can hardly be called by the word "description," such a favourite of early Husserl's.

Before proceeding further I shall go back to discuss the theme of the *visio Dei* in the commentary by Albert the Great. The formula *attingere ipsam substantiam [Dei] sine medio* aims at expressing a kind of immediate relation between the created intellect and God. Yet the opposition of the immediate *vs.* the mediated (and also the concept of relation itself) must be approached with caution. Already the medieval commentators of the *Corpus Areopagiticum* clearly perceived the difficulty lurking here, which we have already touched upon. According to Aristotelian teaching on categories every *relation* must depend, logically as well as *ontically* (i.e. in its definition and in its being), on its own terms which are thought within the category of substance. Yet, as Albert the Great emphasizes in the *Commentary on Book I of the Sentences of Peter Lombard*, the term *substance*, when applied to God, means only that in His being God does not depend on anything else, that He is an *ens a se et per se*. Due to its simplicity and in-finity, God's substance cannot be a subject of accidental de-finitions: God is not composed of *substantia* and *accidentia*. The *absolutely* simple, as Albert states, by definition cannot be combined with (in particular,

---

[65] Cf., E. Tugendhat, *Selbstbewußtsein und Selbstbestimmung. Sprachanalytische Interpretationen* (Frankfurt a. M., 1979), pp. 17f.: "Die Art und Weise, wie uns alle diese Phänomene wie Bewußtsein, Bewußtsein von etwas, Selbstbewußtsein, Ich usw. zunächst gegeben sind, [ist] eben eine sprachliche, und das gilt eigentlich von allen Gegenständen der philosophischen Reflexion; letztlich bedeutet das nichts anderes als Trivialität, daß die Wörter 'Bewusstsein,' 'Selbstbewusstsein,' usw. eben Wörter sind und eine Klärung gar nicht anders beginnen kann, als indem wir nach ihrer Bedeutung fragen. Daß man nicht anders beginnen kann, daran kann eigentlich kein ernsthafter Mensch zweifeln. Die Wege gehen erst dort auseinander, wo man meint, daß wir die Bedeutung von Wörtern dadurch erkennen, dass wir etwas mit einem geistigen Auge erschauen."

relate to) anything else, and God is absolutely simple.[66] Moreover *He Who is* does not enter the category of substance, He is beyond the substance or the essence, as Dionysius says. The name *He Who is,* the name that negates itself through the qualification "superessential essence" (ὑπερούσιος οὐσία) must allow the finite mind to cast a glance beyond beings. If the *vision of God according to His essence without mediation* is understood as an immediate relation, and the term *relation* (*relatio*) is interpreted according to the metaphysical "schemes of category," this inevitably leads to an *objectification* of God as one of the relation's terms. And this danger is to be avoided at any cost.

Starting with quite different premises, phenomenology and onto-theology strive (the former in a universal way and the latter within the context of the *visio Dei*) to reinterpret "the mind's or intellect's relation to ..." in such a way as to avoid a naive objectification of the relation's terms. Yet the theological tradition (despite its huge range and lack of homogeneity) views this problem in a perspective, which is the reverse of the phenomenological perspective. In the "theological reduction" the finite = created mind is traced back or ascends to the Divine intellect as its constituting instance, whereas in phenomenological reduction the whole world is reduced in a sense to the constitutive accomplishments of the pure *ego.* It would seem that such a drastic divergence of the starting points precludes, from the very beginning, any possibility of phenomenological essays on the theme of *visio Dei.*[67] Yet on a closer examination we discover that the difference of *the straight and the reverse perspective* spoken of is based on a fundamental statement of Husserl's phenomenology which does not stand the test of *absence of presuppositions* stipulated by its founder himself.

The original "discovery" of Husserl, which, properly speaking, opened the way to phenomenology, was that of the intermediate field of the *modes of givenness.* This phenomenal field, the primordial dimension of the *appearance of the appeared*, forms a kind of "no man's land" between the "in-itself" of the pure *ego* and the *constituted* exteriority of the worldly objects. Or rather, the distinction under consideration could be termed, in an Aristotelian manner, the distinction between the *energeiai* of manifestation and the manifested forms (εἴδη), so that the starting point of phenomenology is to be considered the field of pure *energeiai.* Yet Husserl interprets this intermediate field, in the Cartesian spirit, as an *immanence of the mind,* and the mind as *the interior*, and in the "mature" period of phenomenology, starting with the *Ideas Concerning Pure Phenomenology and Phenomenological Philosophy* (1913), as "the property" of the executor of the acts of the mind,[68] which is

---

[66] "Simplex dicitur indivisibile omnino divisione sui et alterius non componibile cum alio, non habens hoc et hoc: et hoc modo Deus dicitur simplex."

[67] In § 58 of *Ideen I*, entitled "Die Transzendenz Gottes ausgeschaltet," we read: "Auf dieses 'Absolute' und 'Transzendente' erstrecken wir natürlich die phänomenologische Reduktion. Es soll aus dem neu zu schaffenden Forschungsfelde ausgeschaltet bleiben, sofern dieses ein Feld des reinen Bewußtseins selbst sein soll."

[68] It is necessary to bear in mind that this self-interpretation of phenomenology as *egology* was preceded by attempts to build a *non-egological* phenomenology (in particular, in the *Logical Investigations*). In the lectures delivered in 1907 under the heading "Thing and Space" Husserl says: "The 'I' is inevitably something thing-like which is constituted within the nexus of intentional interconnections, while the

named *ego cogitans*. In this "mature" period phenomenology is interpreted as *egology*, and all its subsequent development becomes an attempt to capture the transcendental *ego* in a net of reflections in order to reach its apodictic certainty. This Cartesian version of phenomenology posits, under the heading of *ego*, a certain centre point of perspective: the no-man's land of the manifestation of pure phenomenality becomes the property (the result of constituting activity, the definition, the attribute, etc.) of the transcendental subject. Yet, as we shall see, such a solution lies outside the scope of phenomenology proper.[69]

Does the phenomenal field possess such a centre; does the manifestation-of-the-manifest (= "consciousness") have an "owner?"[70] Can we actually form sound judgments (according to phenomenological criteria) about the mode of being of such a "performer" of the acts of consciousness (*der Vollzieher der Bewußtseinsakte*), the ultimate functioning instance (*die letztfungierende Instanz*) conceived as the unity which constantly maintains itself throughout all of its acts? And if the answer is yes, what is its or his *name*?

2.3. *The Ego Pursuing Itself*

As already has been said, any phenomenologically reliable state of things must be established in reflection, and reflection is understood as a fulfilled consciousness of consciousness. Besides, for Husserl the term "being" in the proper sense, when it is not seasoned with various prefixes performing the function of a weak negation (like *Ur-sein* or *Vor-sein*), means "positivity," something posed positively. Here consciousness itself plays the role of the subject positing the facts of consciousness. The main difficulty encountered by phenomenology as *egology* is the necessity to explain how the Self as the ultimate functioning instance, the transcendental *ego agens* = *ego cogitans*, can become objective (something positive and posited) for itself. To perform the phenomenological verification of the existence of this ultimate (primordial) instance, it must be turned into an object of reflection. Yet, as it seems, the aim of the phenomenologist, the point towards which the reflection proceeds during the synthesis of fulfilment of the empty intention of the name (and the name at the moment is "the originally functioning, constituting, living subjectivity"),

---

thinking as such is nobody's thinking..." (*Ding und Raum*. Vorlesungen 1907, hrsg. v. U. Claesges, Hua XVI. Den Haag, 1973.) A detailed history of the development of Husserl's concept of pure *ego* can be found in: Alexander Haardt, "Selbstbesinnung und Selbstbewußtsein in Edmund Husserls *Cartesianische Meditationen*," in *Cognitio humana — Dynamik des Wissens und der Werte*, hrsg. v. Ch. Hubig (Berlin: Akademie Verlag, 1997), pp. 433–450.

[69] Concerning the attempts to build a "non-subjective phenomenology" (i.e. to begin with more primordial phenomena than the subject-[intentional]-object distinction) cf. J. Patočka, "Der Subjektivismus der Husserlschen und die Möglichkeit einer 'asubjektiven' Phänomenologie," in *Philosophische Perspektiven*, Bd. 2 (Frankfurt a. M., 1970); H.-U. Hoche, *Handlung, Bewußtsein und Leib. Vorstudien zu einer rein noematischen Phänomenologie* (Freiburg/München: Karl Alber Verlag, 1973); K. Held, "Husserls Rückgang auf das φαινόμενον und die geschichtliche Stellung der Phänomenologie," in *Phänomenologische Forschungen*, **10** (1980), pp. 94f.

[70] Cf. Г. Г. Шпет, "Сознание и его собственник" в: *Георгию Ивановичу Челпанову от участников его семинариев* (G. G. Špet, "Consciousness and its Owner"), Moscow, 1916.

cannot, by definition, become a posited object, cannot be given in a fulfilled consciousness. In fact, all "givenness," all "positivity" is a result of the "giving," positing, constituting activity of consciousness, i. e., of the activity originating in the same functional centre. Such an *ego* cannot "catch itself red-handed," i. e., at the moment of committing the crime (the suicide) of self-positing.

It is necessary here to make a remark concerning terminology. The manner of speaking about the object-beyond-objectivity searched for, adopted by Husserl in his texts, oscillates between two formal alternatives. On the one hand, *ego agens* is thought, from the viewpoint of formal ontology, as something belonging to the category of substance, i. e., as a constant subject, identical to itself, for which various states and functions of the mind serve as determinations or attributes. This tendency is expressed by the substantival pronoun *das Ich* and due to this fact alone refers to the independent possessor of the phenomenal field, to the Cartesian *ego*. On the other hand, in the phenomenological corpus we can find certain indications pointing to the idea that the same "instance" (the same target of phenomenological reflection) must belong to the category of action, i. e., we must anticipate under the heading "*letztfungierende Instanz*" pure activity (*an activity directed towards itself,* according to Fichte) rather than a self-identical agent. In this case Husserl speaks of *cogito* or *das Cogito* (it is to be noted that here, too, the verb is *first person singular* and implies *ego, das Ego*). Sometimes the active living subjectivity is discussed in "topological" terms, i. e., the category of place becomes the most important; the thing sought for is referred to as the focus or the centre of the universal all-encompassing horizon of consciousness.

Yet the formal differences mentioned are not important for us here. Of course, the canon of classical metaphysics, its ruling onto-logic necessarily, one way or another, implies a substance underlying the accidentals: behind the *cogito* the *ego cogitans* is to be revealed. The activity of positing, according to this logic, must be the activity of a self-identical and autonomous (*ontically* self-sufficient) agent, which Husserl identifies with the Cartesian *ego*, the pure *ego*; the horizon of consciousness describes the conscious subject. Nonetheless, Husserl's tendency to call subjectivity *das Ich*, so that this concept formally (according to its functioning in language and in the movement of a text) belongs to the category of substance, represents strictly speaking merely a designation of the place occupied by the transcendental subject within the scope of formal ontology. The substantival *das Ich* refers to a "pole of identity" (*Identitätspol*), and it is nothing but a *result of the synthesis* of identification. The equality sign links together the subjective poles of individual acts of consciousness in their temporal difference. In this synthesis the mind strives to posit as an object precisely that which underlies any synthesis as its necessary condition. Yet *what* this ultimate possibility of all possibilities of the mind is, eludes reflection again and again. The experience of reflection encounters an insurmountable obstacle: it remains ever unclear whether the *ego* actually captures *itself* or just an "imprint," a sign of itself, a falsified copy, a kind of εἴδωλον. Does not a substitution take place in the imaginary identity of the thing reflected and of the reflecting subject, a substitution which cannot be detected, for the gap between the subjective and the objective poles of the reflection act, the gap of reflective intentionality is not seen by reflection at the moment of its performance, and when it becomes the object of reflection on a higher level, the living reflecting instance (the subjective pole of this new act) is, again, an "active agent," not an object posited, so that there is no end to this elusion. The glance of reflection issues forth from the active, "living" Self and overtakes the objectified, "killed" *ego* in its already completed, "past" activity, i. e., in its "already-non-activity." And this movement of reflection cannot be reversed.

Of course, Husserl could not put up with the "invisibility," with the inaccessibility of the "originally functioning instance" in fulfilled consciousness, for this would mean for him a failure of the phenomenological project; the *fundamentum absolutum et inconcussum* of phenomenology would escape all the attempts of phenomenological verification. That is why in his lectures on *First Philosophy* Husserl finds the following solution. There is no doubt that in a particular act of reflection (let us call it reflection$_1$) its subjective pole is not given (positively). Only the objective pole, the object of reflection$_1$, is manifest, and this is the objectified, *patent ego*$_1$, as opposed to the anonymous, *latent ego*$_2$, the performer of the act of reflection$_1$. Yet in the reflection on a higher level (let it be reflection$_2$), for which reflection$_1$ becomes an object,) the subjective pole of the already objectified act of reflection$_1$ (that which *claims to be* the previously anonymous or latent *ego*$_2$) becomes itself an object and with apodictic clarity reveals itself as being identical to its own object at the previous stage = the patent *ego*$_1$, given in reflection$_1$. As a result of this higher level reflection the identity: $ego_1 = ego_2$ is established. And this happens *necessarily* on every new level of reflection:[71] $ego_1 = ego_2 = ego_3 = \ldots$ What has been anonymous gets a name, and it is always the same: *das Ich*. The following was written by Husserl in the beginning of the 30's:[72]

And yet I can direct [the glance of reflection] towards myself. And then my object (*das Gegenüber*) gets split again, and in it the *ego* comes to the fore alongside of that which was its object (*was ihm gegenüber war*), i. e., the *ego* and its [former] object [become a new object]. Here I, the subject, am anonymous with regard to this new object.

That this is how the matter stands concerning this subject, I see precisely by means of this same reflection, for having accomplished it I can at the same time find as an object the *ego* that was anonymous as well as its object a moment ago. Reflecting in this manner and reflecting in this manner again and again, I find again and again a being (*Seiendes*) [as the object] and the *ego*, I find in this reflection the same *ego* [identical to itself], I find also this "again and again" (*das Immerwieder*) of the act of reflection [itself] and of the possibility of going on with my reflection (*Reflektierenkönnen*)... I find, in this permanent self-splitting of the *ego* and in its constant subsequent self-identification, [on the one hand], a kind of primal *ego* (*ein Ur-ich*), which I call the primal pole (*Ur-pol*), the primordially functioning *ego*, [and on the other hand], the *ego* which has become an object for the primal *ego*, which has become a being for it (*das zum **Seienden** gewordene Ich*), and [besides I find] an area of what is present as a non-*ego* for this [objectified *ego*] as well as for me as for the anonymous *ego*...

---

[71] Cf. E. Husserl, *Erste Philosophie*. 2. Teil, *Theorie der phänomenologischen Reduktion*, hrsg. v. R. Boehm, Hua VIII (Den Haag, 1959), pp. 89ff.

[72] Manuscripts C 2 I/3 a. Cited after K. Mertens, *Zwischen Letztbegründung und Skepsis. Kritische Untersuchungen zum Selbstverständnis der Phänomenologie Edmund Husserls* (Freiburg/München: Verlag Karl Alber, 1996), p. 155.

New layers of reflection do not add anything to the image of the *ego* and *necessarily* cannot do so;[73] this means that in this repeated (or rather perpetual) reflection the sought-for *ego* itself, *es selbst*, becomes manifest — since we take the liberty of speaking about it in third person.

However the expression *es selbst* is ambiguous in phenomenology.[74] On the one hand, the "selfhood" of an object may signify its self-identity: *es selbst* = ταὐτό, *idem, dasselbe*. In this sense *selfhood* is identical with the unity of the object, with its *identitas,* which is constituted in the synthesis of identification. The object's "self" is the *noematic* correlate of the noetic concept of *identification*. The identification is an aspect of every perceptual synthesis, more generally of every objectifying act: the object's *selfhood* is posited as the pole of its unity, τὸ ἕν, in the variety of synthesis, in the variety of *noemata* which represent various aspects of the *one* transcendental object. The meaning of objectivity as such becomes manifest to us for the first time in the synthesis of identification.[75]

On the other hand, *es selbst* may also mean αὐτό, *ipsum*. And then, as we have seen, the self refers to the fullness of presence, of the object's "original" self-presentification to cognition. In this sense *es selbst* does not allude to the opposition of the one transcendental object to its "representatives" (aspects, sides, profiles) in the variety of synthesis; it refers to the opposition of the "empty" and "fulfilled," to the original way of self-givenness (in the "now") as distinguished from its intentional modifications.

The experience of continual reflection just described, which strives to pose the *ego* as an object, establishes ("constitutes passively") the *identitas* of the *ego* as a constituted object; but can we say that its *ipseitas* is seen in this kind of reflection?

An evident objection comes to mind here: the necessary identity of the patent and the latent *ego*, established on every level of reflection, is a *result* of synthesis, i. e., of the activity of identification. It follows (if we choose to stick to our premises) that the necessity of identity just spoken of is a necessity *posited* by the sought-for *ego* itself. In other words, a synthesizing *ego* emerges alongside of the reflecting *ego*, and it

---

[73] I am not inclined to underestimate Husserl's achievements in his rigorous and subtle elaboration of the paradoxes of reflection. On the other hand, I am not inclined to overestimate them either. It seems to me that the problem encountered here by phenomenology and especially the method of its analysis is essentially the same as in the passage of Plotinus' *Enneads* quoted above (see sect. 1.1.6 of this chapter). The solution proposed by Husserl in its structural skeleton repeats (is homologous to) the preliminary solution outlined and subsequently rejected by Plotinus. The final solution, as we have seen, is based upon the identity of *energeia* and form in the intellect. The same paradoxes and the same (homo-logical) movement of thought are also found in Fichte's lectures *Wissenschaftslehre nova methodo*; see Kollegnachschrift v. K. Chr. Fr. Krause (Hamburg: F. Meiner Verlag, 1982). The culmination of this movement is the following thesis: "there is such a consciousness in which it is in principle impossible to take apart the subjective and the objective sides, because both are absolutely identical." Fichte writes elsewhere: "The activity of contemplation turning back towards itself and taken as stable and invariable (i. e., *pure noetic energeia – A. Ch.*) ... is nothing but the concept of the *I*."

[74] Cf. E.Tugendhat, *Der Wahrheitsbegriff bei Husserl und Heidegger*, 2. Auflage (Berlin: Walter de Gruyter & Co., 1970), § 3.

[75] Cf. E. Husserl, *Formale und transzendentale Logik* (Halle, 1929), p.139; *Cartesianische Meditationen und Pariser Vorträge*, Hua I, hrsg. v. S. Strasser (Den Haag, 1950), p. 96.

poses the identity of its twins, the latent and the patent one. The acts of the synthesizing *ego* have a different structure, and we need one more "super"-reflection to establish the identity of the synthesizing *ego* with the *egos* already identified. Yet this new identity will be again a result of synthesis, and of a synthesis of a different kind, so that here there is no repetition, no "and so on," but, to the contrary, a necessity of an infinite chain of superimposed reflections having different internal structures becomes apparent. It is extremely difficult to construct such a Babylonian tower even in theory, and quite impossible to experience it in actual self-consciousness.

Only the uninterrupted identity of the separated fragments (ego_1, ego_2, ego_3...) is established in the continual reflexive "self-splitting of the *ego*." Yet the source of the identifying activity remains "behind consciousness' back." This negative statement is a phenomenologically verifiable fact, a conclusion of reflection, and this reflection refutes the results of the previous reflections (the apodictic necessity of the identity between the latent and the patent *ego*). More than that: in the same reflection in which I appear (the *ego* appears) as an object for myself (itself), I become aware of a deficiency, an un-fulfilment, an absence; I become aware of a gap, a distance between the manifest and the non-manifest. "I become aware" does not mean "I posit"; a positive awareness of a deficiency and of an absence is an absurdity. Nonetheless, the positivity of the patent refers in a certain sense to the negativity of the latent; alongside of the recognition: "Ah, it's me," an alienation also takes place: "Is it me?" Here we encounter a paradoxical form of self-consciousness which Husserl's phenomenology (and Cartesian philosophy in general) prefers to pass over in silence: the subjectivity appears not as "the re-establishment of the self-identity in other-being" (Hegel), not as "positing one's Self as positing instance (*sich Setzen als setzend*)" (Fichte),[76] but as an inescapable difference, an absence of coincidence with one's self, a missing of one's self. In reflection the self-identity cannot be re-established. Strictly speaking, I never can say "I am who am" — this is a Divine prerogative. On the contrary, I always am another. The original experience of subjectivity is not the Fichtean "I = I," but an experience of continually losing one's self, an experience of deficiency and lack of self-sufficiency, an experience of substitution, i. e.., formally speaking, an experience of the primordial Difference. Of course, this difference is one of the aspects of time: the reflection is always too late. *Cogito, ergo eram.*

Let us now consider the possibilities of the topological metaphor. The paradoxical absent presence, or presence-as-absence, i. e., a peculiar mode of proximity (which transgresses — *"hebt auf"* — the difference of presence and absence) belonging to the core of subjectivity, is a special kind of "horizon-consciousness" *(Horizontbewußtsein).* Husserl explains the concept of horizon as the contextuality of the phenomenon. The patent phenomenon gets its meaning within the context of what is not yet, but can become, manifest: for instance, the visible side of a spatial object can be understood as such only if those not seen (which, as Husserl says, are present

[76] "Die Anschauung von welcher hier die Rede ist, ist ein Sich-Setzen als setzend [...], keineswegs aber etwa ein bloßes Setzen..." (J. G.Fichte, *Versuch einer neuen Darstellung der Wissenschaftslehre* 1797.)

"in the mode of inactuality") are implied. The functional centre of the acts of consciousness sought for is pres-/abs-ent (*west,* as Heidegger says) within the total scope of the mind. It has *a place* within this scope. This place can be called the centre. It is only in virtue of such a "topological" pre-givenness that reflection becomes possible: to make the *ego* the object of reflection, I must know "*where*" to direct the reflective glance. The ever elusive living *ego* is the "towards-which" of reflection. The empty intention of the *ego* is a choice of direction (a "directional sense," *Richtungssinn*, to use once again Heidegger's terminology), and it supposes an orientation in the topology of the horizon of consciousness, a topological sense of its centre. I know *where* to direct my reflective glance in order to make the hidden latent *ego* enter my field of vision. I always overtake it *here*, in its readiness to move from the "mode of inactuality" to the actuality of presence. Yet, paradoxically, this "latent" pres-/abs-ent thing escapes effort of grasping it, and remains invisible as a *result* of exactly this act of reflection. All positivity supposes a distance between the positing and the posited. But the centre of the horizon of consciousness is a "place" where the glance of reflection proceeds from, and that is why this glance cannot see what is ultimately close (as in Aristotle's theory of vision the eye cannot see itself). When I try to turn the positing into the posited (positive), "*here*" becomes "*there*," and this, let me repeat, as a *result of my effort.* That is why reflection, which is constantly moving around the centre and the source of consciousness, cannot complete its movement, for it finds no term where there is no gap or distance (διάστημα). Reflection, which, according to Husserl, is the only warrant of the positivity of consciousness; reflection which makes the consciousness an entity, *ein Seiendes* (for *per definitionem* entity = positive, posited) repeats the same gesture again and again: it splits the *ego*, and then re-establishes, as it were, its self-identity (which, on examination, proves to be merely the identity of split fragments). The reflection turns around in an infinitesimal circle centred on its source.

When articulated in the text of phenomenological *egology*, this circle seems to bear an unacceptable similarity to a *circulus vitiosus*. That is why Husserl has only the choice to stop the movement in a violent way, by means of an oxymoron: *ungewährte Gegenwart, nicht-gewährende Selbstgegebenheit* (elusive, unnoticeable presence [of self for self], unaccountable self-givenness).

> Any directedness-towards is itself an unnoticed presence, non-thematic self-givenness which is not aware of itself. Yet the transcendental *ego* is always capable of noticing the unnoticed and also of becoming aware of the [corresponding] "I-can" and repeating it again and again.[77]

Yet this absolute immovable self-identity of the Self cannot be termed consciousness in the strict phenomenological sense, for all consciousness supposes a difference between the actuality (*energeia*) of awareness, on the one hand, and the object of awareness, on the other hand, the difference which in phenomenology is called *intentionality.* "The thinking intellect must not remain simple, especially insofar as it thinks itself. For it will duplicate itself by so doing," as we read in Plotinus. The thing Husserl is speaking about is the not-yet-consciousness as the source of con-

[77] MS C 5/3a -3 b. Cited after K. Mertens, *Zwischen Letztbegründung und Skepsis...*, p. 155.

sciousness, the "depth-dimension of the process of constitution," *and it cannot be reached by phenomenological reflection.* When one says, "the transcendental *ego* always can notice the unnoticeable," this is just a way of speaking, not very accurate because it says nothing about the temporal difference. One should say (though this, too, is nothing more than the next step towards an exact description of the matter): "it is possible to be aware, in fulfilled mode, of an object (in a broad sense), which gives itself out to be precisely what was unnoticed." Yet it has become clear that it is impossible to verify phenomenologically whether this claim is well founded: "what was unnoticed" is not a phenomenon to be compared with another phenomenon. The previous description expresses only the dynamics (or rather *kinesis*) of fulfilment.

Further, one can, of course, become in a certain way aware of "I-can," but cannot, in my opinion, be aware of it in an objectified manner ("positively"); i. e., "I-can" as such cannot be given as an object or a fact in a fulfilled consciousness. The "positivity" of "I-can," consists in "I-act," i. e.,, for Husserl, in "I- think," or to be more precise, in "I have just thought" captured by the reflection.

Thoughtlessness is always pres-/abs-ent as a centre of the scope of thinking in a "mode of inactuality." It is impossible to imagine an experience of thoughtlessness as an experience of consciousness.

"That which explicates itself," Plotinus says, "must be many." The necessity of such a self-explication is a categorical imperative of phenomenology. This experience supposes resoluteness and detachment from everyday pursuits. Yet the phenomenologist's will is determined by the search of an object-beyond-objectivity, which has already been named. Naming precedes the will to the fulfilled consciousness of the object named. This "divine name" of Husserl's phenomenology is *transcendental ego;* in pursuit of positivity and in search of the subject lost it initiates an endless movement around a centre (if it is a centre), which remains anonymous, an endless turning in a circle. For, again, "knowledge is a kind of longing [for the absent] (πόθος), and is similar to the discovery made by the seeker. But what is completely other remains itself by itself, and seeks nothing about itself." The difficult experience of a constant reflective longing for self-identification must be gone through to disavow itself. In the place designed for the *ego* one can guess (but not see) "the completely other."

The real parting from Cartesian philosophy would mean for phenomenology a refusal to interpret itself as transcendental *egology.* Yet the "desperate" *apophatic egology* arrives at a conclusion worth careful attention and consideration. It becomes clear that the *ego*, sought for, cannot be overtaken by phenomenological reflection, μᾶλλον δὲ οὔτε ἔστιν — "it is not," if only "to be" means "to be positive."

CHAPTER SIX

SEARCHING FOR THE LOST SUBJECT

1. THE GENEALOGY OF SUBJECTIVITY

> Es gibt immer noch harmlose Selbst-Beobachter welche glauben, daß es "unmittelbare Gewißheit" gebe, zum Beispiel "Ich denke."
>
> Nietzsche, *Jenseits vom Gut und Böse*

1.1. *Predecessors and Heirs of the Transcendental Subject*

We have already seen that in Aristotle's description of the mechanism of the "soul" the metaphor of place, expressed or implied, plays a major role. The soul is a kind of a map of abilities or powers (δυνάμεις), and these powers in action form the system of *energeiai* having various internal forms. The unity of the soul so described is a topological, structural unity. Strictly speaking the soul itself is the principle of unity, since it is "the form (εἶδος) of a natural body, which potentially has life" (*De anima* II 1, 412a20f.). The soul is the internal form or the essence (οὐσία) of a living body as such; it comprises and unites all manifestations or *energeiai* of life (for the soul itself is the *first entelechy* of a living body (412a27), including, if we mean man, the *energeia* of intelligence (νόησις) and the *energeia* of circumspection (φρόνησις), i. e., the *energeia* of good deeds.

Yet after the Cartesian "transcendental turn" *life* becomes a definition of the empirical *Self*, which depends in its being on the "I-am" of the transcendental subject. *Cogito sum* was designed as *fundamentum absolutum et inconcussum veritatis.* But then, as it always happens in the diachronic dimension of philosophy, a definite (delimited, restricted) place was assigned to it within the topography of thought. The claim to absoluteness of the word spoken in the philosophical text is always threatened by the un-spoken, by the surrounding area of silence.

We have seen in the previous chapter that in Husserl's texts the attempt to realize completely the Cartesian positing of an absolute foundation always lapses into a certain lack of foundation. Husserl's "anonymous *proto-ego* (*das Ur-ich*)" is the name of a compromise.

In an article included in an anthology entitled characteristically *Deconstructive subjectivities*, J.-L. Marion asks: "Has phenomenology ever had a more urgent challenge to confront than the determination of what, or possibly who, succeeds to the subject?"[1] This question (to be more precise, the diagnosis disguised as a question) is a result of dramatic changes, which took place in the philosophical consciousness of our century; their origin is to be looked for (among other things) in Heidegger's attempt to resume the "litigation"[2] concerning the meaning of being. "The subject is dead, long live *Dasein*!" Yet there is no doubt that the decisive break in the subject's fate which occurred in *Being and Time*[3] is not merely an "internal affair" of phenomenology, not just the setting of a boundary between Husserl's phenomenology and the new beginning of phenomenology as (fundamental) ontology designed by Heidegger while working on his major work. The de(con)struction of the classical concept of subject, the attempt to go beyond the limits of the "modern ontology of subjectivity" is indicative of a new *topos* of philosophising, or, more precisely, of a certain inevitable *u-topia*, in which the philosophy finds itself after losing the Cartesian *fundamentum absolutum et inconcussum veritatis* but still seeing it in the remote distance.

Hegel says in his lectures on the history of philosophy:

> We are now passing on to the philosophy of the modern age and we shall begin with *Cartesius*. At his side we are entering the realm of autonomous philosophy proper, of a philosophy, which is aware of the fact that in its autonomy it proceeds from reason and that self-consciousness is an essential moment of truth. Here we find ourselves, as one could say, at home, and, like a sailor, we can cry after a long voyage on stormy seas: "Land!" In this new period thinking becomes the principle, the thinking proceeding from itself... [4]

This land of promise rests on the firm foundation of the self-certainty of self-consciousness. More than that, it is the absolute itself, the absolute as such, "the spirit" (*der Geist*), that emerges in Hegel's texts as the absolute foundation. It is on a new soil, i. e., in the new land of philosophising, that Hegel recreates the Aristotelian form of first philosophy as "the science which investigates being as being (entity *qua* entity; that which is as being), and the attributes that belong to it in virtue of its own nature" (*Metaph.* IV 1, 1003a21f.). Now, however, it is said that the genuine philosophical task is "to achieve actual cognition of that which truly is."[5] Heidegger comments on this passage as follows:

> Meanwhile "that-which-truly-is" is established as the actual (*das Wirkliche*), and the actuality of the actual is the spirit. And the essence of the spirit consists in self-consciousness.[6]

---

[1] J.-L. Marion, "The Final Appeal of the Subject," in *Deconstructive subjectivities*, ed. S. Critchley & P. Dews (N.Y.: SUNY Press), 1996, p. 85.

[2] Κρίσις. Cf. Parmenides, fr. 8, l. 15; Plato, *The Sophist*, 242c5.

[3] Cf. A. G. Chernyakov, "Das Schicksal des Subjektbegriffs in *Sein und Zeit*," in *Reihe der Österreichischen Gesellschaft für Phänomenologie*, Bd. 3 ("Siebzig Jahre *Sein und Zeit*"), hrsg. v. H. Vetter, 1997, pp. 175–188.

[4] Cited after M. Heidegger, "Hegels Begriff der Erfahrung," in *Holzwege*, 7. Aufl. (Frankfurt a. M.: V. Klostermann, 1994), pp. 128f.

[5] "das wirkliche Erkennen dessen, was in Wahrheit ist" (PhG, "Einleitung," p. 63).

[6] M. Heidegger, *op. cit.*, p. 128.

The spirit is thought as the foundation of the being of entity, the entity's "beingness," its essence, οὐσία, *substantia*. Besides, it is important that this "essence" is not *just* a substance, even the "absolute substance"; it is, as Hegel puts it, a *living* substance. Further, "the living substance is being (*das Sein*) which is in truth a *subject*, or (which means the same) which truly is actual being insofar only as it (sc. the substance) is the movement of self-positing (*Sichselbstsetzen*) or, in other words, insofar as it is a mediation of its becoming-another-for-itself."[7] Let us remark that here the subjectivity of the subject consists, according to this passage, in the possibility of a special movement or gesture which has a double nature: first, of alienation (of itself) from itself, of positing itself as an object, and, second, of recognizing itself in the other (or being-other, *Anderssein*, as Hegel expresses himself), in the posited, alienated, foreign; of re-establishing its self-identity and its totality. And it is precisely the subjectivity of the subject, which determines the climate of the promised land of philosophy.

To understand the essential meaning of the break (περιπέτεια), which took place in the subject's fate in *Being and Time*, we have to answer the following questions:

1. What does the "classical character" of the classical subject of modern metaphysics consist in? What are the principal attributes of the said "classical character?"
2. What is essential in the "transgression" or "destruction" of the subject in *Being and Time*? What are its motives, intentions and results?
3. What or who succeeds to the subject in *Being and Time*? Can we speak of a successor at all? And if so, what kind of heritage does it get?

Of course, these questions are very closely interrelated and have to be answered simultaneously. Ultimately we are trying to understand whether first philosophy, whether phenomenology, which sees itself as first philosophy, must (can) abolish definitively the concept of subject and, recognizing the absence of an heir, embark again on a voyage "on stormy seas," or pause to reproduce in its constructions again and again (of course, in a new shape and within a new system of concepts every time) the functions of subjectivity, for the nostalgia for the promised land of philosophy proves to be irresistible. If philosophy has considered a delusion its autonomy in which it "rests on reason" and "establishes that self-consciousness is an essential moment of the truth," *who* or *what* is today writing the text of philosophy? And, most important, *how* is this text being written?

1.2. *Negative Analogy*

In § 6 of *Being and Time* Heidegger speaks of the task of destruction of the history of ontology. And, of course, we must try to understand the metamorphoses undergone in SZ by the concept of subjectivity, within the framework of this general task, for the "subject" is a key concept, a "concentrated term" of metaphysics, compris-

[7] "Die lebendige Substanz ist ferner das Sein, welches in Wahrheit *Subjekt*, oder was dasselbe heißt, welches in Wahrheit wirklich ist, nur insofern sie Bewegung des Sichselbstsetzens, oder der Vermittlung des Sichanderswerdens mit sich selbst ist." (PhG 20)

ing the most important events of its history. And here we must be aware of the fact that in SZ *destruction* does not mean *destroying* or *annihilating.*

Destruction ... is far from having the *negative* sense of shaking off the ontological tradition. We must, on the contrary, stake out the positive possibilities of that tradition, and this always means keeping it within its *limits*; these in turn are given factically in the way the question is formulated at the time, and in the way the possible field of investigation is bounded off. (SZ 22)

More than that, the "staking out the tradition within its limits" is connected in Heidegger's text with the designation of the place or the point where he himself goes beyond them. The metaphor of the limits is ambiguous: it not only indicates the limited character of the previous ways of inquiry, but also (by this very fact!) limits, i. e., outlines, the new field of research. We must not undervalue the role of the "destruction of the history of ontology" in SZ. Many of Heidegger's constructions can be understood only if their *apophatic* aspect, a kind of a *via negativa* which contours the metaphysical tradition in its limits, is retained. The *sigetic* spoken of by Heidegger in his later period does not so much conquer (i. e., make sound) the area of silence as guess its limits. To guess, to see the limits of what seems limitless is the task of philosophy-as-*sigetics*, and then "logic" must take the next step. In the text of SZ there are many passages — and these are very important — where the meaning of the basic concepts of existential analytic (ontology) of *Dasein* is formed by means of *opposing* them to the fundamental concepts of classical ontology (*the ontology of the present-at-hand*, as Heidegger terms it). These oppositions are not just illustrations; they perform an important function of meaning-formation. The opposition points to a special kind of connection between concepts, which I would like to call *negative analogy.* In what follows I shall try to ascribe a more ontologically relevant meaning to this term.

1.3. *Categories and Existentialia*

In the previous chapters we have already touched upon one of the most important instances of negative analogy. Heidegger writes that the *existentialia* as characteristics of existence "are to be sharply distinguished from what we call categories" (SZ 44). Here the concept of *existentialia* and that of *category* are distinguished and thereby connected in a relation of negative analogy (existentialia *are not* categories). We read further:

Existentialia and categories are the two basic possibilities for characters of being (*Seinscharaktere*). The entities which correspond to them require different kinds of primary interrogation: every entity is either a "who" (existence) or a "what" (presence-at-hand in the broadest sense). (SZ 45)

The "existential question of the *who* of Dasein" is the subject matter of sections 25 and 64 of *Being and Time*, and Heidegger's approach to this question takes its starting point in the analysis of "saying *I*" (*Ich-sagen)*, by means of which *Dasein* expresses itself about itself in everyday self-interpretation (SZ 318). The fact that the categories of language cease to be categories of thought in *saying I*, was demonstrated by Kant in his teaching on *Paralogisms of pure reason.* But here the transcendental subject is referred to under the heading *das Ich,* and "saying *I*" means "forming assertions about the transcendental subject." When we say "I," meaning "the thinking being as the absolute subject of all my possible reasoning" (A 348), such an "I" *(das Ich)*, according to its functioning in language and thought, cannot be anything but a substance. The *paralogism*, the causes of which shall be discussed later, consists in the fact that the "I" as a transcendental subject cannot, according to Kant, be thought within the category of substance. Such an "I" cannot be inscribed into the system of categories because it is itself an ultimate basis ("categoriality") of the categories. And yet I intend to show that we are fully entitled to speak of analogy between *existentialia*, and *categories*, and this precisely when we understand categories in the Kantian sense, i. e., if we take into account the results of the *transcendental deduction of the pure concepts of understanding.*

It seems at first sight that Kant shares Thomas' viewpoint (or rather the viewpoint of the Thomist School) and considers existence (in the scholastic sense) a simple concept.[8] In his essay entitled *Die einzig mögliche Beweisgrund zu einer Demonstration des Daseins Gottes — The Sole Possible Argument for a Demonstration of God's Existence* (1763) he says in connection with the concept of existence: "This concept (i. e., *Existenz* = *Dasein* in Kant's terminology) is so simple that nothing can be said in explanation of it, except to take careful note that it must not be confused with the relationships things have with their distinctive marks."[9] It follows, as it seems, that for Kant also existence is a simple *yes*, a simple positing of a thing alongside of all its meaningful definitions. In the first section of the *Beweisgrund* we find indeed the following thesis: "Existence is the absolute position of a thing and thereby differs from any sort of predicate..."[10] The *Critique of Pure Reason* develops the same thesis (albeit in a more concise treatment) in connection with the same problem:

Now, if I take the subject (God) with all its predicates (omnipotence being one) and say: "God is," or, "There is a God" *(es ist ein Gott)*, I add no new predicate to the concept of God, I merely posit the subject itself with all its predicates, and I posit it as an *object* in relation to my *concept*. (B 627)

Existence, according to Kant, is "the absolute position." And yet it is precisely in The *Critique of Pure Reason* that we find a development and "explication" of the concept of existence, for within the framework of transcendental method, existence (*position*) is interpreted as a certain relation of the thing posited to the faculty of cognition, a relation which has a complex *internal structure* (e.g., of an intrinsically coherent and harmonious perceptual synthesis). That is why the discussion of the *internal form of existence* becomes for Kant not only possible, but even necessary within the context of the "transcendental interpretation" of the concept of *Existenz* = *Dasein*. We shall see further that Kant's categories, like Heidegger's *existentialia*, are characteristics of the internal structure of *existence*, although the term *existence* is used in both contexts (that of Kant's *Transcendental Logic* and that of the *Existential Analytic of Dasein*) neither univocally nor simply equivocally in the sense of ὁμωνυμία ἀπὸ τύχης, but expresses a pair of concepts connected and unified by the

---

[8] Cf. chapter 3 of this book.

[9] *Immanuel Kants Werke*, hrsg v. E. Cassirer, Bd. 2, pp. 77f.

[10] *Ibid.*, p. 77.

relation of negative analogy. The analysis of this analogy will allow us to answer, to a certain extent, the question about the "heir" to the transcendental subject in SZ.

1.4. Ὑποκείμενον *and Subjectum*

Yet first certain precautions against possible conceptual confusion are to be taken. When Hegel depicts the edifice of philosophy as resting on a firm absolute foundation of the "living substance" or subject of self-consciousness, he already relies on the results of a series of major steps taken by modern age metaphysics. These results consist, besides other things, in positing the concept of subject within the configuration of the concepts which have, strictly speaking, quite a different origin: *ego cogitans*, "the I," "consciousness," "reason," "spirit," "personality." Originally there is, however, no necessary conceptual connection between "subjectivity" and the "I-ness" (*Ichheit*, to use Fichte's favourite term) or the Selfhood of a "man" existing for himself one way or another. If we connect from the outset the question of subjectivity of the subject with Descartes' exclamation *Sed quid igitur sum?* ("What, then, am I?"),[11] and with the question of "my authentic nature," we shall bar the way to correct investigation.

Sub-jectum is a morphologically exact (or almost exact, if certain linguistic subtleties are to be taken into account) Latin translation of the Greek ὑποκείμενον. In Aristotle's non- or pre-terminological usage (in which the rigorous terminology originated) τὰ ὑποκείμενα has the meaning of the totality of things that can be subjected to cognition, perception, etc., or (one way or another) to everyday dealing-with.[12] Yet as a *terminus technicus* or "in the formal *apophantic* sense" τὸ ὑποκείμενον means, as we have seen in chapter 2, *the one* (τὸ ἕν) or *something* (τὸ τί) underlying the variety of qualities and characteristics which the thing shows to us "proximally and for the most part." In an *apophantic assertion* the predicate "shows" *something as something*, i. e., something becomes manifest — disclosed and ex-pressed ("said"), — strictly speaking, through the predicate (κατηγορούμενον), and that which underlies the variety of predication and shows itself by means of predication is the logical subject (τὸ ὑποκείμενον). Thus the formal concept of subject implies: *to be one in a plurality;* the subject is the *unifying unity* which as an underlying sub-ject makes possible the thing's presence in the variety of its determinations and thereby allows the thing to be present to knowledge as a definite something.

In the Middle Ages the object of cognition as distinguished from cognition was called *conceptus objectivus* or simply *objectum*. In Scholasticism (and in Descartes' writings too) the characteristic "objective" has a meaning opposed to today's average usage, in which "objective" means "independent of cognition," "existing in itself." *Ob-jectum* signifies in Scholasticism, on the contrary, the thing which is put *before* the intellect as something conceivable and already conceived in a conception, as something "thrown before" (from *ob-jicio*) the intellect as something proximate

[11] *Meditationes de prima philosophia...*, *Meditatio* II.

[12] Cf. E. Tugendhat, ΤΙ ΚΑΤΑ ΤΙΝΟΣ. *Eine Untersuchung zur Struktur und Ursprung aristotelischer Grundbegriffe* (Freiburg/München: Karl Alber Verlag, 1968), pp. 14ff.

and exposed to the cognition, as distinct from what is "thrown under" (*sub-jectum*) and shows itself only through a mediation.[13] "Objective existence" means, accordingly, an existence *for* the intellect, but by no means *in* the intellect. The form, which exists *in* the intellect and represents "formally" a thing to the mind (*representat menti rem cognitam*), is a form of the mind itself (*forma ultima mentis*), an act of the intellect, "its offspring as it were,"[14] and it is called *conceptus formalis*. *Conceptus objectivus*, on the contrary, is not a form determining internally the act of cognition (*conceptio*), but a matter of a kind, to which this act resorts and with regard to which it performs, as a *forma formans*, its form-shaping task. It is towards the *conceptus objectivus* that the sharp point of the intellect is immediately directed (*directe tendit*), and that is why it is also called *intentio intellecta*. On the other hand, *conceptus objectivus* and *res concipienda* are by no means be confused; the object is not always a "positively existing" thing, for one way or another we can also understand negative, abstract, universal or, on the other hand, obscure and confused things. Such objects of the intellect can be called *entia rationis*, for they have *objective being* in the intellect only (*esse objective in intellectu*). The last sentence is to be understood as follows: objects are not located in the intellect in the literal sense as its actual definition, but can be termed entities because they are *represented* in the intellect by means of a formal concept (*conceptus formalis*). For this latter the "object" serves as a foundation or a "target" of intellectual intention. Besides, the *conceptus formalis*, an act of understanding actually taking place in the intellect, is always a "real (concrete) and positive thing," and a unique thing as well, whereas the object of the corresponding act can be universal and/or imaginary. And finally, in contrast to both the *conceptus formalis* and the *conceptus objectivus* (*objectum*), the *subjectum* is the object's foundation existing in itself, having, generally speaking, no relation to the mind except the possibility of "throwing" (*objicere*) an ob-ject before it. This possibility, as an effort or a tension, was called *intentio rei*, the intention of the thing (having its origin in the thing itself!), as distinguished from the intention of the intellect.[15]

Thus the second formal characteristic of the concept of subject consists in the fact that the subject is defined *with regard to the object* as a condition of the possibility of the latter.

---

[13] In what follows I use a survey on the Scholastic distinction between *conceptus objectivus* and *conceptus formalis* to be found in Suarez' *Disputationes metaphysicae*.

[14] "... dicitur conceptus, quia est veluti proles mentis..." F. Suarez, *Disputationes metaphysicae*, disp. II, sect.I, 1.

[15] I would like to cite in this connection several passages from Suarez' *Disputationes metaphysicae* (disp. II, sect. IV, 1): "Conceptus formalis dicitur actus ipse, seu (quod idem est) verbum quo intellectus rem aliquam seu communem rationem concipit; qui dicitur conceptus, quia est veluti proles mentis; formalis autem appelatur, vel quia est ultima forma mentis, vel quia formaliter representat menti rem cognitam [...] Conceptus objectivus [...] non est conceptus ut forma intrinsece terninans conceptionem, sed ut objectum et materia circa quam versatur formalis conceptio, et ad quam mentis acies directe tendit, propter quod ab aliquibus, ex Averroe, *intentio intellecta* appellatur, et ab aliis dicitur *ratio objectiva*. Unde colligitur differentia inter conceptum formalem et objectivum, quod formalis semper est vera ac positiva res et in creaturis qualitas menti inhaerens, objectivus vero non semper est vera res positiva; concipimus enim interdum privationes, et alia, quae vocantur entia rationis, quia solum habent esse objective in intellectu."

We have seen already in chapter 2 that there is a specially marked category in the Aristotelian list, which no longer manifests further definitions in a different "subject" (ὑποκείμενον), but defines or forms the "subject" in itself in a primordial way. This is the category of substance (οὐσία). Substance is the absolute subject (as long as it is a definite something, i. e., on the condition that we do not speak merely of an indefinite possibility of definition, called "matter" in Aristotle). Let us now read the last sentence in a somewhat different way: the absolute subject is the substance or the essence, i. e., the concept of subject determines the fundamental meaning of the entity's essentiality — "being-ness." The Aristotelian term — οὐσία — is originally an indication of a problem raised or a research project, a designation of the entity's being-ness; for "the question 'What is an entity [as entity]?' is the question 'What is the essence?'" (*Metaph.* VI 1, 1028b4). The answer given in the *Categories* and reproduced later, with many important correctives, in the *Metaphysics*, is: *essence is the absolute subject,* τὸ ὑποκείμενον πρῶτον (*Metaph.* VI 3, 1029a1f.).

1.5. *The Subject and the Ego*

In his essay *The European Nihilism* Heidegger describes thus the first steps made by philosophy on the firm ground of self-consciousness:

> Thanks to Descartes and after Descartes the "subject" of metaphysics becomes mainly *man*, the human *ego*. How does man assume the role of the proper subject and of the only subject? Why is this human subject superimposed on the ego, so that subjectivity is here equal to the sphere of the ego? Is subjectivity defined by the *ego* or, vice versa, the *ego* by subjectivity?

It seems that both statements are true. On the one hand, the formal concept of the subject spoken of before is interpreted in the sphere of the human *ego* (in the field of reflection) according to the structures and relations discovered here. On the other hand, "the authentic or pure *ego*," as a projected ontological foundation, or, more precisely, the foundation of the *certainty* of the entity's being, is separated from the "holistic" self-interpretation of man and is constructed after the pattern of the metaphysical concept of *subjectum*, ὑποκείμενον. From the very beginning Descartes' question — *Sed quid igitur sum?* — follows, within the promised land of metaphysics, the beaten track in search of an answer. The self-oblivious catharsis effected on this way leads to the pre-defined answer: *Sum subjectum ultimum* (*cogitationum*). My thoughts are my *determinations*, and I myself am the subject of these determinations. I can doubt the "*formal* being" of the object of my knowledge (in Descartes' terminology this means to doubt the existence of a certain "subject" in the classical, scholastic sense, i. e., of a "thing outside of my mind"), but the existence of the content of my thoughts as such ("objective being" in Descartes' *and* scholastic terminology as well) is certain, though this being must be characterized as derivative. Descartes says: "The reality which I apprehend in my thoughts is *only objective.*" The way of being proper to representations or "ideas" is nothing but *objective being* (*modus essendi objectivus*). For Descartes (as for the scholastics), the latter means, contrary to what the "philosophical ear" hears today in the word "objective," quite a trivial thing: ideas are simply modes of my thought (*modi cogitandi*).

I can — more than that, *I must* (following the onto-logical scheme pre-defined by tradition) — deduce hence the existence of the corresponding *substance*, a subject of thoughts, existing in itself though not as yet determined and named. For this substance the ideas are accidental properties or modes, the "I-think" as such is its *proprium*, its attribute. It remains then only to recognize in this subject — since representations are mine — my "proper" *ego*. The *ego* as the absolute subject, the subject in the modern sense, appears in Descartes' works *alongside of other* subjects, understood in the classical way — e.g., a tree or a stone. Yet this subject has, if not an ontological (this is a later interpretation of Descartes' conclusions), at least a "logical" superiority, the superiority of its certainty for knowledge, for it represents the *fundamentum absolutum **veritatis***, the absolute foundation of the truth, of true judgments about the being of other things (including God, the only, strictly speaking, *ens a se, ens absolutum*, on which the human subject depends in its being, though, paradoxically, not in its certainty for itself).

1.5.1. *The Fundamental Axiom of Classical Ontology. Existence and Positivity*

If we now ask ourselves what ontological solution has led to the identification of the *ego* with the subjectivity and try to represent this solution by means of one axiom, this axiom will be: "to be" means "to be able to become an object." Being is being-a-(possible)-object. This axiom follows from the basic principle of Greek ontology, *being* = *presence*, in virtue of a further chain of identifications: presence is "visibility"; to be present is to be exposed to "the soul's eye," i. e., to the mind (ὄμμα τῆς ψυχῆς = νοῦς). What-is, the present, testifies to its being through the *conceptus objectivus*. A being is always a possible object. According to this view, *objectivitas*, objectivity, is a fundamental, and at bottom the only, way of being-present (of ἀ-λήθεια), which philosophy can take into account, the only way the entity manifests itself as entity.

In Kant's writings the concept of object does not simply refer to a presence for knowledge and in knowledge; it is connected with the possibility of experience, i. e., relates, directly or indirectly, to *intuition*. "The possibility of experience," Kant writes, "is what gives objective reality to all our knowledge *a priori*" (B 195). In this sentence it is the adjective "objective" that designates entity as existing (*existens*), whereas "reality" refers to the *whatness*, the form, the essence (*essentia*) of what is present. The essence, *essentia*, is the totality of real predicates, i. e., of meaningful determinations proper to the thing, *res*, whereas existence (*Existenz, Dasein*), as we have already seen, is not, according to Kant, a real predicate: it is the *absolute position*. Let us analyse this thesis more deeply.

Kant writes:

The characteristic of a thing's existence (*Dasein*) cannot be found in the thing's concept alone. For, even if the concept is so complete that it has not the smallest defect preventing us from thinking the thing with all its inherent determinations, existence has all the same nothing to do with all this; it is connected with the question whether this thing is given to us, and given in such a way that its perception can in any case precede its concept. For the fact that the concept precedes perception means the bare possibility of concept, and perception, providing the concept with matter, is the only characteristic of actuality. (B 272f.)[16]

Essentially Kant does not say anything new here. As we have seen, Leibniz in his work *On the Method of Distinguishing Real Phenomena from Imaginary Ones* proposes exactly this method of distinguishing between essence and existence: the entity (*ens*) as something possible (*possibile*) is comprehended by means of an adequate knowledge — *cognitio adaequata*, and to know in this way is to have carried out a final analysis of a concept back to its irresolvable constituents, whereas that which exists (*existens*) is understood by means of clear perception; and the clarity of perception implies a constant harmony between its "living" moments. According to Kant, "by means of the concept the object is thought only as being in accordance with the general conditions of possible empirical knowledge in general, and by means of existence it is thought as contained within the whole context of experience" (B 628). Of course, Kant "re-interprets Leibniz transcendentally": the coherent perception is not any longer a method of *verification* of existence, it is the only possible way of understanding a thing's actuality as such. Later Husserl will repeat this definition of existence by saying that an X, thinkable in concept, exists only when X is retained as the *invariable* objective meaning in the course of a coherent perceptual synthesis alongside of other *perpetually changing* moments of the integral *noema*.[17] To be sure, in such a synthesis one can never exclude the possibility of a certain *plus ultra*, including a violation of regularity and destruction of coherence, so that, instead of the existing X, the perception suddenly discovers an existing Y, which, on the initial stages of the perceptual synthesis, "was pretending to be X." Since this *plus ultra* makes any verification of existence inconclusive, Husserl speaks in this connection of an "approximation" or *idea* in the Kantian sense.[18]

In order to be able to appear as a being (a thing in its formal = essential thing-determinateness) and in its claim to existence, an object must appear as *one*, as a synthetic unity. The manifold content of intuition must be combined *in* and *by* the concept in order to become present. It would be superfluous to explain here that perception cannot precede *every* concept in general. And this is not Kant's statement. "The content of intuition" before there is any concept, a kind of primal matter, can be thought only as an abstract moment of the appearance, the moment which is never manifest itself. Every object is a synthetic unity, and an object (an entity showing itself) is possible only as a unity of a certain manifold of intuition (cf. A 158/B 197).

1.5.2. *Man Assumes the Role of Subject*

Let us now go back to our question: how does the "I" become the absolute subject? For Kant, the *ego* in its first and basic sense is nothing but the *ego* of the representation of "I think," which makes possible all representations as such (B 131f.). The "I think" means, first of all, "I combine" (*Ich verbinde*); therefore the *ego* as

[16] "Actuality" (*Wirklichkeit*) signifies here the same thing as "existence" (*Dasein, Existenz*).

[17] See the previous chapter for the details.

[18] See chapter 4, sect. 2.1; cf. *Ideen I*, § 143.

the transcendental unity of self-consciousness, as the unifying one, is the necessary condition of combining and the integrity of the manifold given in intuition, i. e., the necessary condition of the presence of what is present, of the ob-ject. Thus the *ego* in the sense of *personalitas transcendentalis* is the *subject as the possibility of the object.*

Besides, I become (the *ego* becomes) the absolute subject (ὑποκείμενον) since all my representations (and all representations are "mine") are my determinations. "Consciousness in itself is not so much a representation," says Kant, "as it is a form of representation in general" (B 404). The concrete content of a representation is only an accidental determination in relation to this essential form. To *possess* determinations or predicates means for the mind to *be conscious* of them. And vice versa, all that is represented in a representation (the content of the representation, its object *qua cogitatum*) is necessarily related to the "I think" (B 132) and, as a determination of the representation, is also a determination of the thinking *ego*. Since in any *apophantic* statement "S is P" the logical subject S as my representation is a determination of the *ego*, the *ego* serves as "the absolute subject of all my possible statements, and this representation of myself cannot be used as a predicate of any other thing" (A 348). This argument (removed, by the way, from the second edition of the *Critique*) is to be accepted with all possible caution. Of course, within the framework of Kant's theory, the logical subject S is, in the sense indicated above, a determination of the "I," but, obviously, S cannot be *predicated* of the *ego* in the same sense in which P is predicated of S.

Nevertheless, according to this line of arguments, the *ego* appears to be the absolute subject, for both the structural moments of the classical concept of the subject are present here: (1) the *ego* is one in a plurality, a unifying unity; and (2) the *ego* is the source of the possibility of any object.

On the other hand, the search for the *ego* was carried out according to the pre-defined structure of the formal concept of the subject. And so it is that the identification of the subject and the *ego* comes to an end, and so man assumes the role of the subject.

1.6. *The Category as the Internal Structure of Existence*

We have already seen that the *existence* of a transcendental thing, apprehended according to a concept X, means (for Kant as well as for Husserl) its ability to become manifest in a developing coherent perceptual synthesis. Here the belonging of the concept X to a particular category determines the most general *rule or law* of the synthetic activity in question. If we want now to understand existence as distinguished from essence, we need to speak of the *activity* of combining (*Verbindung*), of its pure "kinetic form" as distinguished from the structure of *what is combined*, of a *forma formans*, not of a *forma formata*. Kant calls this "active form" a *function* (of the synthetic unity); a function is "the unity of action subordinating various representations to one general representation" (B 93). The system of categories is an exhaustive list of such most general kinetic forms. From the point of view of transcendental logic, categories are not simply the most universal *genera* or types of predication, i. e., determinations of the thing's whatness (*essentia*), as is the case in formal logic, but the most universal *rules of synthetic activity*, independent of the given *hyletic* content provided by intuition. As forms (functions) of *positing* a thing distinguished from the form of what is posited, from *quidditas, essentia*, the categories are, in Kant's transcendental logic, determinations of the thing's existence (*existentia*), for existence is nothing but the *absolute position*. Thereby, as has already been said, an "explication" of the concept of existence takes place, and a discussion of its *internal structure* becomes possible.

1.7. *Whatness and Whoness*

So the categories in Kant's transcendental logic are as much fundamental *characters of being* (*Seinscharaktere*) as the *existentialia* in Heidegger's existential analytic. Yet categories are definitions of *being-what* or of *being-present-at-hand* (*Vorhandensein* = *das Sein des Vorhandenen*), whereas *existentialia* are definitions of *being-who*, of *Dasein's* being (= existence in the Heideggerian sense). However on a more attentive examination it becomes clear that a simple reference to the distinction between the "what" and the "who" does not allow us to tell apart categories and *existentialia*.

1.7.1. *Category as a Determination of the Transcendental Subject*

It is true that in a certain (and quite strict) sense Kant's categories are determinations of the transcendental subject itself. The representation "I think" is included as a necessary formal invariant in any representation, i. e., it determines the representation itself as a *form* irrespective of its content. The form spoken of is nothing but the pure representedness of what is represented, the objectivity of the object, which in its turn is already *represented as a relation to* the unifying unity of the *ego* (*das Ich des "ich denke"*). This self-identical *ego* is thereby already co-represented (*ad* + *perceptum* = *apperceptum*) in any representation in pure apperception. It is pure apperception that is, according to Kant, *self-consciousness* (B 132).

The following concepts reveal themselves as equally primordial, i. e., paraphrasing the same first principle: (1) (pure, transcendental, primordial) *unity* of apperception, i. e., self-identity of the co-thought "I think," (2) *unity* or "constant *(durchgängige)* identity" of self-consciousness, (3) *unity* or self-identity of the transcendental subject, (4) isomorphism (formal *unity*) of all representations generally as representations, (5) synthetic *unity* of any object as object, (6) *unity* of the action or function of combining the manifold of intuition. Some of these equiprimordial moments are identified in the following passage from the first edition *Critique*:

> The original and necessary consciousness of the identity of the self is thus at the same time a consciousness of an equally necessary unity of the synthesis of all appearances according to concepts, that is, according to rules, which not only make them necessarily reproducible but also in so doing determine an object for their intuition, that is, the concept of something wherein they are necessarily interconnected. (A 108)

The *represented* combining subject, the "I" of the *Ich-verbinde*, and the *represented* combined object (synthetic unity of the manifold of intuition) define each other in an equally primordial way. There is no need to say that in Kant's writings the term *self-consciousness* does not mean that the transcendental subject can, in a certain way, be given to the *intuition* (by means of the "inner sense"). On the contrary, such a possibility, as we shall soon see, is rejected in a marked manner. Kant distinguishes sharply between the *personalitas transcendentalis* and the *personalitas psychologica* or, in other words, between the *ego* of apperception and the *ego* of apprehension. In the *Anthropology* it is said that *apperception* is divided into self-consciousness in the sense of *reflection* (pure apperception) and self-consciousness in the sense of *apprehension* (empirical apperception). Reflection is *the representation of the internal activity (Spontaneität)*, through which conceptualisation (thought) is possible; apprehension is the representation of receptivity, through which *perception* is possible, i. e., the empirical intuition. The representation of receptivity is the "appropriation" of what has been perceived, its representation (interpretation) as "mine," as "these states of my Self," manifest to the inner sense. My states are interconnected by associations and form a certain unity — the concrete history of a mind's life. Kant calls this concrete "narrative" unity of the mind's "biography," the *subjective or empirical unity* of consciousness (B 139). It is "the determination of the inner sense," whereas the *objective unity* of self-consciousness is that by means of which the manifold of intuition, including the content of the inner sense, is united in the concept of object. The latter unity is totally "original" and hence not explicable or intuitable, but is necessarily presupposed when any appeal is made to empirical consciousness itself.

Thus the *ego* of apprehension is the extant integrity of the extant manifold of my psychic life, the *personalitas psychologica*. The *ego* of reflection (the transcendental subject), on the contrary, is a functional centre (never present for the inner sense) in virtue of which the combination (*Verbindung*) as such is possible. It follows that the combining activity itself (whatever the content being combined may be) is every time a determination of *the transcendental subject*, and the most general kinetic forms of synthesis, the categories, are its necessary or "essential" definitions. If we have recourse to the classical division of determination into modes and attributes (which Kant makes use of too, e.g., in his lectures on metaphysics), then concrete forms of the (empirical) synthesis must be called *modes*, and the categories must be called *attributes of the transcendental ego itself.* Thus Kant would be fully entitled to say that categories are, in the first place, determinations of who-ness, i. e.,, of the transcendental *subject* as the basic forms of its synthetic acts, and only in the second place (as reflected in the internal structure of the synthesized) are they determinations of the whatness of its objects.

1.7.2. *The Limits of Analogy*

Yet it is impossible to continue to move further on along the way of similarity, of homo-logy between the *categories* in the sense of transcendental logic, on the one hand, and the *existentialia*: the categories are *attributes* of the transcendental subject, determinations of the subject's "essence," but they can by no means be thought as determinations of its *existence*, whereas the *existentialia* are characters of *Dasein*'s *being*, or of existence in the sense of SZ.[19] Indeed, while existence, according to Kant, is the absolute position, and the absolute position means coherent perception, we can *experience* only the existence of the *personalitas psychologica,* of a synthetic object of the inner sense, but by no means that of the transcendental subject.

In the transcendental synthesis of the manifold of representations in general, and therefore in the synthetic original unity of apperception, I am conscious of myself, not as I appear to myself, nor as I am in myself, but only that I am. This representation is a thought, not an intuition. (B 157)

This unity of consciousness would be impossible if the mind in knowledge of the manifold could not become conscious of the identity of function whereby it synthetically combines it in one knowledge. (A 108)

The passage suggests that we are aware of our own mental activity as activity that in some way transcends its mere "reflection" in the intentional objects of that activity. This implies for Kant that the awareness in question could not be intuitive, though it needs, as he also insists, to be brought into connection with intuition in order to constitute awareness of an individual person, *qua* individual. The *personalitas transcendentalis* is (is represented as) an absolute functional centre of positing, which can never be posited as an object of intuition. As the ultimate condition of any object, the transcendental subject itself can never become an object (in the Kantian sense, i. e., an object of experience). It is important to repeat here once more that in Kant's terminology *the ego of reflection is not an object* (of experience), for reflection does not include intuition; it is only the necessary representation of spontaneity, of autonomous activity of understanding, which makes possible the pres-ence of an object given in internal or external contemplation.

The *ego* of reflection can be thought only as an absolute subject, "to which thoughts belong as determinations only, and this *ego* cannot be a determination of any other thing. It follows that everybody must necessarily regard himself as a substance, and his thinking as accidental determinations of his existence (*Dasein*) and of his states" (A 349). Here Kant reproduces Aristotle's formal definition of the first οὐσία (*substance* in Latin rendering) as absolute subject (ὑποκείμενον), as well as the Cartesian thesis stating that *cogitatio* is the only attribute of the thinking *ego.* Yet after that, serious divergence between Descartes and Kant's viewpoints begins to become apparent. The conclusion that the transcendental subject *exists* as a substance cannot be deduced from the premise stating that it can be *thought* only as a substance, for "the categories (including the category of substance) do not have any objective meaning in themselves unless there is a certain intuition behind them, to the manifold of which they can be applied as functions of the synthetic unity" (A 348f.). Since the *personalitas transcendentalis* cannot become an object, any dis-

[19] It is necessary to keep in sight that the term "existence" is used equivocally while referring to Kant, on the one hand, and Heidegger, on the other; though the equivocality is not merely accidental. The relation between the two meanings is the relation of negative analogy.

cussion of the substantiality of the transcendental *ego* is deprived of any objective meaning.

The consequences of this thesis are dramatic. The *ego sum*, the *ego existo qua substantia cogitans*, does not follow from the *ego cogito* any more. It is impossible to give an answer to the question of the transcendental subject's way of being, if we want our answer to have an objective meaning.

In the previous chapter we have seen that Husserl's attempts to resume the discussion of the being of the pure *ego* without going beyond the classical interpretation of being as objective being or positivity only revive, in a modified form, an old paradox, which Kant rejects at a high cost.

2. CARE AS THE SUCCESSOR OF SUBJECTIVITY (ONTOLOGY OF BEING-READY-TO-HAND)

2.1. *Being as "having-to-be"*

If we are trying to build an ontology which no longer identifies, from the outset, the entity and what is posited, being (*esse*) and positivity, then "the distinction between positivity and transcendentality" can no longer be regarded as the answer to the question of the meaning of being. Heidegger wrote to Husserl:

> What constitutes is not nothing at all, but also something and an entity, though not in the positive sense. The question of the way of being characteristic of the constituting instance cannot be eluded. The problem of being has a universal character, and it bears equally on the constituting instance and on what is constituted. [20]

The strategies of reflection that have led to the modern European understanding of subject, the play of mutual definitions ("dialectics") of objective and subjective being, begin earlier than required with the question what "to be" means generally. Heidegger considers this question "in a reversed perspective" as compared to the Cartesian paradigm: if the way of being (existence) of the pure *ego* (or of transcendental subjectivity) turns into a fundamental *aporia* of modern ontology, and if we do not want nonetheless to give up, like Kant, any "objectively meaningful" discussion of it, could we not make this being our *starting point*? Besides, must all meaningful discussions be necessarily *objectively* meaningful? In other words, could we not take being as the starting point of phenomenological analysis — the being which is always "mine," which is already laid open for me in a certain way, and in which I already find myself?

And if language in its facticity says "I find myself..." or even "I catch myself in the midst of my being," this does not mean at all that the philosopher has to correct it, purge it, harmonize it with the *fundamentum absolutum et inconcussum veritatis* and to transform the phrase into the third person sentence: "The ego (*das Ich*) grasps (conceives, perceives, apperceives) itself." Besides, is it quite necessary to presuppose (as the only possible "philosophical" viewpoint) that "catches" inevitably

[20] See "Supplements" to Husserl's *Phänomenologische Psychologie*, hrsg. v. W. Biemel, Hua IX (Den Haag, 1962), p. 602.

means "forces to appear," "represents," "thinks" – "the *ego* thinks, the *ego* poses itself in thinking as the positing instance," etc.? Could we not go the other way round and begin with clarifying the meaning of this *being*, which "is every time mine," and then ask *who* its "possessor" is and what kind of *ego* "catches itself" in this being, which is always mine?

And that is how Heidegger begins: the *sum* precedes the *cogito*, "having-to-be"(in the sense of Heidegger "I am and have to be"), being in my being, precedes the intentional relation with the object and therefore the objectivity of the object and the subjectivity of the subject.

2.2. *The Existential Structure of Care*

Now we can make use of the negative analogy between *categories* and *existentialia* in all its layers of meaning. According to Kant, the transcendental subject is an active or functional centre, which makes the system of categories whole and one. The *ego* belongs to the representation "I combine" and is the necessary condition of the system of categories as functions of the synthetic unity.

Heidegger's attempt in SZ to restate the question about the meaning of being by starting with the entity which does not just occur among other entities, but is "ontically distinguished by the fact that in its very being, that being is an issue for it" (SZ 12) – by starting, in a word, with the being of *Dasein* or existence, leads him to the concept of being-in-the-world. Being-in-the-world is one of the attributes of the existence itself, i. e., the way of being of *Dasein* itself. At a certain stage of development of the existential analytic, being-in-the-world appears as a structure "primordially and constantly whole," which at the same time is "phenomenally so manifold" that it can easily "screen off (*verstellen*) the unified phenomenological view of the whole as such" (SZ 180). The *existentialia* are the basic structural moments of the complex system just mentioned. Let us ask now, what collects and unites the variegated aspects of existence into a structured system, and in particular, what makes the variety of *existentialia* a single whole. What must the "unified phenomenological glance" be concentrated on? Heidegger's question is: "How is the totality of that structural whole which we have pointed out to be defined in an existential-ontological manner?"[21] Some lines further Heidegger writes:

> The being of Dasein, on which the structural whole leans as such, becomes accessible in a kind of a glance going through this whole and directed towards a single primordially unitary phenomenon, which is already in this whole in such a way that it provides the ontological foundation for each structural item in its structural possibility.

This unitary phenomenon, the "one," which unites in itself and carries as a primary foundation the existential "multiplicity," is called *care* (*Sorge)* in ***Being and Time***. Now it can be easily seen that along the lines of the interpretation that I have been offering, within the context of the analogy discussed, it is "care" that must serve, within the framework of the new ontological beginning, as a counterpart of

[21] SZ 181.

the modern concept of transcendental subject (or, to speak more cautiously, of transcendental subjectivity).

We shall further make an attempt to understand in detail this correspondence, making use of the work done hitherto. It has been said many times that the transcendental *subject* of modern European metaphysics is, first, an ontological foundation that exists *for itself*. Second, the transcendental subject is the fundamental condition of the presence of any *object*, the condition of the presence (positivity) of what is present, and this means of being too, for being has already been interpreted as the presence of what is present at hand, as a presence-at-hand. This definition, very general and therefore not very precise, is further specified in various ways in different versions of transcendental philosophy. Yet since "to be" means "to be (potentially) an object (of thinking)," the figure of being-for-itself turns into a figure of self-consciousness. We shall devote the remainder of this chapter to re-interpreting within the context of fundamental ontology the second characteristic property of the subject — the subject as a possibility of object or objectivity. In the next chapter the topic of the fundamental ontological "successor" to self-consciousness (*cogito me cogitare*), already touched on in various contexts, will be investigated.

So let us ask ourselves: is it true (and if so, in what sense) that the concept of care expresses in SZ the primordial and basic condition of the possibility of encountering beings in their being, of the openness (*Offenheit*) or disclosedness (*Erschlossenheit*) of an entity in its being? In SZ Heidegger remains within the transcendental formulation of the problem of being, in the sense that to ask about being means for him to ask about the ways of accessing beings, about the basic conditions in which an entity lets itself be encountered. Yet the event of this encounter is not at all reduced now to a representation of what exposes itself to the gaze of mind, to an intentional relation of consciousness to its intentional objects. Heidegger asks about the hidden *ontic* mechanism of the intentional relation.

2.3. *Entity-within-the-world and Its Being*

In what sense is care a condition of encountering an entity? Let us narrow this question: In what sense does the existential structure of care allow us to encounter the *entities within the world*? In SZ entity-within-the-world (*das innerweltliche Seiende*) means, formally speaking, any entity, which is not itself "worldly" (*weltlich*), i. e., the way of being of such an entity differs from being-in-the-world understood existentially. Although Heidegger's terminology seems paradoxical, the mode of being which the *entity-within-the-world* possesses cannot be termed *being-in-the-world*, for the latter is a characteristic of existence only, of the being of *Dasein*. As we have seen, Heidegger tries to get beyond the intentional relation, stating that any "theoretical" (in the original sense of the Greek verb θεωρεῖν — "contemplate distantly," "observe," "scrutinize") relation to an entity, which from the very beginning treats an entity as an object, is founded in the relation of *care*. Accordingly *entity* in its first and basic sense must have the original meaning of the Greek τὸ πρᾶγμα — *equipment* or *tool* (*das Zeug*). And this is the starting point of the version of "phenomenological" analysis developed in SZ. Heidegger rejects the basic procedures of Husserl's phenomenology, ἐποχή and various models of phenomenological reduction, for all of these begin with *the object* and its *ontic* claims and at the same time clarify the structures of the object's objectivity by means of thematization of the ways of (objective) presence = presence-at-hand. Nonetheless Heidegger's research, according to the canon of transcendental philosophy, proceeds from entities we encounter within the world, which are taken as the basic examples for the interpretation of being, towards their being-ness, that is, towards the fundamental conditions of our encountering them.

2.4. *The Sphere of Equipment*

A thing, according to Heidegger, is first and above all a means of ... The being of equipment, the being-ready-to-hand as distinguished from the being-present-at-hand, is not the object's presence in its semantic determinations (whatness) for the intellect, the soul's eye; it is the entity's inclusion in the structure of non-indifference as of something *assigned-to or assigned-for.* Ontologically the first determination of a thing, according to Heidegger, is not its "what" (*essentia*), but its "in-order-to." The thing as equipment is a bundle of references or assignments having the character of *in-order-to.*

If being is understood as presence, then the essence = *beingness* of a thing amounts to its *form*, from which we read off *what* it is, in its semantic appearance or "look" (εἶδος),[22] which the thing shows to the mind's eye. It is in this ability to appear, to be present as this or that (in its "what," in its *whatness*) that the thing *is.* Essentially *form* is a synonym of *presence.* A thing present-at-hand is present-at-hand because it shows itself *as something.* Accordingly the predicate shows the subject in an *apophantic* statement. The extreme (and most important ontologically) case of this *apophansis* is:

> THIS (matter) is X (form)[23]

Yet, Heidegger insists, "the kind of dealing [in the world and with entities within the world] which is closest to us is [...] not a bare perceptual cognition, but rather that kind of concern which manipulates things and puts them to use; and this has its own kind of 'knowledge'" (SZ 67). Equipment as something ready-to-hand discloses itself in a different way: to understand a thing ready-to-hand in its readiness-to-hand means to understand its being-assigned-to or its being-in-order-to.

> THIS is in order to X, but also Y and Z... In its turn X is in order to A, B and C, and Y is in order to D...

Even this last formula is only a method of description. The completeness of understanding of what is ready-to-hand is not achieved when something objectively posited gets further definitions of a teleological nature; it is achieved in our con-

[22] Since the time of lecturing on the *Basic Problems of Phenomenology* the expression *das Aussehen* ("the look") becomes for Heidegger the regular German rendering of the Greek term εἶδος.

[23] Cf. Aristotle, *Metaph.* 1041b5ff.

cerned dealing with it, when *the equipment is skillfully used* to such and such a purpose.

"The world" appears not as a totality of objects, but as a totality of involvements *(Bewandtnisganzheit),* a field of forces or, to speak more precisely, a field of references or assignments: *in order to this and in order to that.*

Each individual piece is by its own nature equipment-for... — for travelling, for writing, for flying, etc. Each one has immanent reference to that *for which* it is what it is. It is always something *for* (*etwas um-zu*), pointing to a *for-which (Wozu)*. The specific structure of equipment is constituted by a contexture of what-for, in-order-to. Each particular item of equipment has as such a specific reference to another particular item of equipment. We can formulate this reference even more clearly. With any entity that we uncover as equipment there is a specific involvement.[24] The contexture of what-for or in-order-to is a whole of involvement relations (*Bewandtnisbezüge*). This involvement that each entity carries with it within the totality of involvements (*Bewandtnisganzheit*) is not a property adhering to the thing, and it is also not a relation, which the thing has only on account of something other being present-at-hand. Rather, the involvement that goes with chair, blackboard, window is exactly that which makes the thing what it is. The contexture of involvement is not a relational whole in the sense of a product that emerges only from the conjoint occurrence of a number of things. The totality of involvements, narrower or broader — room, house, neighbourhood, town, city — is the primal [structure], within which specific entity, as the entity of this or that character, is as it is and exhibits itself correspondingly. (GP 233)

Thus an entity as something ready-to-hand shows itself within a system of references or assignments having the character of *in-order-to*. Assignment, in fact, is a *power* or *force* of a kind since it can generate and direct a concernful dealing of *Dasein. Dasein* is always already included in this field of forces. The latter is a certain field of *indifference* with regard to the classical difference subject/object; on the one hand, it is the medium of the *ontic* meaning of the thing ready-to-hand, and on the other hand, an aspect of the being of *Dasein* itself. And the "in-order-to" reference, drawing what is near into its sphere, reveals what is farther and gets lost in what is the farthest. The limits of the scope of assignments, as far as one can speak of them at all, are vague. The completeness of definition of equipment in its "in-order-to" cannot be achieved, because the variety of references seems to be open as a matter of principle: assignment refers to a new assignment. "Dealings with equipment," writes Heidegger, "subordinate themselves to the manifold assignments of the 'in-order-to'. And the sight with which they thus accommodate themselves is circumspection."[25] This "sight" of concernful circumspection never reaches the boundaries, the end points, the dead ends of referencing.

---

[24] "Mit jedem Seienden, das wir als Zeug entdecken, hat es eine bestimmte *Bewandtnis*." Albert Hofstadter translates this passage as follows: "Every entity that we uncover as equipment has with it a *specific functionality, Bewandtnis [an in-order-to-ness, a way of being functionally deployed]*." See M. Heidegger, *The Basic Problems of Phenomenology*, trans. A. Hofstadter, (Bloomington & Indianapolis: Indiana Univ. Press, 1988), p. 164. Because throughout this text I was trying to keep to John Macquarrie's principles of translation and his system of terminology, Hofstadter's translation has been slightly changed. Macquarrie's renders the German expression *Bewandtnis* as "involvement." The detailed discussion of merits and demerits of this translation can be found on p. 115 of the English *Being and Time*, seventh edition (Oxford: Basil Blackwell, 1987), ftn. 2.

[25] "Der Umgang mit Zeug unterstellt sich der Verweisungsmannigfaltigkeit des 'Um-zu'. Die Sicht eines solchen Sichfügens ist die *Umsicht*." (SZ 69)

Nonetheless the incompleteness of equipmental determinations of a thing being drawn into a net of manifold assignments does not mean that the completeness of being-ready-to-hand, its ἐντελέχεια, can never be achieved. On the contrary, it is achieved every time the thing ready-to-hand is skilfully used.[26]

2.4.1. *The Totality of Involvement*

Let us repeat it once more: in SZ Heidegger understands the *being* of entity phenomenologically, i. e., proceeding from the *ways we encounter* the entity. Therefore the being of a thing ready-to-hand is determined not by its ability to be present to consciousness as a *what* (or, as Husserl puts it, by means of self-representation, self-givenness of an object as *cogitatum*), but by its *ability* (*power*) to draw *Dasein* itself into the open variety of "in-order-to" references, to assign to *Dasein* a certain structure of activity. The fundamental condition of this ability belonging to the ready-to-hand lies in the structure of existence itself, which can, as being-in-the-world, be drawn and is drawn every time into the variety of references, into the totality of involvements (*Bewandtnisganzheit*). How is this possible? The answer given by Heidegger is articulated by means of substantival conjunctions, and before reproducing it we have to say some words about the meaning of the word *Bewandtnis* and the government of the verb *bewenden.*

The German idiomatic expression *damit (mit einem X) hat es folgende Bewandtnis* means "the matter stands with it (with the thing X) as follows; *es hat mit dem X eine eigene (seine besondere) Bewandtnis* means "there is something particular about X."

On the other hand, *es dabei (bei Y) bewenden lassen* means "to leave the matter as it stands (concerning Y)," "to leave it at that," "to let it go at that," "to let it rest there (within the Y-circumstances)."

Finally, *dabei hat es sein Bewenden* means approximately "there are [some proper] reasons for that," though the English equivalent of the expression *dabei muß es sein Bewenden haben* would be, perhaps, "here the matter must rest."

Heidegger writes:

Es hat *mit* ihm [sc. mit dem Zuhandenen] *bei* etwas sein Bewenden. Der Seinscharakter des Zuhandenen ist die *Bewandtnis.* In Bewandtnis liegt: bewenden lassen mit etwas bei etwas. Der Bezug [...] des "mit... bei..." soll durch Terminus Verweisung angezeigt werden. (SZ 84)

With any such entity there is an involvement which it has *in* something. The character of being which belongs to the ready-to-hand is just an involvement. If something has an involvement, this implies letting it be involved in something. The relationship of the "with...in..." must be indicated by the term "assignment" or "reference."

The matter stands so and so with (*mit*) this or that item of equipment, there is a certain involvement with it. The "in" (*bei)* (when the circumstances, the functional

[26] Aristotle remarks in *De anima* II 1, 412b28 that the *entelecheia* of an axe is its "to chop," but, Heidegger adds, this fullness of meaning is understood not by someone who observes the action at a distance but by the person who performs it.

context are such and such) refers to the horizon of non-indifferent beings and concerned dealings[27] into which a particular equipmental thing is getting involved and in which it becomes fit for such and such use — properly speaking, this is what the thing ready-to-hand is *used for* (*das Wofür*).[28] The relationship "with...in...," singling out as it were a specific knot of assignments within a functional whole, is "ontologically definitive for the being of such an entity, and is not an ontical assertion about it" (*ibid.*).

Yet the totality of involvements itself goes back finally to such a *what-for* (*das Wozu*) in which there is no further involvement, which cannot be used for anything else, be fit for anything else. And this ultimate and primordial *what-for* must be called *das Worumwillen*, "for-the-sake-of-which," and such a "for-the-sake-of-which" concerns only the being of *Dasein*, a being, for which, in its being, this being itself "is an issue."

It is to be remarked that in his attempt to articulate the shades of meaning by means of substantival prepositions and pronouns, as well as in the method itself of making semantic distinctions, Heidegger imitates Aristotle. Using the formula τὸ οὗ ἕνεκα ("the for-the-sake-of-which," *das Worumwillen*) Aristotle refers to one of the meanings of the cause or principle, *the causa finalis* (the aim). In *De anima* II 4 415b2f.; 20f. he introduces the following distinction: τὸ δ' οὗ ἕνεκα δίττον, τὸ μὲν οὗ τὸ δὲ ᾧ — "For-the-sake-of-which can be understood in two ways: (1) what the action is directed towards, what it aims at (*terminus ad quem operatio derigitur*), (2) for whom, for whose sake it is carried out (*subjectum cui hic terminus procurandus sit*)." The soul is the "for-the-sake-of-which" of the living body in both senses: on the one hand, nature generates (ποιεῖ) for the sake of the soul; on the other hand, any living body performs its functions so that the soul as the form (εἶδος) of the living organism may sustain its being in the individual and reproduce it in the species. A living being may pursue certain particular aims, but the ultimate aim in the sense of τὸ ᾧ is the soul as such, and in the sense of τὸ οὗ — "the participation in the eternal and in the divine," i. e., in the εἶδος, by means of its completion (completeness and perfection) in the individual and the continuation of the individual in the species; i. e., the ultimate aim in the second sense is also the soul, as the form-eidos-species, about which it is said that every living being, its ancestors and descendants, remain one in species (εἴδει ἕν).[29] In any case the soul as a whole is not any more equipment or a tool, but the ultimate "for-the-sake-of-which;" the living body that Aristotle calls also σῶμα τὸ ὀργανικόν exists, as a complex and integral "tool" (ὄργανον), for the sake of the soul (415b18ff.).

---

[27] J. Macquarrie and E. Robinson emphasize the active character of involvement: "The reader must bear in mind that the kind of 'involvement' with which we are here concerned is always an involvement in some activity, which one is performing, not an involvement in circumstances in which one is 'caught' or 'entangled'"(*Being and Time*, p. 115, ftn. 2). As it was already mentioned, A. Hofstadter translates *Bewandnis* as "functionality."

[28] "Das Wobei es die Bewandtnis hat, ist das Wozu der Dienlichkeit, das Wofür der Verwendbarkeit." (SZ 84)

[29] Cf. *De anima* 415b3–7.

2.4.2. *For-the-sake-of-which and For-whose-sake*

Going back to the text of SZ, we must say that for-the-sake-of-which (*das Worumwillen*) is no longer an "internal" structural element of the manifold of references or assignments, but the foundation itself of the possibility of any such assignment understood as the thing's instrumental intention, its fitness-for, its designedness-for. Heidegger understands "for-the-sake-of-which" in two ways: as τὸ ᾧ — it is *Dasein* itself, and as τὸ οὗ — it is its ownmost ability-to-be (*das eigenste Seinkönnen*).[30] In both cases "for-the-sake-of-which" is a definition of the existence of *Dasein*, which, in its being, is concerned with this being itself. It is precisely due to this determination that *Dasein can* let there be an involvement *with* something *in* something (*bewenden lassen mit... bei...*) and every time already (or even beforehand) has let there be an involvement (*apriorisches Perfekt*). Yet "to let there be an involvement..." means to disclose in understanding the structure or the field of the assignments. To *disclose* the being of an entity in a specific understanding or skilfulness means to enter in an understanding way the manifold of references *for-the-sake-of-which*, and in such an understanding an encounter with the entity-within-the-world is becoming possible, i. e., the ready-to-hand as such can disclose itself in its being. "For-the-sake-of-which" is an *a priori* condition of encountering the entity-within-the-world as something ready-to-hand.

2.4.3. *The Ontological Concept of Care*

It is precisely this structure of primordial un-indifference, making the "for-the-sake-of-which" possible and expressed by means of the phrase *es geht um*, that forms the core of the ontological concept of *care*. Only the entity to which its being is "entrusted" as the first and last "for-the-sake-of-which" of caring (the openness of the "for-what" can certainly appear *ontically* as not-caring, negligence, foolery, self-destruction), only the "caring" entity can be concerned with the question "what for?" and understand (accomplish in an understanding way) the answer: "for this or that."

Yet it is fundamentally important not to confuse the ontological concept of care with a certain *ontic* definition of human existence. Of course, after the existential analytic of *Dasein* has made a series of important steps, the concept of care in SZ acquires a more definite and varied content; care is the common root of three constitutive structures of *Dasein*'s being, existentiality (*Existenzialität*), facticity (*Faktizität*) and fallenness (*Verfallenheit*).[31] At the same time it cannot be ignored that *care,* as an *ontological* structural concept in the pay of the transcendental method in the Heideggerian version, is developed from a corresponding *ontic* concept (a generic term for a range of phenomenologically captured emotional states, "inner circumstances," motives and drives) by means of *an a priori ontological ge-*

[30] J. Macquarrie translates Heidegger's term *das Seinkönnen* as "potentiality-for-Being." I prefer to render it as "ability-to-be," because "potentiality" as a *terminus technicus* of classical metaphysics generates a lot of misleading allusions.

[31] Cf. SZ 316.

neralization or emptying. The connotations of *care* which become manifest in the naive interpretation of the world and in everyday talk, when we say "he is concerned with this or that (is sad, depressed, despairs, or, on the contrary, is taken up, intent), he cares for this or that," these connotations must be passed, i. e., used only as a support or a medium of interpretation moving towards the ontological foundations of the *ontic* definitions listed above. The question our investigation is guided by is the classical question of transcendental philosophy: "How is it possible that ... ?" How is it possible to encounter the entity understood not as present-at-hand, but as ready-to-hand?

The emptying mentioned is for Heidegger a special procedure, which has succeeded to the phenomenological reduction:

As compared with the *ontical* interpretation (*Auslegung*), the existential-ontological interpretation (*Interpretation*) is not, let us say, merely a theoretical *ontical* generalization. That would just mean that *ontically* any man's ways of behaviour are "full of care" and are guided by his "devotedness to" something. The "generalization" is rather one that is ontological and *a priori*. What it has in view is not a set of *ontical* properties which constantly keep emergent, but a state of being which is already underlying in every case... (SZ 199)

What underlies (*das Zugrundeliegende*) is τὸ ὑποκείμενον in Greek and *subjectum* in Latin. Despite all his caution Heidegger hardly avoids lapsing into terminological heresy. We must be aware, however, that he always speaks about determinations of being, of existence. The *a priori* ontological concept of care expresses that which under-lies all other existential structures, the constitution of *Dasein's being*. The term "care," after undergoing the *a priori* ontological emptying, refers to the formal structure *es geht um*, which Heidegger, using his method of gluing words together by means of hyphens and of turning them into morphemes of a compound term, expresses as "to-be-ahead-of-itself-in-already-being-in-the-world as being-alongside (the entity-within-the-world) and in co-being-with-others" (cf. SZ 192).[32] The meaning this linguistic monster refers to is usually rendered, within the context of existential analytic, by the term "care."

Yet the choice of this name is not an accident. The *ontic* connotations of care are by no means completely lost in the ontological concept: they are somehow kept in the background (or put between parentheses, though in a sense different from that of Husserl's phenomenology). It is the way traversed in the *a priori ontological emptying*, the ascension from the *ontic* to the *ontological* (this means from the present-at-hand, proximate definitions of entity to the conditions and the structures of its being) that allow bringing to light the unitary and unifying phenomenon of care. The special *ontic-ontological* homonymy bearing on many central terms of the existential analytic of *Dasein* is inevitable and exercises a definite meaning- and concept-generating function. We encounter the same kind of inevitability in Husserl's phenomenology too: the majority of its main terms can be interpreted in the sense of psychology as well as in that of transcendental phenomenology; this homonymy of terms expresses the most important fact that the phenomena of transcendental

[32] „Sich-vorweg-schon-sein-in-der-Welt als Sein-bei-innerweltlich-begegnendem-Seienden im Mitsein-mit-anderen."

phenomenology are the objects of "psychological" reflection put between parentheses; the language of phenomenology is impossible without this ambiguity.

To sum up: first, it is *care* as an *a priori* ontological concept that is the condition of unity of existential structures and plays the same role with regard to the system of *existentialia* as the transcendental unity of apperception with regard to the system of categories in the *Critique of Pure Reason*. Second, as the transcendental subject is the possibility of the object's presence (appearance), so care is the most fundamental condition of the possibility of encountering the entity-within-the-world, the ontological foundation of being-ready-to-hand (*Zuhandensein*), on which, according to Heidegger, the being of the present-at-hand (*Vorhandensein*), the objective being, depends. *Thus within the scope of our negative analogy it is care that must succeed to the transcendental subject.*

3. THE UNITY OF THE THING AND THE UNITY OF CARE

3.1. *Negative Analogy as a Distinction*

Up to this point, speaking about the negative analogy, we have been discussing the concepts of transcendental subject and of care with regard to their similarity or, to be more precise, with regard to their structural (formal), functional and methodological homo-logy. Yet the task of the destruction of the history of ontology as understood by Heidegger (i. e., the hermeneutical strategy adopted by him) demands that their hetero-logy as well be brought to light and pointed out.

The fact that the concepts discussed (*subject, care*) are results of different ontological projects pre-defines the fundamental differences between them. These differences characterize our analogy as negative. I would like to discuss only one (though most important) of the differences mentioned.

3.2. *Entity as "One"*

Both the concept of subject and the concept of care, as we have seen, play the role of unifying unity. Both the subject and care are conditions of encountering the entity in its being *because* they play the role of such a unifying unity. As early as Parmenides' epoch the formula τὸ ὄν = τὸ ἕν becomes one of the basic ontological principles. This formula appears again in Plato's *The Sophist* (237d6f.). Plato confirms that τὸ ἕν, one of the "signs" discovered by Parmenides on the "way of being," is not encountered by accident: that, which is, essentially is *one*. Yet in *The Sophist* we find an important difference as compared with Parmenides' viewpoint (which finally leads to the "parricide"): not *the* entity as such is one, but *every* entity is one (which, of course, does not mean, for Plato, *every individual sensible thing in the world*). The "one" has a double meaning: first, it is the basis of identity, the unifying one; second, it is the excluding one, "this and not the other." This distinction of meanings becomes particularly clear in the following discussion of the "largest genera" (μέγιστα γένη). After "entity" (or "being," τὸ ὄν), "movement" and "rest," it is "the same" and "the other" that make their appearance. The necessity to take into account the last two genera is imposed by the following reasoning: each one of the first three genera is something other than the remaining two, but the same as itself.[33] The genus "entity" (or "being") coincides neither with "the same" nor with "the other," but partakes of both, since it is *the same* with regard to itself and *other with regard to other*, and "other" is always said in a relative sense (πρὸς ἄλλα), and to be what it is [sc. "the other"] it needs to be other for something other (255d6f.).[34] These dialectical exercises, added to the preceding reasoning, show that τὸ ἕν, the foundation of self-identity of entities, must appear also as the principle of its difference ("repulsion") from any *other*, and in this sense we can speak of the "excluding one."

3.2.1. *Aristotle: the "One" as Essence*

In the *Metaphysics* Aristotle confirms the ontological equation τὸ ὄν = τὸ ἕν with the reservation that its members are one and the same "by nature" but different by definition (*Metaph.* IV 2, 1003b24ff.). To be *one* (τὸ ἕν) means, in the most general sense, to be indivisible or inseparable (ἀδιαίρετον) with regard to itself (*Metaph.* VII 17, 1041a18f.). Such is *any* entity: it is itself, αὐτό ἐστιν αὐτό, the same, it is identical to itself, X is always equal to X. And since in the internal differentiating of being according to the "figures of predication,"[35] τὸ ὄν in the first and basic sense is the entity of the first category, οὐσία, *substantia* in the classical Latin translation, substances are what can be called "one" in the primordial sense.

In *Metaph.* V 1 it is said:

> And in general those among the entities whose *noesis*, which conceives what they are according to their essence (τὸ τί ἦν εἶναι), is indivisible and cannot be separated either in time or in place or according to its *logos*, are in the truest sense one; such are — in the first place — substances. (1016b1ff.)

We see that the meaning of the *unifying one* is retained and differentiated in these definitions. First, the substance is one with regard to its accidental determinations, and in this sense (as the principle of unity of the variety of *accidental properties*) it is the "subject," ὑποκείμενον, *subjectum*. Here the substance is understood as a composite whole consisting of matter and form. But a composite whole needs in its turn a principle of unity which makes the indefinite manifold of matter *one* and *something*, and this is the internal form of the thing, its τὸ τί ἦν εἶναι as a form which defines and unifies matter and turns a "heap" (σωρός) of elements into one whole (1041b11f.).

Aristotle says that, in the first and basic sense, that is one *of which the thought is one*. In Kant's *Critique of Pure Reason* the entity (object) is one *because* thought of it is one, i. e., because the understanding (*Verstand*) combines the variety of sensa-

---

[33] 254d14f.: αὐτῶν ἕκαστον τοῖν μὲν δυοῖν ἕτερόν ἐστιν, αὐτὸ δ' ἑαυτῷ ταὐτόν.

[34] ὅτιπερ ἂν ἕτερον ᾖ, συμβέβηκεν ἐξ ἀνάγκης ἑτέρου τοῦτο ὅπερ ἐστὶν εἶναι.

[35] Cf. chap. 2, sect 3.

tions according to a concept. The ultimate condition of unity or integrity is the transcendental unity of apperception, the transcendental subject.

3.3. *The Unity of the Ready-to-hand as a Topological Unity*

However applied to the being of the ready-to-hand the classical scheme just explained loses its validity. The unity of an entity ready-to-hand (*eines Zuhandenen*) or of equipment cannot be understood according to the Aristotelian definition of the "one." Since an item of equipment is a bundle of references, a *topos* within the open variety of assignments, the encounter with an entity ready-to-hand, contrary to what Aristotle says about the *noesis* of the form, is always divided (distributed) in time, and according to its meaning (*logos*, *Bedeutung*) or rather according to its significance (*Bedeutsamkeit*). Heidegger says: "Taken strictly, there 'is' no such thing as *an* (one) equipment (***ein*** *Zeug*). To the being of any equipment belongs a totality of equipment (*ein Zeugganzes*), in which it can be the equipment that it is. Equipment is essentially something-in-order-to..."[36] What equipment is in its "essence" is not disclosed in a simple grasping of the whatness of an object present-at-hand (an act "indivisible in time"); it is understood in skilful use or dealing-with it. A thing ready-to-hand always outlines some "in-order-to." Actively planning and consequently carrying out something (the future result), we "hold in our hands" the "by-means-of-which," and the latter can be understood as such only in the wholeness of the already present (already fulfilled or achieved) "having-been-involved" (*Bewandtnisganzheit*). The present is adjusted to the fulfilment of the forthcoming, and in this adjustment *Dasein* projects its possibilities from their achieved facticity, which has the character of *having-been* (*Gewesenheit*), towards some "what-for" (*das Wozu*). In this open stretching temporal topos a thing is encountered as ready-to-hand; it is in this topos that appears the gap in which the entity shows itself as it comes to be before-hand, at hand, to be handled.

> Expecting the "for-which," we retain the "with-which" in our view; keeping it in view, we first understand equipment as equipment in its specific involvement relation. The letting-be-involved (*das Bewendenlassen*), that is the understanding of the involvement, which makes possible an equipmental use at all, is retentive expectance, in which the equipment is made present (*gegenwärtigt*) as this specific equipment. In expectant-retentive making-present (*im gewärtigend-behaltenden Gegenwärtigen*), the equipment comes into play, becomes present, enters into a present (*Gegen-wart*). (GP 415f.)

3.3.1. *The Tool as "One"*

Our goal is to understand in what sense a tool or a piece of equipment nonetheless possesses "unity" and an instrumental individuality of a kind. Then we proceed to the question, what the condition of *Dasein*'s unity is.

Heidegger writes:

> Equipment is encountered always within an equipmental contexture (*Zeugzusammenhang*). Each single piece (or item) of equipment carries this contexture along with it, and it is *this* equipment only with regard to that contexture. The specific *thingness* of a piece of equipment, its individuation, if we take the

[36] SZ 68.

word in a completely formal sense, is not determined primarily by space and time in the sense that it appears in a determinate space- and time-position. Instead, what determines a piece of equipment as an individual is in each instance its equipmental character and contexture. (GP 414f.)

As we already know, equipmental character and contexture are constituted by what Heidegger calls *Bewandtnis*, *involvement*:

The being of something we use... is characterized by a specific way of being put to use, of being involved [in the functional whole of dealings]. (*Ibid.*)

We are searching for such a concept of unity that would allow us to speak about "*one* piece or item of equipment," "this *particular* tool" and which would not depend on the *objective* unity of the thing present-at-hand (*das Vorhandene*). The unity of something ready-to-hand means nothing but a unity of a special *topos* in the variety of references or assignments, within the contexture of involvement.

3.3.2. *Interpretation and Freeing of the Thing*

Circumspective understanding interprets something as something-in-order-to; and "the circumspective question as to what this particular thing that is ready-to-hand may be, receives the circumspectively interpretative answer that it is for such and such a purpose (*es ist zum...*)" (SZ 149). Heidegger uses the term "understanding" in an extremely broad sense: "In the 'for-the-sake-of-which,' existing being-in-the-world is disclosed as such, and this disclosedness we have called 'understanding'" (SZ 182). To understand means primarily to understand the "for-the-sake-of-which." This ultimate mode of understanding underlies the entire totality of understanding acts.

As understanding *Dasein* projects its being upon possibilities.[37] [...] The projecting of understanding has its own possibility — that of developing itself. This development of the understanding we call interpretation (*Auslegung*). (SZ 148)

The bundle of references, which *Dasein* discovers, one way or another, as being-in-the-world and in which it becomes involved in skilful dealing-with, plays the role (in the sense of negative analogy) of the classical concept of forming form. The interpretation (*Auslegung*) articulates (*legt auseinander*) — i. e., *takes apart* in its "in-order-to" — the totality of involvements, which at first is an indefinite medium of designedness-for and assignment-to, and divides it into "towards-which" or "for-which," on the one hand, and "by-means-of-which," on the other. Heidegger distinguishes these meanings by means of different prepositions (*das Wobei/das Womit des Bewendenlassen*) and writes: *Aus dem Wobei des Bewendenlassen her ist das Womit der Bewandtnis freigegeben.*— From the world already understood, from the medium of references and assignments already powerful, from the contexture of involvements in which we already catch ourselves, the interpretation singles out and frees *that-by-means-of-which* we can fulfil what has been planned ("projected") within the all-embracing structure of care. When an entity within-the-world has al-

[37] "...entwirft sein Sein auf Möglichkeiten."

ready been freed for its being-ready-to-hand, this is an essential counterpart, or perhaps an important constituent, of *Dasein*'s "projecting its being upon possibilities," i. e., of Dasein's understanding. But then the understanding develops itself in interpretation, and this means, "we take apart in its 'in-order-to' that which is circumspectively ready to hand, and we concern ourselves with it in accordance with what becomes visible through this process" (SZ 149).

We see that the principle of unity appears for the entity ready-to-hand not as a unity of synthetic (gathering, combining) activity (form gathers matter, the transcendental subject as a "living" concept combines the manifold of intuition), but as an explanation, which singles out an entity through interpretative articulation (*Auseinanderlegung*). "To say that 'circumspection discovers' means that the world which has already been understood comes to be interpreted" (SZ 148). The interpretation defines the boundaries of a certain topos within the variety of references, and the topos thus delimited is the specific "what" belonging to a thing ready-to-hand.

In interpreting (Heidegger enumerates: preparing, putting to rights, repairing, improving, rounding-out) the ready-to-hand comes to be ex-plained, taken apart (*auseinandergelegt*); the circumstances of the matter become articulate, visible to the circumspection. And then an individual piece of equipment is freed from the undistinguished totality of involvement (cf. SZ 84; 149).

Here Plato's negative sense of the "one" as "one *and not* the other," as "being what it is through the compulsion of some other" is more relevant than the positive sense of self-identity or indivisibility. In *The Sophist* the "genus" receives and manifests its definiteness, its semantic appearance, its "look" exposed to the "eye of the soul," when its limits are defined dialectically "through the compulsion of the other." The genera participate or do not participate in each other; they are connected by the relation of *participation* or *communication* (μέθεξις, κοινωνία). The clarification of meaning takes place in the dialectical movement, which sets limits within the initially undivided medium. Understanding as contemplation (θεωρία) de-fines the genera and interprets the "participation," which at first was just the guideline of the dialectical movement, as a *relation* between the genera. This relation can be expressed or "manifested" (in the sense of Plato's λόγος δηλῶν) in the statement "S is P."

Something like this happens in the originally undivided (though always already understood one way or another) totality of involvements. Only the understanding is not now intelligent contemplation; it is caring circumspection which lives not in the element of "I think," "I know," but in the element of "I need," "I can," "I must." Heidegger says: "The being-towards-possibilities as understanding is itself an ability-to-be (*ein Seinkönnen*). And it is so because of the way these possibilities, as disclosed, exert their counter-thrust (*Rückschlag*) upon *Dasein*" (SZ 148). Hegel's "Reflection of consciousness into itself" turns here into "counter-thrust" of the disclosed possibilities and capabilities into *Dasein*'s being (existence).

3.4. *The Topos of Da*

We have already ascertained that understanding and interpretation in their turn are rooted in the structure of care proper to existence itself, which allows the "for-the-sake-of-which" to be disclosed ("understood") in *Dasein*'s being. This disclosedness of *Dasein*'s being does not at all belong to a certain *second ontic sphere* or region delimited "alongside of" the "region" of the entities-within-the-world. In Heidegger's ontology there are no such "regions" of being. On the contrary, as we have seen, the unveiling of the entity-within-the-world is founded in the disclosedness of existence as being-in-the-world. And it is not true that the latter just *precedes* the former in ontological hierarchy. The unveiling of the entity-within-the-world is "woven" into the disclosedness of *Dasein*'s being-in-the-world; both are aspects, interpreted by means of each other, of the same gap, of the same *topos* designated by the German vocable *Da*.

Da can mean "here" as well as "there," and also "now," "then,"etc., i. e., by itself it means neither "here" nor "there" nor "then." *Da* is indifferent to the distinction of time and space, of the proximate and distant. It is this invariant of meaning that I want to designate by the Greek word *topos*.

Dasein means *das Sein des Da*: the being of *Da* itself. To be more precise: *Dasein* means being (of something) *Da*, being (something) *Da*, being (somewhere, at some time, somehow) *da*. The root of this being is care = *the principle of unity of the topos of Da*.

CHAPTER SEVEN

PRIMORDIAL TEMPORALITY AND ONTOLOGICAL DIFFERENCE

1. CARE AS PRIMORDIAL TEMPORALITY

Time present and time past
Are both perhaps present in time future,
And time future contained in time past.
If all time is eternally present
All time is unredeemable.

T. S. Eliot, *Four Quarters, Burnt Norton.*

1.1. *Being-ready-to-hand and the Time of Dealing*

It has become clearer for us, that it is a certain "image" (ἔδωλον) of Chronos, which governs both everyday language, and the history of ontology from Parmenides to Bergson. The task of "destroying the history of ontology" can be accomplished, therefore, only by repeating and interrogating its relation to the problem of time. The meaning of being is understood in the tradition as *permanence* and *presence.* Heidegger asks: "What project lies at the basis of this comprehension of being?" And answers: "The project relative to time, the projection towards time (*der Entwurf auf die Zeit*), for even eternity, taken as *nunc stans...* is conceivable as 'now' and 'persistent' only on the basis of time" (KPM 233). But the image of time as it was fixed in the tradition corresponds to the primordial project of being as *presence-at-hand.* Or perhaps the primordial "project of time" and the "primordial project of being" are, ultimately, one and the same project. Now, after Heidegger's "Copernican revolution" within the hierarchy of the ways of our encounter with beings, according to the new arrangement of the ontological scene where "I care" ontologically precedes "I think," the task to disclose the mode of temporality underlying the structures of care is urgent. The traditional treatment of the meaning of being as παρουσία or οὐσία, is only an "*outward evidence*" (SZ 25) of the fact that within the scope of the ontological tradition, "entities are grasped in their being as *presence*; [and] this means that they are understood with regard to a definite mode of time — the *Present*" (*ibid.*). Now it is necessary to proceed from "outward evidence" to a thorough analysis and cogent demonstration.

First we need to investigate the temporality of "concernful dealings" with πράγματα (SZ 68) and to show, that it is indeed a peculiar *structure of time* that underlies the *comprehensibility*, the disclosedness of the ready-to-hand, and it is different from the temporality of presence-at-hand, i. e., from the "present" as "now."

The event in which the ready-to-hand first manifests itself cannot be thought as a simple immediate presence in the "now," as Husserl's *primal impression*, as the *primordial self-presentation* of the entity which undergoes "subsequently" a succession of retentional modifications following one another in some pre-temporal sense.

To understand something as something ready-to-hand in a primordial manner means, as we have seen, to be drawn into the totality of involvements (*Bewandtnisganzheit) and to move within it following the "in-order-to" references.*

In understanding them I dwell with the equipmental contexture which is ready-to-hand. I stand neither with the one nor with the other but move in the *in-order-to*. It is for this reason that we proceed *in order* in dealing with things (*haben wir Umgang*) — we do not merely approach them (*nicht einen bloßen Zugang*) as they lie before us but deal with them, have commerce with them as they exhibit themselves as equipment in equipmental contexture. Letting-be-involved (*Bewendenlassen*) as understanding of involvement is that projection (*Entwurf*) which first of all gives to *Dasein* the light in whose luminosity things of the nature of equipment are encountered. (GP 416)

The "area" of luminosity spoken of is not one of the immediate presence reduced to the "point" of the "now" (*Momentanjetzt*), but rather an open and stretched "topos of proximity" to which *Dasein* in its ability-to-be *(Seinkönnen)* can reach out. When care is considered the ontological condition of being-ready-to-hand, *possum* precedes *cogito. Dasein* primordially "discloses" the entity (ready-to-hand) not as an object of *cogito, but as a counterpart of possum*, of "I can," i. e., in its ability to respond to the original *ontic* effort of *Dasein* (entity understood as equipment, as something familiar and obedient) or to resist this effort (entity as something infinitely alien, as a burden, as a resisting medium without which a tool ready-to-hand cannot be even considered).

Of course, the central term of Husserl's phenomenology, *intentio*, also means *stretching* and *being stretched*, but according to its original definition it points rather to a "vertical" span between the *ego* and its intentional objects. It is the inescapable difference between being-manifested (*presence*) and the manifested object (*that which is present*) that lies at the foundation of the intentional relation. Intentionality is a condition of *ego cogitans'* self-distinction from its object (*cogitatum qua cogitatum*).

1.2. *"Tendency" before the* Cogito

Remaining within the limits of Husserl's phenomenology, we can hardly ask further about the "conditions of possibility" in relation to this primordial form of intentionality, for it itself serves as the first foundation of all phenomenological constructions. It is true that in his work entitled *Experience and Judgment* Husserl speaks of a certain affective *tendency* belonging to the transcendental subject, which *precedes* in a sense the completed form of intentional relation: *ego — cogito —*

cogitatum, of a *Tendenz vor dem Cogito.*[1] This affective tendency has a double nature: that of excitement *(Reiz)*, by means of which a certain object imposes itself on consciousness, and that of the consciousness tending to the object, of the desire to give in to excitement, to let the object be present. This objective *eros* of consciousness generates a turning of the transcendental *ego* to the object (*Ichzuwendung*), of which the "intentional relation" is the result. Yet, as it seems to me, all this reasoning refers to the appearance of the object as a definite "what" in the focus of attention, to the object's passing from the mode "of inactuality to the mode of actuality."[2] Yet strictly speaking consciousness never remains without object; it is always living through the experience of an "intentional background" (*intentionales Hintergrunderlebnis*). In this intentional experience the objectivity is already there as "thisness" (*haecceitas*) preceding the "whatness" (*quidditas*), as an indefinite horizon of objectivity as such. If according to Husserl intentionality can be thought *in its preliminary or undeveloped form* as an affective *energeia* or "longing" (*Streben*) of consciousness, then the fulfilment of this desire is the objective intentional relation: *cogito cogitatum. The "affective tendency" of the consciousness is nothing but the only good eros*, the *eros* of intelligent contemplation, which Aristotle discusses in *Metaphysics* XII 7. In the actual possession of the object the intellect is manifested to itself, and this is "*energeia*," the actuality of the intellect, which is identical with the highest form of life.[3]

Heidegger wants to "deduce" the intentionality of consciousness from a more deep ontological foundation, from the existence of *Dasein*. Here is one more excerpt from Heidegger's letter to Husserl, written while preparing the article for the *Encyclopaedia Britannica:*

> It is necessary to show that the way of being of human Dasein is completely different from the way of being of any other entity, and that this way, precisely as what it is, contains in itself the possibility of transcendental constitution. Transcendental constitution is one of the main abilities of existence, belonging to the factual self (*Existenz des faktischen Selbst*). And this "self," i. e., a concrete person as such, is never only present as an entity, but exists (*existiert*). And the most wonderful thing is that the existential structure of Dasein makes possible transcendental constitution of everything positive (*alles Positiven*).[4]

Thus constitution, and therefore intentionality also, is one of the existential possibilities of *Dasein*. According to Heidegger understanding as one of the *existentialia* precedes intentionality, if the latter is interpreted formally as "being-directed-towards-an-object." In order to be directed, says Heidegger, *Dasein* must somehow

[1] E. Husserl, *Erfahrung und Urteil* (Hamburg: F. Meiner, 1985), § 17, pp. 79ff.

[2] Cf. *Ideen* I, § 35, p. 62: "Es liegt aber darin, daß gewisse Modifikationen des ursprünglichen Erlebnisses möglich sind, die wir bezeichnen als freie Wendung des „Blickes" – nicht gerade und bloß des psychischen, sondern des '*geistigen Blickes*'..." "Zum Wesen des Erlebnisstromes eines wachen Ich gehört es aber..., daß die kontinuierlich fortlaufende Kette von *cogitationes* beständig von einem Medium der Aktualität umgeben ist, diese immer bereit, in den Modus der Aktualität überzugehen, wie umgekehrt die Aktualität in die Inaktualität." (p. 64)

[3] ἡ γὰρ νοῦ ἐνέργεια ζωή. *Metaph.* 1072b26f.

[4] In: E. Husserl, *Phänomenologische Psychologie*, Vorlesungen Sommersemester 1925, hrsg. v. W. Biemel, Hua IX (Den Haag, 1962), Anhang, pp. 601f.

antecedently understand the correspondent "direction," and this means – the objective being of the objects of consciousness.

Not only *intentio* and *intentum* belong to intentionality but also each *intentio* has a directional sense (*Richtungssinn*), which must be interpreted with reference to perception as follows. Presence-at-hand must be antecedently understood if an entity present-at-hand is to be uncoverable as such; in the perceivedness of the perceived there is already present an understanding of the presence-at-hand of what is present-at-hand (*Vorhandenheit des Vorhandenen*). (GP 446)

1.2. *Understanding, Interpretation and the Meaning of Meaning*

The directional sense of the *presence-at-hand* as a way of being precedes the whatness, the meaning (meaningful appearance-image) of the *present-at-hand*. What is directional sense? What is *meaning* in general in the context of *Being and Time?*

First of all, "meaning is an *existentiale of Dasein*, not a property attaching to entities, lying 'behind' them, or floating somewhere as an 'intermediate domain'" (SZ 151). In other words, meaning is a certain structural moment of existence, i. e., of such being for which this being itself "is an issue." Only *Dasein has* meaning and can be meaningful or meaningless (*ibid.*). It follows that meaning is an aspect of the way the being of *Dasein* is an issue for itself. Further, meaning belongs to the structure of understanding-interpretation, and as understanding *Dasein* projects its being upon possibilities (SZ 148); this means that as understanding it plans and calculates in the sense of the Greek verb λογίζομαι. As we might remember, in the *Nicomachean Ethics* Aristotle uses this verb to designate that "part of the soul" to which the truth of moral action and the truth of "making" (ποίησις) are revealed in contradistinction to the truth of knowledge. Heidegger's plan to appeal against the verdict of classical ontology and to take away the ontological primogeniture from *sophia* bears on the concept of meaning also. For him, meaning understood existentially is primordially connected with projecting, outlining, calculating, and not with contemplating an εἶδος. Understanding always has in itself the existential structure of *projection* (*Entwurf*); yet besides that, it is always constructed and developed in interpretation. Interpretation is the working-out of the possibilities projected in understanding and thereby the development of meaning (SZ 148).

The projection always proceeds from the totality of involvements already understood in some preliminary way (though perhaps, as Heidegger puts it, the ready-to-hand has not yet explicitly come into the "field of vision" of the understanding). Within this medium of preliminary comprehensibility, interpretation (*Auslegung) performs the operations of taking-apart (Auseinanderlegung) and articulation. In this process understanding becomes explicit* and the entity visible. That which has been circumspectively taken apart, i. e., that which is interpreted and thus *explicitly* understood, has the structure of *something as something [in order to...]*.

In connection with this hermeneutical scheme Heidegger speaks about "fore-having" (*Vorhabe*). There can be no unprejudiced interpretation and no construction of meaning without premises. The mastering-constructing of meaning takes place in a certain pre-defined perspective: something must be first taken into account in order to become the guideline of interpretation. This pre-defined perspective in which meaning is seen is called "fore-sight" (*Vorsicht*). Finally what has been interpreted in a manifest and clear way requires a certain preliminary set of concepts (*Be-griffe*); interpretation has already decided on a definite way of conceiving the entity, "either with finality or with reservations; it is grounded in something we grasp in advance — in a fore-conception (*Vor-griff*)" (SZ 150).

Meaning is the "upon-which" of a projection in terms of which something becomes intelligible as something; it gets its structure from a fore-having, a fore-sight, and a fore-conception. (SZ 151)

It follows that it is *Dasein* itself that *has* meaning or *is meaningful* in the primordial sense, and when one speaks of "the meaning of Dasein" (*der Sinn des Daseins*) then the genitive case here (*whose meaning?*) must be understood as *genetivus subjectivus*. It follows also from such a definition of meaning that the complex existential structure of *Dasein* (*different* meanings of the concern, which is for Dasein its own being) must generate manifold *meanings of the existential meaning*. And since the "primordially one phenomenon" of care underlies the plurality of *existentialia*, the primordial meaning of meaning (*genetivus objecti*) is the meaning of care (*genetivus subjecti*).

1.3.1. *The Meaning of Care*

Since existential meaning is the *upon-which* of a project, we must ask first of all: what project is care as such connected with? To be sure, if we are interested not in the *ontic*, but in the ontological description of care, if the procedure of *a priori* ontological generalization or emptying has already been carried out, we find ourselves in a situation where it is impossible to *name* (for the name supposes τὸ τί, "whatness") and to enumerate *that which care cares for*. One can say, of course, that we are dealing here with the being of *Dasein* as such, but this will not bring us a step forward.

Meaning "lives" in the medium of what has been understood beforehand, of the fore-having, in the sphere in which I catch myself, where I happen to be. Yet meaning is constructed and clarified-articulated in the operations of taking apart being accomplished by interpretation. We cannot posit beforehand a general object of care having a universal significance, because the first principles of the entity which bares its being in *Dasein*'s concerned and "sympathetic" disclosing of the truth, are in each instance different, as Aristotle already knew. And yet it is possible to discuss these principles. We must not ask *what* they are, because this question bears on the epistemic part of the soul; it designates one of the epistemic strategies listed by Aristotle in the *Analytica posteriora* II 1 and aims at the manifestation of the form-eidos (which always remains identical to itself), and the answer is given in an apophantic assertion, by means of proposition "S is P." However we can ask: to whom are these principles, "different in each instance," revealed, and what does this revelation mean? Aristotle's answer is: to the prudent person in that special way of disclosing the truth which belongs to prudence as to the virtue of the calculating, projecting "part" (faculty) of the soul.

And "it is held to be a mark of a prudent man to be able to make correct decisions concerning what is good and advantageous for himself, and this not in particular (for instance, as regards what is good for his health or strength), but generally with regard to a good life as a whole (πρὸς τὸ εὖ ζῆν ὅλως)."[5] (Heidegger translates: ...*was zuträglich ist für die rechte Weise des Seins des Daseins als solchen im ganzen* – "that which is advantageous for the correct way of being of Dasein as such and as a whole.")

To set forth the meaning of care means, then, to follow up the projection which guides and underlies the primordial existential interpretation of *Dasein*, and to follow it up in such a way that in what is here projected, its "upon-which" may be seen. And what has been projected is the being of *Dasein*, disclosed precisely in that which constitutes that being as an authentic ability-to-be-a-whole (*Ganzseinkönnen*). (SZ 324)

Care is the ontological hidden background of every project. "The first or primordial existential projection" *(der primäre/ursprüngliche existenziale Entwurf)*, to use Heidegger's phrase, is connected with it. In this projection I always outline myself, my *Self*, in a fundamental way; not the substantial *ego*, not *what I am* in the sense of the Cartesian "*sed quid igitur sum?*" but *how I am*, since I am delivered (as an entity) to my being. The being in its "how?" is projected first, and the "I" in its "what?" or "who?" is constituted as a superstructure.

As something that keeps silent, authentic being-one's-self is just the sort of thing that does not keep on saying "I"; but in its reticence it "is" that thrown entity as which it can authentically be. (SZ 323)

Projection always discloses possibilities-powers underlying my own authentic ability-to-be, known only to myself as to a witness bearing witness to himself. The search for one's ownmost is connected with a refusal to be absorbed and dissolved in the generally accepted, and universally recognized as in something secure and "tranquillising" (SZ 177), with a rejection of the public "ready-made" interpretedness of the world, of the temptation "to get lost in the publicness of the 'they' (*in die Öffentlichkeit des Man*)"[6] and thus *partake* of the secure and self-evident common opinions and common goals. *Dasein* must be resolved to bear the burden of its own being and of its own decisions. This primordial resoluteness, preceding any decision of mine, is precisely the way of disclosing the truth in which the primordial existential projection reveals itself. Heidegger calls it *die vorlaufende Entschlossenheit*, anticipatory resoluteness.[7]

That which was projected in the primordial existential projection of existence has revealed itself as anticipatory resoluteness... [And it is] being-towards one's ownmost, distinctive possibilities. (SZ 325)

In primordial projection the "for-the-sake-of-which" and the "for-the-sake-of-whom" (τὸ οὗ and τὸ ᾧ of Aristotle) are united; more than that, the question about

---

[5] *Eth. Nic.* VI 4, 1140a25ff.

[6] SZ 174.

[7] Unfortunately, in English translation the connection between "disclosedness" (*Erschlossenheit*) and "resoluteness" (*Entschlossenheit*) disappears, as well as the relation of the latter term to the Aristotelian ἀληθεύειν.

the "who" of existence (SZ §25) is clarified only *after* the question about the "for-the-sake-of-which."

1.4. *The Ecstases of Temporality*

1.4.1. *The Forthcoming*

The fact that care "has" meaning implies that *Dasein* in the disclosing of its ownmost possibilities can arrive at itself, access itself, *come towards itself*, and this is, Heidegger says, the primordial *phenomenon* (SZ 325) of the *forthcoming* future and its primordial *concept*:

> *Dasein* understands itself by way of its ownmost capacity to be (*das eigenste Seinkönnen*), of which it is expectant (*gewärtig ist*). In thus comporting towards its ownmost ability to be, *it is ahead of itself*. Expecting a possibility, I come from this possibility towards that which I myself am. *Dasein*, expecting its ability to be, comes towards itself. In this coming-towards itself, expectant of a possibility, *Dasein* is futural in an original sense. This coming-towards-oneself from one's ownmost possibility, a coming-towards which is implicit in *Dasein's* existence and of which all expecting is a specific mode, is the *primary concept of the future*. (GP 375)

1.4.2. *The Past*

Yet when we are discussing meaning in general (and the meaning of care in particular), we must not forget that meaning as the "upon-which" *of a projection* is structured by *a fore-having, a fore-sight, and a fore-conception*. For *Dasein*, coming towards itself, towards its ownmost possibilities, the *fore-having* means the necessity to proceed from the facticity of its own self-representation, from *what I represent myself to be and who I am represented as*. Yet only that which already has been can be represented in representation. The projection upon the future presupposes a retro-jection into the past as a necessary condition. The primordial phenomenon and the primordial concept of the *forth-coming* presupposes coming-back to *Dasein's* having-been-ness. "As authentically futural, *Dasein is*, properly speaking, as '*having-been*'." [8]

Things may have gone by, become past, but "that which we *are* as having been has not gone by, passed away, in the sense in which we say that we could shuffle off our past like a garment. *Dasein* can as little get rid of its past as escape its death" (GP 375). The *having-been-ness* (*Gewesenheit*) as distinguished from *bygoneness* (*Vergangenheit*) is a mode of being proper to *Dasein,* the point "from which on" existence exists (*ex-sistit*). Without *what-has-been* there is no "forthcoming" for *Dasein*; care is the primordial "stretchedness" or passage in which ἐκ τινός — "where from" (the meaning of what has been) and εἰς τί — "where to" (the that-towards-which of the projection, the meaning of the future), are *articulated* in interpreta-

---

[8] "Eigentlich zukünftig *ist* das Dasein eigentlich *gewesen*." (SZ 326) Macquarrie and Robinson translate: "...is *authentically* as *having-been*." It seems to me however that the word "eigentlich" in the second case does not refer to the authenticity (*Eigentlichkeit*) of *Dasein*. The German phrase is deliberately ambiguous.

tion, and this means — are simultaneously taken apart and joined together. Care is the primordial articulation of time; it is *that from which on* time temporalizes itself. Yet as we know already the ultimate thing (τὸ ἔσχατον) care cares for, is *Dasein*'s most authentic ability to be. Resolving on such a way to be I reject a flight into the "tranquillising" community of the "they," structured by idle talk, curiosity, and ambiguity; I expose myself to extreme solitude:[9] I am the only witness of my authentic self, i. e., the witness of the meaning of meanings.

Aristotle says that the intellect is εἶδος εἰδῶν (the form of forms). Care understood ontologically is the meaning of meanings, Heidegger could say. In the imperfect and doubtful acoustics of my being I am always in danger of hearing the call of care, i. e., my conscience, improperly; yet having taken upon itself the burden of witnessing, *Dasein is* (exists) nonetheless in the mode of the proper or authentic understanding. It is to such understanding that the *forth-coming* as "the time" (the vector) of the projection belongs; it is that into which *Dasein* throws itself, projecting its own ability to be. To such understanding belongs also as its integral part the character of "having-been" — that is the mode of *Dasein*'s being, in which it takes itself over (*übernimmt*) and brings itself back again or repeats itself (*holt sich wieder*).

Resoluteness temporalizes itself as repetitive coming-back-towards-itself from a chosen possibility to which *Dasein*, coming-towards-itself, has run out in front of itself [preceded itself]. (GP 407)

1.4.3. *The Present as the "Twinkling of an Eye"*[10]

Yet both "provinces of time" (the *forthcoming* and the *having-been*) are thinkable only in a unity with a certain mode of the present. We shall approach it within the framework of the primordial projection towards my authentic Self, as a form of the "present" that is held in resoluteness and springs from it. I overtake and repeat myself, the one-who-has-been, and throw myself in the moment of choice or decision (Aristotle's προαίρεσις), which must be accomplished "now or never." I find myself within the domain where the ethical determination of the principle of action overcomes its physical determination. It is impossible to give up choice by avoiding this or that action *ad extra*, for such avoidance is a certain voluntary action, the result of my decision. And it is the preferential choice that according to Aristotle makes human action worthy of praise or blame, inasmuch as it allows us "to discriminate characters better than actions do."[11] There is only one way of giving up

[9] "Because *Dasein*, as being-in-the-world, is at the same time being-with other Dasein, authentically existent being-with-one-another must also determine itself primarily by way of the individual's resoluteness. Only from and in its resolute individuation is *Dasein* authentically free and open for the 'thou.' Being-with-one-another is not a tenacious intrusion of the I upon the thou, derived from their common concealed helplessness; instead, existence as together and with one another is founded on the genuine individuation of the individual... [which] does not mean clinging obstinately to one's own private wishes but being free for the factical possibilities of the current existence." (GP 408)

[10] I refer here to Heidegger's "Augenblick." Macquarrie and Robinson translate this word as "moment" or "moment of vision," Hofstadter as "instant." Neither version seems to be satisfactory. We shall combine different ways of rendering this important term.

[11] *Eth. Nic.* III 2, 1111b5. Cf. P. Ricœur, *Oneself as Another*, Study 4, § 1.

responsible action: the tranquillising way "to say as *they* say," to do "as *people* do." But this means giving up resoluteness as the disclosure of my being, of my Self. The time of choice, the time of action is not the moment of "now"; it is a "twinkling of an eye," a blink of an instant (GP 409ff.), not τὸ νῦν, but καιρός, about which Pindar says: καιρὸς πρὸς ἀνθρώπων βραχὺ μέτρον ἔχει[12] — it is not made to human measure, it is too brief, too narrow, too acute; καιρός is the only appropriate moment, the fleeting moment of opportunity.

It is in this moment that resolute *Dasein,* stretching from what has been to the forthcoming, reaches out to the things in the world and to those who are near, discovering them in the optics of a responsible action.

The *Augenblick is making-present (Gegenwärtigen)* of something present *(von Anwesendem)* which, as belonging to resolve, discloses the situation, upon which the resoluteness has resolved. In the *Augenblick* ... Dasein is carried away, as resolved, into the current factually determined possibilities, circumstances, contingencies of the situation of its action. (GP 407f.)

According to Aristotle, the being of "the first principles, which admit of variation," is revealed to the deliberating-projecting part of the soul; these principles cannot be stopped, retained in the *epistemic element*, gone back to again and again like something which is always accessible to the glance of contemplation (theory). The first principle disclosed in action does not "wait," for time does not wait, and καιρός is not made to human measure.

The "twinkling of an eye" (*Augenblick), the moment of vision is proper to the primordial and authentic temporality of Dasein* and is the primordial and authentic mode of the present (*Gegenwart*) as letting-be-present (*Gegenwärtigen*).[13] The "now" of the objective presence depends on it. At this stage of our discussion the most important characteristics of temporality can be introduced.

The essence of the forthcoming [the future] lies in *coming-towards-oneself*; that of the having-been-ness [the past] lies in *going-back-to*; and that of the present in *staying-with, dwelling-with*, that is, being-with. These characters of the *towards, back-to* and *with* reveal the basic constitution of temporality. As determined by this *towards, back-to*, and *with*, temporality is outside itself. [...] Moreover, it is itself *the original outside-itself (das ursprüngliche Aussersich)*, the ἐκστατικόν. For this character of carrying-away we employ the expression the *ecstatic character* of time. (GP 377)

1.5. *Ecstatic Tripartite Unity of Time and Topological Unity of Dasein*

Thus the ontological meaning of care is the primordial temporality that is nothing else but a triad of *ecstases* of time, every one of which is unthinkable by itself and constitutes just an aspect of a tripartite unity. Care as the "root" of the being of *Dasein* means, in temporal interpretation, *ecstasis* before *stasis*, the primordial opening of the topos of *Da, in which time temporalizes itself in the primordial way.*

In the tripartite unity of time lies the foundation of the unifying unity of the *existentiale of care, and therefore the ultimate foundation of all unities lies in it also, as*

---

[12] *Pythian Odes*, IV 286. The verse has become a proverb analogous to the English saying: "time and tide wait for no man."

[13] The English translations of this term are also varied and numerous: "letting-be-present," "making-present," "enpresenting," "presentification," etc.

the ultimate foundation of all differences lies in the difference of ecstases. Yet let us ask ourselves once more what meaning of unity is manifested in the unity of temporal *ecstases.*

1.5.1. *The Point of the "Now" and the Unity of the Transcendental Subject*

We shall go back again to the negative analogy between transcendental subjectivity and care, so that the concept of ecstatic temporality becomes clearer for us in this comparison-opposition.

Kant's transcendental subject (as well as Husserl's "polar ego") is *one* in the strict Aristotelian sense: the "I" (belonging to the representation "I think") is something simple and self-identical ("one and the same in every consciousness"),[14] something which is co-represented in every representation. The act of apperception is indivisible in itself and cannot be divided ("either in time or in place or according to its *logos*"), because division, analysis, implies synthesis, and the unity of apperception is the condition of every synthesis, and cannot therefore be thought of as a synthetic unity. "This self-consciousness cannot be accompanied by any further representation 'I think'"[15] and that is precisely why it is called primordial apperception.

In Husserl's phenomenology the pure or polar *ego* always occupies the position of "absolute now," and the temporal depth of the phase "passing away into the past" is *constituted* with regard to the temporal position (*Zeitstelle*) of the *ego.* This position itself is a *nunc stans* of a special kind, an universal frame of reference, in relation to which the *nunc fluens* of the primordial impression, undergoing retentional modifications, is defined. This pre-temporal or super-temporal position of the pure *ego* serves, according to Husserl, as a "source" of temporal modifications and therefore as the foundation of the unity of the temporal stream, which in its turn is the ultimate foundation of all objective unity. The unity of the temporal stream, in its turn, is "an accomplishment of the ego in its innermost dynamo."[16] The stream of inner time, according to Husserl, "flows out of" the primal "now" (*Ur-Jetzt*). In time the "life of the *ego*" temporalizes itself. The moment of "now," the "now-phase" (*nunc fluens*) is the proximate "point" within the inner time-flow in relation to the pre-temporal position of the pure *ego* (*nunc stans*). As the primordial "overturning" of the *ego* into time, retained and modified in the temporal flow, it "puts the stamp of existence" (to use already cited Husserl's phrase) on the object of consciousness, which is constituted as having a certain duration in time — and not only the stamp of existence, but also the stamp of *unity.* The unity of an object is due to the fact that each piece of the manifold of synthesis, each fragment of the "internal horizon" of the intentional object preserves its bond with the point-like "now": it is either the retained or the anticipated content of the primal impression. The original principle of unity is always "now." Husserl would agree with Plato saying that the verb "is" in the present tense (as distinguished from "was" and "will be") expresses the *One* as it

[14] I. Kant, *Critique of Pure Reason,* B 132.

[15] *Ibid.*

[16] MS C 10, p. 23 (1931), after Landgrebe, *The Problem of Passive Constitution,* p. 54.

is in itself, the foundation of all unity as an "eternal essence" in an appropriate way (*Tim.* 37d–38b).[17]

1.5.2. *Involvement and the Horizon of an Object*

We have established earlier that the entity ready-to-hand "is" only in an extended temporal horizon; the unity of an item of equipment is topological. The topos of the thing ready-to-hand is brought to light as a *meaning* in practical understanding-interpretation (which includes, as we have seen, preparing, putting to rights, repairing, improving, etc.) and this means that it is de-termined, de-limited out of, or as Heidegger puts it, "cut out from" (SZ 150), the implicit comprehensibility (*fore-having*) of the totality of involvements. The *fore-sight* "takes the first cut" with a view to a definite way in which the entity can be interpreted. Finally, something ready-to-hand which is freed for its being in such a way can be articulated in the pre-solved conceptuality, which is already here as something we grasp in advance, as a fore-conception.

Yet do we not encounter exactly the same scheme in Husserl's phenomenology, which Heidegger would have called phenomenology of the present-at-hand? For after the appearance on the phenomenological stage of the concept of *horizon* (*Horizont, Horizontbewußtsein*) Husserl also could have said: taken strictly, there "is" no such thing as *one* object. To the being of a thing there always belongs its external and internal horizon;[18] i. e., its thingness, its objective meaning is constituted only in relation to these horizontal structures, in particular, it is constituted within the context of its surroundings, and every horizon presupposes the temporal horizon as its ultimate condition.

And yet we can say, remaining within the framework of Husserl's phenomenology: the thesis "there is no such thing as one object outside of the horizon," is true only if we mean the thing's *realitas*, thingness, its semantic definiteness. The "stamp" of the primordial impression already marks the object as "this entity" independently of its "what" (*quidditas*). The *ego* is co-present with the instantly exploding "thisness" (*haecceitas*) of the object. And though the absolute temporal flow is to be interpreted as the *horizon* of the *ego itself*, i. e., of the *ego* objectifying itself in its reflective pursuit of itself, Husserl insists that the self-recognition of the *ego* in its "life" or primal being (*Ur-sein*) does not require temporal duration. Husserl writes (arguing implicitly against Kant): "The *ego* does not arise originally out of experience — in the sense of the associative apperception, in which the unities of the manifold of the nexus constitute themselves — but rather out of its own life (it is, what it is, not *for* the *ego* but, rather, it is itself the *ego*)."[19] The *ego* has already recognized itself in the special pre-intentional form of self-consciousness,

[17] Cf. chap. 5, sect. 1.

[18] Cf. SZ 68.

[19] E. Husserl, *Ideen zu einer reinen Phänomenologie und phänomenologischen Philosophie*. Zweites Buch: *Phänomenologische Untersuchungen zur Konstitution*, hrsg. v. M. Biemel, Hua IV (Den Haag, 1952), p. 252.

termed *ungewährte Gegenwart* or *nicht-gewährende Selbstgegebenheit.*[20] However this positing *ego* possesses the Aristotelian point-like, polar unity, which corresponds to the "point of the 'now'." The unity of the thing and the unity of the temporal stream as its fundamental condition are derivative in relation to the point-like unity of "now" (or "proto-now," the source of the temporal flow). The horizon of an object given in perception is centred on the point of the primal impression, and this makes it one, while the unity of the topos of "this here" entity ready-to-hand within the variety of "in-order-to" references is determined by its de-finition in practical understanding-interpretation "spanned over" the three *ecstases* of temporality. The topos of the ready-to-hand in contradistinction to the present-at-hand *has no distinguishable centre.*

1.6. *Understanding of Being and Transcendence of Dasein*

Thus according to Heidegger the primordial temporality as the unity of the *forthcoming*, the *having-been-ness* and the *present-being-with* is the original "outside of self," τὸ ἐκστατικόν.

We have seen in the previous chapter that the unifying unity of care is not "point-like" or polar; it is *topological.* Care means *to-be-ahead-of-itself-in-already-being-in-the-world as being-alongside (the entity-within-the-world) and in being-with-others. Dasein* must be already ahead of itself so that the presence of the "for-the-sake-of-which" and of the "in-order-to" becomes possible. The referential totality of significance does not originally come apart into a multiplicity of ob-jects, and therefore that which fills in the signification of the term "care" in its formal structure of "ahead-of-oneself" is not a kind of protention, an anticipation of the object's presence. Therefore "ahead-of-itself" is not a special intentionality belonging to *Dasein*, but only a "vector" of projection of one's own ability-to-be, either disclosed in resoluteness or abandoned to the disposal of the "they" (cf. SZ 193).

Still the expression *ahead-of-oneself is to be dealt with very cautiously. What is this* "Self" ahead of which *Dasein* may find itself? Is this not a mere awkward play with pronouns generating a semblance of meaning? Out of what *stasis* does the *ec-stasis* go, and where to? Is not this "stopping" (*stasis*) identical with the temporal position (*Zeitstelle*) of the transcendental subject, the *nunc stans* spoken of above? The last admission would mean that there is no new "topological" unity of care and that the unity of the topos of *Da appears on closer examination to be determined by its being centred on the point (pole) of the ego.*

Yet as a matter of fact "ahead-of-oneself" does not signify that a kind of preliminary immanence, out of which the *ecstasis* goes, has been already posited. The *ecstasis* as original "outside-itself" is not defined in relation to the preceding "inside-itself."

We must not think that some Self, already possessing a point-like unity, already self-identical in its autonomy, steps aside away from itself, goes out of itself, gives itself up in order to establish after that a relation of identity with itself, recognize it-

---

[20] Cf. chap. 5, sect. 2.3.

self, acquire possession of itself, in a word, to perform the Hegelian *reflection* or "mediation of becoming other [for itself]."[21] The self (*ipseitas*) as a possibility and as an actual effort ("being-able-to") of caring does not require a hidden background in the form of unity (*identitas*) of the subject. The unifying unity of the self is *the unity of temporal ecstases, a seam of time.*

According to Aristotle, time "follows" movement, and the latter is a transition ἔκ τινος εἴς τι, from one thing to another, from one entity to another entity. Accordingly we discover in time the "before" and the "after" (τὸ πρότερον καὶ ὕστερον). Yet could not the transition be conceived *before* its "where-from" and *before* its "where-to" — as something constituting the "where-from" and the "where-to"? Could not time be conceived not as alienation-holding back (*retention*), not as anticipation-drawing into (*protention*), but as a unity of temporal *ecstases constituting* the "now," the "before" and the "after"? Could not relation be conceived before what is related, and *ecstasis* before *stasis*? Of course, the answer has to be in the negative within the framework of the ontology where "being" signifies *positivity*, presence of the present.

Yet *Dasein* is (exists) differently. The being of *Dasein* is *transcendence*. Heidegger rejects the classical understanding of transcendence as of something other with regard to the immanence of consciousness. Transcendence means a transition, a stepping over (preceding any where-from and where-to), a passage, which serves as a foundation for positing the "where-from" and the "where-to."

> *Transcendere* means to step over; the *trancendens*, the transcendent, is that which steps over as such and not that towards which I step over. The world is transcendent because, belonging to the structure of being-in-the-world, it constitutes stepping-over-to... as such. *Dasein* itself steps over in its being and thus is exactly not the immanent. (GP 425)

Only that entity is transcendent of which being has a character of transition, stepping over, projecting a meaning, and projecting itself.

Ontology as phenomenology searches for the ultimate conditions of the possibility of encountering entity as entity. This encounter can take place only because the entity goes out to meet us as something already understood: the entity is "freed" out of the undistinguished totality of involvements as "that here" in the interpretation-articulation of meaning. Yet the initial *letting-be-involved (Bewendenlassen)* and the understanding ex-planation itself are possible insofar as *Dasein* itself "has" meaning, i. e., projects itself, transcends itself. It follows that the ultimate condition of the possibility of encountering entity as entity must be understood phenomenologically as a condition of transcendence. And this condition has already been discovered; it remains now just to restore some missing links in the chain of reasoning.

---

[21] "die Vermittlung des Sichanderswerden mit sich selbst" (PhG 20). Cf. also: "[Die lebendige Substanz] als Subjekt ist die reine e*infache Negativität*, eben dadurch die Entzweiung des Einfachen; oder die entgegensetzte Verdopplung, welche wieder die Negation dieser gleichgültigen Verschiedenheit und ihrer Gegensatzes ist: nur diese *sich wiederherstellende* Gleichheit oder die *Reflexion* im Anderssein in sich selbst... ist das Wahre" (*ibid.*).

Intentionality is called by Heidegger "ontic transcendence," and as such "it is itself only possible on the basis of original transcendence, on the basis of being-in-the-world. This primal transcendence makes possible every intentional relation to entities."[22]

1.7. *The Concept of Schematism*

1.7.1. *Directional Sense and Horizontal Schema*

Existence viewed ontologically, says Heidegger, is "the original unity of being-outside-itself that comes-towards-self, comes-back-to-self, and makes-present" (GP 378). As existent (*ex-sistens*) *Dasein* is carried away towards something, projects itself towards something. As we can see the concepts of *ex-sistence, transcendence, ec-stasis* are firmly interrelated, moreover they have a common root, and this is primordial temporality. *Ecstasis* is a vector of projection, which points in one of the three primordial "directions of projecting," the three "shifts" of the primordial temporality: towards the forthcoming, towards the having-been-ness, towards the presence in the present. As the basis of any definite meaning, i. e., of any "upon-which" of the projection, *ecstasis* does not point to some "what." More than that, it does not point to the interpreted or to-be-interpreted "for-the-sake-of-which" or "in-order-to either." And nevertheless we have to admit, after closer examination of the phenomenon, that the *ecstasis* of the having-been-ness points *in its own* "direction," different from that pointed to by the *ecstasis* of the forthcoming or the present. Heidegger articulates this phenomenon of direction proper to *ecstasis* by saying that every *ecstasis* has its own *Richtungssinn*, "directional sense" or "directional meaning." In the concept of *Richtungssinn* Heidegger tries to grasp a certain formal moment of any projection, a certain *character* of the "letting-be-present" which is also called *horizontal schema*. As we have mentioned in the beginning of this paragraph, each *ecstasis* contains within its own nature a *carrying-away towards something* in a formal sense. "Every such re-motion or shift (*Entrückung*) is intrinsically *open*. A particular openness, which is given with the *outside-itself* belongs to *ecstasis*. That towards which each *ecstasis* is intrinsically open in a specific way we call the *horizon of the ecstasis*" (*ibid.*). The directional meaning is the primordial *schematisation* of meaning in general. "Of itself," writes Heidegger, "the *ecstasis* [of the future] does not produce a definite possible, but it does produce the horizon of possibility in general, within which a definite possible can be experienced" (MAL 209).

[22] M. Heidegger, *Metaphysische Anfangsgründe der Logik im Ausgang von Leibniz*, GA 26 (Frankfurt a. M.: V. Klostermann, 1978), p. 170. Cited as MAL.

1.7.2. *Kant's Transcendental Schema*

The term "schema" and the idea of "schematisation" or "schematism" in its most general form is borrowed by Heidegger from Kant.[23] In the lectures on *The Basic Problems of Phenomenology*, however, these terms acquire quite a different meaning.

In the *Critique of Pure Reason* the concept of transcendental schema makes its appearance in connection with the problem of "subsumption" of an intuitively apprehended object under a concept (this is, of course, the source of the faculty of judgment or predication as such). To make conceptualisation possible and expressible in a judgment (predication), Kant says, such features as are *thought* in the concept (and expressed through the predicate) must necessarily be present in the intuitive representation of a thing (performing the role of the subject of the judgment in question). When one says: "The plate is round," this means that something of the same kind (*Gleichartigkeit*) as the pure geometrical concept of circle ("the roundness of the plate") is present (can be given in the intuition) in the empirical concept of a plate (A 137/B 176).

But then the question presents itself: *how can categories be applied to appearances?* For in the objects of intuition there can be no "features" to which categories, being the *pure* concepts of understanding, could refer. What intuitively apprehended "features" could correspond, say, to the category of substance or the category of cause?

That is why Kant needs a mediation between pure concepts of understanding and appearances. The transcendental schema becomes such a mediator. We have already said that Kant's categories are the most general kinetic forms of synthesis, the forms of the synthesizing activities as activities. Yet the mind is aware of the *form* of its activity on the basis of its results; *a priori* it possesses only a representation "of the identity of a function," whereby it combines the manifold of (sensuous) intuition into one object, and this is ultimately the representation "I think." The mind is conscious of the kinetic form of its synthesizing activities insofar as this form is "reflected" or "reproduced" in the objects synthesized. To leave its trace, i. e., *to determine*, as Kant puts it, *the inner sense*, the pure concept, the pure kinetic form needs an indefinite ὑποκείμενον to be defined, i. e., it needs matter. The mind can in a certain way provide itself with "intuitive matter," because it possesses the faculty of imagination; and "the faculty of imagination is the ability to represent an object in intuition without its presence (*ohne dessen Gegenwart*) also" (B 151).

"The schema itself is always a product of the faculty of imagination..." (A 140 / B 179). Yet it must not be confused with *what* is imagined as an image. An image is not important in itself but only insofar as it "reflects" the kinetic form of a concept: what matters, is the activity of the understanding recorded in the image. A pure concept as *forma formans* must be recognized in what is formed (*forma formata*),

[23] An interpretation of Kant's doctrine of schematism as a "preliminary stage in a problematic of temporality" was planned as the content of the Second Part of *Being and Time, which has been announced but never written.* (Cf. SZ 40.) The plan was only partially carried out in the lecture courses *Kant und das Problem der Metaphysik* and *Die Grundprobleme der Phänomenologie.*

and for this the concept has to be "imagined"[24] (and that means – "embodied in an image"); it has to form the matter of intuition turning it into an image by imagining the image-object. Kant says: when I put side by side several points – ..... – this is the *image* of the number five. And when I grasp the method or the rule of com-posing (synthesizing), the form of *activity underlying the creation of this image, as a form of movement, I grasp the schema of number as such or the number as a schema (ibid.)*.[25] The *schema*, according to Kant, is a representation of a general method of the faculty of imagination, by means of which it creates an image for the concept (A 140 / B 179 f.).[26]

In this sense every category has its own transcendental schema, which portrays (in the sense indicated above) the corresponding pure concept as one of the basic pure forms (the most general rules) of *activity*, by means of the transcendental faculty of imagination and a so-called "pure image."[27] Yet the imagination, connected with quasi-sensibility, "endogenous" (given in, and by means of, the productive imagination) sensuous matter, always performs its operations depending on the form of time as the *a priori* form of inner sense. Kant says: the faculty of imagination de-termines the inner sense, and this always means that it determines time (is a kind of superstructure built on top of the most fundamental form of time). "It follows that the schemata are nothing but a priori determinations of time according to rules..." (A 145 / B 184). It is of special interest for us in connection with the preceding analysis of the Aristotelian time-concept that according to Kant, the schema of quantity (*quantitas*), as of a pure concept of understanding, is the *number*-as-schema, i. e., the rule of com-position, articulation of something homogenous:

That is why the number is nothing but the unity of synthesis of the manifold of a homogenous intuition (*einer gleichartigen Anschauung*) in general, due to the fact that in apprehension of this intuition I produce time itself. (A 142 f./B 182)

1.7.3. *Schematism of Temporal Ecstases and the Unity of Horizontal Schemata*

For Heidegger, as well as for Kant, schemata are *a priori* determinations of time; only time is understood now as "primordial temporality," the tripartite unity of temporal *ecstases*.

The *ecstases* of temporality (the forthcoming, the having-been-ness, the present) are not simply removals to... (*Entrückungen zu...*), not removals as it were to nothing. Rather, as removals to... and thus because of the ecstatic character of each of them, they each have a *horizon*, which is prescribed by the mode of removal, the carrying-away, the mode of the forthcoming, having-been, and present, and *which belongs to the ecstasis itself*. Each *ecstasis*, as removal to..., has at the same time within itself and belonging to it a predelineation of the formal structure of the *whereto of the removal*. We call the *whither* of the *ecstasis* the horizon or, more precisely, the *horizontal schema of the ecstasis*. (GP 428f., cf. SZ 360)

---

[24] In the sense of the German "eingebildet werden."

[25] In chap. 2, sect. 2 we have called this form of synthesis "articulation" in connection with Aristotle's concept of number.

[26] "Diese Vorstellung nun von einem allgemeinen Verfahren der Einbildungskraft, einem Begriff sein Bild zu verschaffen, nenne ich das Schema zu diesem Begriffe."

[27] The list of these "pure images" is given in chapter I of *The Transcendental Doctrine of the Faculty of Judgment*.

The "non-indifference" underlying the being of *Dasein*, the *existentiale of care, is the condition or possibility of any directional sense. Dasein* is the directedness, the projectivity, in which three formal moments coincide — the "that-upon-which" of the projection (the pure meaning of care), the "for-the-sake-of-which" and the "in-order-to" of existence. In other words, ecstatic temporality is *the primordial self-projecting as such* (GP 436), the underlying structure of transcendence.

Of course, all the preceding explanations cannot pretend to "deduce" anything in a strict sense, as categories are "deduced" in Kant's *Critique of Pure Reason.* We must not forget that the concept of ecstatic-horizontal unity of temporality as a fundamental condition of the possibility of "understanding" being in general is a result of phenomenological clarification. We are dealing here with a phenomenon, which we have to let "come to light" in the logos of phenomenology, "to exhibit it directly and to demonstrate it directly" (cf. SZ § 7). In order to accomplish this it is necessary to actually carry out the procedure of "*a priori* ontological generalization or emptying" in relation to the most familiar sphere of everydayness, in a way similar to that in which the phenomenological reduction performs its procedures in relation to the natural attitude of consciousness. And then a location is found (posited) for the concept of primordial temporality in the area of silence; the boundaries of this location are gradually outlined and marked more distinctly in "the work of the negative," to use Hegel's phrase,— in the negativity of negative analogy among other things, in the destruction of the opacity of the most familiar. In this process the topos of meaning, indefinite at first, gets its definition as something "other in relation to the other." Philosophy is a clarification of meaning, and it is possible because *Dasein* as the only temporal entity "has" meaning and bears witness to the meaning.

We cannot ask about the conditions of temporality as such any further. The way of questioning as outlined by Heidegger in his fundamental ontology stops (ends before the boundary of the surrounding area of silence) after bringing to light the phenomenon of primordial temporality and articulating the concept of the *ecstases* and horizontal schemata pertaining to them. Heidegger's texts record a certain procedure of clarification of meaning. If we examine this procedure from inside the topos of meaning, already explored by us in understanding and interpretation, we shall see that we are dealing with a series of mutually dependent projections "as it were inserted one before the other — understanding of entity, projection upon being, understanding of being, projection upon time. [And this sequence of projections] has its end at the horizon of the ecstatic unity of temporality. [...] At this horizon each *ecstasis* of time, hence temporality itself, has its end. But this end is nothing but the beginning and starting point for the possibility of all projecting" (GP 437).

As Plato's "idea of the good" has no eidetic determinations of its own but serves as the ultimate condition of "eidetic character of any *eidos*," as the "idea" of "eideticity" itself; as Aristotle's "intellect" has no form of its own in order to be receptive of any intelligible form, so the ecstatic temporality is also a self-projecting as such, deprived of any other determinations. It is the projectivity itself of a projecting of meaning. The only "internal structure" we can find in it is an "empty" schematisation of three temporal horizons corresponding to the three *ecstases* of temporality in their primal unity and primal difference.

We must keep in mind, however, that the *ecstasis* surpasses every being and the horizon is not located, say, in the sphere of subject. Hence this horizon is also nowhere, since it presents no determinate being: it is neither spatially nor temporally located, in the usual sense. It "is" not as such, but it temporalizes itself. The horizon manifests itself in and with the *ecstasis*; it is *ecstema*, ἔκστημα (formed analogically as, say, σύστημα is to σύστασις or σύνθημα to σύνθεσις). And, corresponding to the unity of ecstases in their temporalization, the unity of horizons is a primordial unity. (MAL 269)

As we shall see in the next section, the difference of the temporal horizons is the primordial difference as well, the *differentia differens*.

2. THE ONTOLOGICAL DIFFERENCE

2.1. *Two Forms of Ontological Difference*

We are now going back to the problem of ontological difference, the problem of the distinction between entity and being: "*In der ontologischen Differenz handelt es sich um den Unterschied von Seiendem und Sein*" (GP 109). The entity (according to its concept and also as a condition of encounter with it) always implies a certain constitution of being, and being itself is not entity.

The problem of distinction between entity and being belongs to *ontology*, and ontology strives to solve it *according to the truth*, i. e., to un-cover, to unveil this distinction, to let it come to light in word. Heidegger writes: "We characterized the unveiledness of something as truth... Truth is a determination (a warranty or responsibility) of *Dasein*, that is, a free and freely seized possibility of its existence" (GP 455). This possibility is chosen freely, but it is chosen as a *possibility* proper to factical existence. This possibility can be freely chosen insofar only as it "is already there" *(ist da)* in some way in the existential constitution of *Dasein*. Ontology is possible because *Dasein* is (*exists*) in the understanding of being (SZ 12).

The fundamental act of constituting ontology is a *thematization* of being, i. e., *the projection of being upon the horizon of comprehensibility* or understandability (GP 459). The articulation of the meaning of ontological difference is its ex-planation in interpretation, i. e., disclosure, de-limitation, de-finition, demonstration in language. Yet originally, meaning, according to Heidegger, is "the 'upon-which' of a projection structured by a fore-having, a fore-sight, and a fore-conception." The answer must come from where the question already points to, the interpretation (*Auslegung) implies an operation of "taking apart" (Auseinanderlegung) the milieu of the inherited and constructing anew* ontological conceptuality according to a certain fore-sight guiding our efforts. The fundamental ontology is a definite *projection* of being in which being is "outlined" towards *primordial temporality*. Yet ontology is possible as phenomenology only (SZ 35). And being phenomenology it must pose the problem of ontological difference in a certain definite way. It must proceed from the fact that originally being and entity are given in their indistinctness: entity in its being, being as being of entity. If it is possible to understand the difference between being and entity, *how* is it possible? What phenomenon, "which proximally and for the most part does not show itself,"[28] must be brought to light and exhibited as the ultimate condition of such understanding?

What does this difference looks like in the light of the logical *apophansis? How does language express entity in its being? We say: "X* (the name of a thing) *is*."[29] The name X refers to the entity as to something already possessing a certain "countenance of a thing," a certain "thingness"; the "is" tells about the being of the entity called X. It would seem that language, by distinguishing between parts of the sentence, already provides us with a basis of ontological difference: X, on the one hand, and "is," on the other. Yet an essential ambiguity is hidden there. What is distinguished in the existential proposition "X is"? We could ask using the terminology of Aristotelian metaphysics: does the name X, as one of the parts of the sentence distinguished, signify the entity itself, i. e., the composite whole (*compositum*), or does it express only the form, the whatness, the essence (understood as τὸ τί ἦν εἶναι)? Of course, according to Aristotle (and to everyday common sense as well), it is the compositum that *is*, but this does not remove the ambiguity of the attempt to distinguish "is" from *what-is*.

We have seen that medieval ontology interprets the distinction testified to by different parts of an existential proposition as a distinction (and composition) of essence (*essentia, realitas*) and existence (*existentia*). It is for this formal reason that Heidegger, leaving aside more serious meaningful points of divergence, distinguishes the traditional distinction of essence and existence from his "ontological difference."

Thus the distinction between reality and *existentia*, or between *essentia* and *existentia*, does not coincide with the ontological difference but belongs on the side of one member of the ontological difference. That is to say, neither *realitas* nor *existentia* is an entity; rather, it is precisely the two of them that make up the structure of being. The distinction between *realitas* and *existentia* articulates being more particularly in its essential constitution. (GP 109)

Yet how essential is the difference between the two kinds of distinction mentioned? If "is" and *what-is* have already been distinguished, what else can *entity* mean but "what" in which we find "is," or "is" in which we find "what?" Taking into account not only the *distinctio*, but also the *compositio essentiae et existentiae*, we by no means overlook the *compositum*, the entity itself, *that-which-is* (or, as Suarez puts it,— *actualis essentia existens*).[30] Yet here the *compositum* is already dis-

[28] Cf. SZ, § 7.

[29] As we have seen, Aristotle considered this "existential" proposition to be an elliptic form of a correct *apophantic* assertion: "X is Y" (*Metaph*. VII 17). The latter construction is a genuine logical counterpart of the fundamental ontological principle – the "internal duality of being" (see chap. 1, sect. 3). For Leibniz (and for Fichte) the identity principle "X is X" performed the role of the deepest ontological foundation.

[30] As we have already mentioned in chapter 3, one of the most important sources of the medieval concept of *distinctio et compositio essentiae et existentiae* had been Boethius' treatise "Quomodo substantiae in eo quod sint bonae sint..." We read in this text: "Diversum est esse et id quod est; ipsum enim esse nondum est; at vero quod est, accepta essendi forma, est atque consistit." And further on: "Omni *composito* aliud est esse, aliud *ipsum* est." Thus, as it seems, the distinction posited here, formally speaking, is completely identical with Heidegger's "ontological difference" – the difference drawn between being and entity, and not between essence and existence.

tinguished from the *componenta*. The distinction of *ens/esse* is inevitably posited *simultaneously, together with* the distinction of *essentia/existentia*, and depends closely on the latter. It is not without good reason that Heidegger says (changing, in a manner which may seem innocent at first sight, his mode of expression) that the medieval distinction distinguishes two *meanings of being:* the *being-what* and the *being-how*, i. e., in this dividing-joining we are dealing with *the fundamental articulation of being*.

Thus, formally speaking, the distinction between *being-what* and *being-how* also "belongs on the side of one member of the ontological difference." Nevertheless, the only possible way to point to the other side is to consider the combination of what has been distinguished: from the phenomenological point of view a being (*ein Seiendes*) is nothing but some "what" that *is* (some*how*). "Only as phenomenology, is ontology possible" (SZ 35), and "with regard to its subject matter, phenomenology is the science of the being of beings — ontology" (*ibid.*, p. 37). The precise phenomenological formulation of the problem of ontological difference reads as follows:

We must now manage to exhibit more precisely the interconnection between the uncoveredness (*Entdecktheit*) of a being and the disclosedness (*Erschlossenheit*) of its being and to show how the disclosedness or unveiledness (*Enthüllung*) of being founds, that is to say, gives the ground, the foundation for the possibility of the uncoveredness of the entity. In other words, we must manage to conceptualise the distinction between uncoveredness and disclosedness, its possibility and necessity, but likewise also to comprehend the possible unity of the two. (GP 102)

Thus, we find ourselves in the face of a difficult task: to bring to light the distinction between two forms of ἀ-λήθεια — the *uncoveredness* of a being (*ens*) and *unveiledness* of its being (*esse*). This state of affairs means that, as phenomenologists, we always remain encapsulated within being (*das Sein*) phenomenologically understood; we are doomed to take side of one member of the ontological difference, and to point to the other member only indirectly. An attempt to overcome this insularity, to penetrate to the "being's other," "beyond the essence-beingness"[31] would mean simultaneously parting from phenomenology.

The term *entity* itself acquires various meanings in the version of ontology we are dealing with (and the list of these meanings differs considerably from the one medieval metaphysics borrowed from Aristotle): first, it is *Dasein* itself, and, second,— the entity-in-the-world. The latter can in its turn have the meaning of *ready-to-hand* or *present-at-hand*, and the primary topic of ontology, according to *Being and Time*, is entity as ready-to-hand.

For the entity-in-the-world Heidegger introduces a distinction similar to that of essence and existence. He says:

To the being of this entity (sc. of something ready-to-hand) belongs its inherent content (*Sachgehalt*), the specific whatness, and a way of being. The whatness of the entity confronting us every day is defined

[31] I have in mind the title of E. Levinas' treatise *Autrement qu'être ou au-delà de l'essence*. Levinas sees clearly that an effort to substitute "ontological separation" for Heidegger's "ontological difference" and to force his way through being towards the beings (*de l'existence à l'existant*) requires departure from phenomenology. Cf. Th. de Boer, *The Rationality of Transcendence. Studies in Philosophy of Emmanuel Levinas*, chap. 6, "Ontological difference (Heidegger) and ontological separation (Levinas)" (Amsterdam: J. C. Gieben, 1997).

by its equipmental character. The way an entity with this essential character, equipment, is, we call being-ready-to-hand or readiness-to-hand, which we distinguish from being-present-to-hand, extantness. (GP 432)

In my opinion, it is precisely the distinction between being-what (*Wassein*) and being-how (*Wiesein*) that Heidegger discusses in his lectures on *The Basic Problems of Phenomenology* under the heading of "ontological difference." As we have already remarked, this distinction does not coincide formally with the ontological difference but is closely connected with it (also in a formal way): any one of these two distinctions is easily "converted" into the other. This does not remove the possible subtle content-related difficulties of such a conversion; however Heidegger is concerned precisely with the difference of *being-what* and *being-how*, i. e., with *the fundamental articulation of* being. "What" and "how" are essentially interconnected here: being is not a simple *yes* answered to the question, whether something is *(an sit)*. We know already that being as such has its own internal constitution (*Seinsverfassung*), which is related in a certain way to the "thing-content" (*realitas*) of the entity. Further we shall complete the task of distinguishing *being-what* and *being-how* for the entity ready-to-hand in its being-ready-to-hand, for the entity present-at-hand in its being-at-hand, and finally for *Dasein* itself as the existing "who."

2.2. *The Ready-to-hand, the Present-at-hand and the Existential Ontological Modification*

Yet before proceeding any further we must take into account a very important circumstance. When Heidegger speaks about the entity ready-to-hand as distinguished from the present-at-hand, about equipment (*Zeug*) as distinguished from the thing-object (of perception, knowledge, etc.— *Ding*), he does not mean something like "a classification of entities," a division of the common genus "entity" into species or classes (so that a hammer, for instance, may belong to the species of tools, and a stone to the species of objects present at hand). Heidegger's idea is not at all identical with the Aristotelian division of entities into *natural entities* (φύσει ὄντα) and *artificial entities* (τέχνῃ ὄντα) either. First, after Aristotle, a philosopher does not easily venture to call "entity" a common genus, and second, Heidegger does not at all try to propose a new method of *diairesis* (division of genus into species), still less to add two new items to the Aristotelian list of meanings of "entity," the "entity-ready-to-hand" and the "entity-present-at-hand." And nonetheless he solves an essentially Aristotelian problem, that of distinguishing the meanings of "being" (and the corresponding meanings of "entity").

Aristotle's meanings of being are connected by the unity of analogy, and the *analogia entis* traces all meanings of entity back to the one (πρὸς ἕν), first and initial meaning, that of substance (οὐσία). Entity is united as πρὸς ἓν λεγόμενον. Yet Heidegger also asks the question about the unity of the distinguished meanings of entity. The terms "ready-to-hand" and "present-at-hand" refer to two ways of *Dasein*'s understanding of entity, i. e., to two ways of disclosing the being of the entity-within-the-world. These two ways of understanding are hierarchically connected. Heidegger asserts that being-present-at-hand is a derivative way of being in relation to being-ready-to-hand. When speaking about the meaning of being, derivative or secondary does not refer, of course, to the traditional logical dependence; it designates a change of attitude in understanding-interpretation (i. e., in the existence of *Dasein*). This change of attitude is called "hermeneutical change-over" (*hermeneutischer Umschlag*), "change-over in the fore-having" (*Umschlag in der Vorhabe*) or "existential ontological modification" (*existential-ontologische Modifikation*). Later we shall describe this change-over in more detail. What is relevant for us now is the circumstance that *Dasein* becomes related to entity as entity and interprets entity in a certain way depending on the original projection of the entity's meaning. Being and entity are understood simultaneously, and the inherent content of the entity (*Sachgehalt*), to which *Dasein* relates one way or another, depends on the way of understanding being, on the projection of the entity towards a certain mode of being. But how can being be *distinguished* from entity?

2.3. *Being-what and Being-how of the Ready-to-hand*

We have discussed at length what the specific equipmental content of an entity-ready-to-hand means and how it is de-fined-"ex-plained" or, as Heidegger puts it, "freed for its being." The equipment in its specific instrumental whatness is a topos in the variety of references or assignments, which has a topological unity of its own. The phenomenon which now plays a decisive role for us is that this topos may be *empty*: it may happen that the thing needed is *not* ready to hand, is not handy (*ist abhanden*).

Let us recall the argument of Thomas Aquinas in favour of the distinction between essence and existence: we can understand (*intelligere*) what a man or a phoenix is (*quid est*), ignoring whether they have being in the nature of things. In terms of Husserl's phenomenology this means that the *distinctio essentiae et existentiae* is posed here as a distinction of empty and fulfilled. When the empty intention of meaning, initiated by the name *man*, is fulfilled, we say: "man is or exists (*in rerum natura*)."

A similar interplay of the empty and the fulfilled allows us to distinguish between the equipmental content of the ready-to-hand and its *being*-ready-to-hand. Only the meaning of "empty" and "fulfilled" is quite different here (and more fundamental ontologically, according to Heidegger). The *letting-be-involved* (*das Bewendenlassen*), in which the ready-to-hand is already understood as having a place of its own, i. e., having its topos in the totality of involvements, succeeds in Heidegger's scheme to the empty intention.

We begin with an almost trivial statement: the topos of a thing ready-to-hand is fulfilled when the thing is actually ready to hand (which does not at all mean immediately present in perception). A thing ready-to-hand is understood to the extent that it is usual and familiar. This "confident closeness," familiarity (*Vertrautheit*) of the thing is its being included into the dealing-with (*Umgang*) and consists in an appropriate *use* of it. A thing is ready to hand if, without any delay, I can use it in order to... *Dasein* as being-in-the-world in its inclusion in dealing-with distinguishes "ready-to-hand" and "not-ready-to-hand" as the positive and the negative modes of one fundamental phenomenon which has a different meaning than pres-/absence (*An- /Ab-wesenheit*) in perception or, in a broader sense, *being for consciousness*. Let me repeat: it does not matter at all *what kind (species) of things* is spoken of here. To be sure, the hammer may not be ready-to-hand when I want to drive a nail home, yet a *word* may also not be ready-to-hand, ready-to-tongue when I want to say something. The phonetic signs as such (as signs) are absent *for me* when I am speaking quickly and eagerly; they *are*-in-order-to allow me to say something necessary in the situation of talking, they are ready-to-my-tongue in their transparency, inconspicuousness (*Unauffälligkeit*), and unobtrusiveness. Yet it may happen (particularly if I am speaking a foreign language) that I lack words, and then I discover, first, a lack, a shortage, a void in the totality of my linguistic tools, and, second, immediately after that words cease to be for me unnoticed "equipment," and become objects (of an empty intention). When I "find the right word," the empty intention is fulfilled, and then the words as signifiers, as objects of my reflection, are dissolved again in the inconspicuousness of dealing with the language, involvement in the language, in the necessity to say what I want. "One and the same thing" can uncover its being for *Dasein* in *Dasein*'s different hypostases ("reflective" positions) as the ready-to-hand and also as the present-at-hand. Still the meaning of the phrase "one and the same thing" needs to be defined more precisely.

Thus *Dasein* distinguishes in its dealing-with the ready-to-hand and the present-at-hand of "this here" thing (*Zeug*). Let us render this in the language of ontology: in the situation of identity of being-what *Dasein* discovers the difference between the positive and the negative mode of being-how. By answering the question of how this is possible, we shall clarify the foundation of the difference between being-what and being-how of the ready-to-hand.

Not-ready-to-hand is a lacuna, a hiatus that we encounter in our confident and customary progress along the lines of force in the variety of references. Absence in this sense can be encountered because, first, the totality of the circumstances of the matter has already been understood; second, the location in the variety of references (the thing content of such and such an entity ready-to-hand) has already been determined or delimited and is found to be empty; third, the way itself of "fulfilling the location" and of being-ready-to-hand has been outlined.

2.4. *Praesens as the Horizon of the Present*

Being-ready-to-hand in the immediately surrounding world and being "not-handy," being unavailable or distant are, respectively, the positive and the negative mode of one phenomenon, called *Praesenz* by Heidegger. We use its genuine Latin equivalent — *praesens* — in our text. The *praesens* is the *a priori* schematised horizon of the *ecstasis* of the present (or, to be more precise,— of making-present).

2.4.1. *Praesens as Being-ready-to-hand*

The entity ready-to-hand is understood (disclosed in its being) in concernful dealing with it. The latter "constitutes itself with respect to its temporality in a retentive-expectant making-present of the equipmental contexture" (GP 432).[32] The way of being of the ready-to-hand is thus understood beforehand and projected towards time. This means, first, that the being of the ready-to-hand or the being-ready-to-hand (*Zuhandensein*) in the sense of its specific equipmental "what" (*Wassein*) is freed out of the original indistinctness of involvements by means of delimitation of a certain topos within the referential manifold, and the *unity* of the three temporal horizons serves as a condition of this delimitation (as was sufficiently explained earlier). Second, the being-how (*Wiesein*) of the ready-to-hand is connected with *singling-out* of the horizontal schema belonging to the *ecstasis of the present.* The horizon of the present is outlined beforehand (*a priori*) in understanding: *Dasein* can detect that a piece of equipment is not ready to hand, can detect an absence, a loss, since it has understood beforehand and outlined beforehand a project of being-ready-to-hand. The *praesens* as the schematised present allows Heidegger to understand a thing in its being-ready-to-hand and (in a secondary way, as will be explained further) in its being-present-at-hand.

Thus the difference of the temporal ecstases and of the horizons belonging to them serves as the fundamental condition of the difference between being-what and being-how.

Let us go back for a while to our negative analogy. For Kant the schematism of the understanding (*Verstand*) is a condition of an "objective application" of categories. The deduction of categories in the *Critique of Pure Reason* begins with an enumeration of the types of judgments, i. e., ways of predication. That is to say, in language the categories correspond to the different meanings in which the copula "is" tells about being, showing the subject *as* the predicate (here Kant loyally follows Aristotle), and in the synthesizing activity of the understanding (in its *objective application*) the categories correspond to certain kinetic forms or functions of combining. Thus, the schematism of pure understanding is *the condition of the possibility* to assert "is" so that the corresponding statement might have an "objective meaning," i. e., might be a statement concerning an object *given in the intuition*. And this possibility exists *a priori* as a "concealed art lying in the depth of the human mind,"[33] as a fundamental condition, I daresay, of encountering the entity in its being through logical *apophansis.*

The *praesens,* the schema of the *ecstasis* of the present, is a condition of encountering the entity-within-the-world. However, the concept of *praesens* is considerably broader and, as Heidegger asserts, is more primordial ontologically than the concept of *positivity* of the object of consciousness. One of the meanings of *praesens* is being-ready-to-hand (in the sense of its "how"), which differs dramatically from being for the *now-consciousness.*

---

[32] "...in einem behaltend-gewärtigendenden Gegenwärtigen des Zeugzusammenhanges."

[33] "Dieser Schematismus unseres Verstandes, in Ansehung der Erscheinungen und ihrer bloßen Form, ist eine verborgene Kunst in den Tiefen der menschlichen Seele, deren wahre Handgriffe wir der Natur schwerlich jemals abraten, und sie unverdeckt vor Augen legen werden." (A 141/B 180f.)

2.4.2. *Praesens in the "Twinkling of an Eye"*

Let us repeat once more: the *praesens* is a horizontal schema of the *ecstasis* of the present. Every horizontal schema pre-defines (schematises) a specific character of the disclosure, not the inherent content (*Sachgehalt*) of an entity disclosed, neither its specific equipmental "whatness," nor its extant form. The phenomenon of horizontal schematism must be "explicitly shown and exhibited." That means that the corresponding ways of being-disclosed are to be somehow indicated in phenomenological analysis. The ontological tradition concentrates exclusively on objective presence or positivity. Heidegger refers to the being-ready-to-hand as the most important way of encounter with the entity-within-the-world. Yet the *praesens* as the horizontal schema of the present can be exhibited phenomenologically not only in the "presence" as readiness-to-hand or in the presence-at-hand of an extant thing, it manifests itself (and in the most primordial sense, according to Heidegger) in the moment of vision (*Augenblick*), in the "now-or-never" of human action, when the world, the things-within-the-world and the other people (my near and my far neighbours with whom I co-exist in my being-in-the-world) enter the situation of the action and disclose their relevance, their inclusion in the ultimate "for-the-sake-of-which," in the most complete way. People and things have been and will be that which they are in the sense of the Aristotelian τί ἦν εἶναι; they retain and reproduce their "form," but they have never had before and will never have any more the same meaning and the same relevance for me as those which they disclose in the twinkling of an instant. In the *praesens* I disclose the truth (ἀληθεύω) of the action, the first principles of what-must-be-done, and these are always "different again and again."

"*Praesens,*" Heidegger says, "is a more original phenomenon than the 'now.' The instant (*Augenblick*) is more original than the 'now' for the reason that the instant is a mode of making-present (*Gegenwärtigen*) of a particular disclosure" (GP 434). It can *express* itself by a "now," but by saying "now," it loses its specific character, for what is disclosed in this "now" is at that instant included in my existence, not in my knowledge; what is disclosed in the *Augenblick* will never be the same after the closing of that instant "eye-blink."

2.5. *Being-what (essence) and Being-how (existence) of the Present-at-hand*

Let us recall Husserl's definition of a "truly existing" object:

> As a matter of principle, to every "truly existing" object (*wahrhaft seiendem*) corresponds (in an *a priori* unconditional essential universality) the idea of a possible consciousness, for which this object itself is originally graspable (*originär erfassbar*) in a *perfectly adequate* way. And vice versa, if this possibility is granted, the thing itself is *eo ipso* truly existing.[34]

It is important for us now that the *essence* (*Wesen, Essenz*) of the object (this or that *objective category* in the technical sense of Husserl's phenomenology), as well as Kant's "concept" (*Begriff*), determines the *rule* of synthesis; and *existence* corre-

[34] *Ideen* I, § 142; cf. chap. 5, sect. 2 of this book.

sponds to an adequate self-evidence. The latter implies a *fulfilled* consciousness of the object, and particularly its presence *now.* The primal impression (*Urimpression*) as a moment of the "now"-consciousness, Husserl says, "marks the object with the stamp of existence."

Every objective category ... is a universal essence, which is to be brought, as a matter of principle (if we want to speak about its existence — *A. Ch.*), to adequate givenness. [...] It (sc. the objective category — *A. Ch.*) prescribes the rule for the way in which the object subordinate to it should, according to its meaning, be brought to complete determination and an adequate original givenness.[35]

The adequate givenness of the transcendent (in the classical sense) object, as we already said, cannot be attained in a finite act of consciousness and is "an *a priori* determined continuum of appearances, possessing different but definite dimensions."[36] Yet the endless synthesis of the object's aspects is "governed by the strict lawfulness of its essence (*Wesensgesetzlichkeit*)," and this law or rule can be apprehended in the *finite* act as a *representation* of an essentially motivated and at the same time endless and continuous synthesis of fulfilled consciousness.[37]

In Heidegger's terms this means that the "truly existing" thing present-at-hand is already being outlined as a project of "adequate givenness." The present-at-hand is projected, first, in its being-what, and this is the "essence" as the *a priori* determined rule of synthesis (*forma formans*), defining the "what" (*quidditas, forma formata*) of the object synthesized, and, second, in its being-how (presence-at-hand), and this is nothing but the adequate givenness. We should now ask: *what underlies the distinction of essence and existence thus understood, what underlies the distinction of being-what and being-how of the present-at-hand?*

Let us remark first of all that the rule of synthesis, the rule determining in what way the object must, according to its meaning, be brought to adequate givenness in order to be "visible" (*einsichtig*), requires an extended temporal field including the horizon of retentions and protentions.

This is particularly clear in the act of perception of a spatial thing, in which the *kinaesthesis* of the organ of perception plays a major role. Husserl contrasts his own notion of sensory receptivity to that of Kant. Kant speaks of receptivity as of "the ability to receive percepts through the manner in which we are affected by the objects" (A 385). For Husserl it is not only a passive undergoing of affections but likewise a consciousness of activity, and between these two there is a causal relation. Let us consider visual perception as an example. The spatial "meaning" of *res extensa*, that is to say,— *what* it is as a spatial object, its geometric form and location — are constituted not only in passive receptivity, but in a special correlation of the movements of the eyes (of the head, and of the body in general) and of that which comes

[35] "Jede Gegenstandkategorie... ist ein allgemeines Wesen, das selbst prinzipiell zu adäquater Gegebenheit zu bringen ist. [...] Sie schreibt die Regel vor für die Art, wie ein ihr unterstehender Gegenstand nach Sinn und Gegebenheitsweise zu voller Bestimmtheit, zu adäquater originärer Gegebenheit zu bringen wäre..." (*Ideen* I, § 142, p. 296)

[36] *Ideen* I, § 143, p. 297.

[37] "Die Idee einer wesensmäßig motivierten Unendlichkeit ist nicht selbst eine Unendlichkeit: die Einsicht, daß diese Unendlichkeit prinzipiell nicht gegeben sein kann, schließt nicht aus, sondern fordert vielmehr die einsichtige Gegebenheit der *Idee* dieser Unendlichkeit." (*Ibid.*, p. 298)

as *a response* to these movements: "because I have moved my eyes, head, have walked around the object, I have beheld the object in these perspectives."[38] The form of movement[39] of the sense organ must be projected beforehand and given beforehand in some way as an outline within the scope of the "I can," of the system of my capabilities. The looking at the object from various sides is a realization of a rule, which governs the "bringing of a *res extensa* to adequate givenness." This "because-thus," the *kinaesthetic* causality, must somehow become beforehand my "possession" as a *habitus* of the faculty of sense perception; and while this rule is being realized in a concrete situation of perception, the protentional field of the corresponding act (anticipation of *my activity as of a perceiving agent, my kinaesthesia* conforming to the rule) is projecting depending on what has been already perceived and held back in retention. It is thereby evident that the perceptual meaning of the spatial thing "lives" in a full temporal horizon. The act of perception, one could say imitating Heidegger, is a retaining-anticipating-exposedness-to-"now"-impression. The being-what of the thing present-at-hand requires a whole structure of the temporal field.

The idea of original and adequate givenness as such is, on the contrary, connected with fulfilled consciousness as distinguished from "empty positing." Heidegger would say that adequateness and fulfilment, as the meaning of being in the sense of *Wiesein* or *existentia* of a "material spatial thing," are projected antecedently *towards the praesens*. And the *praesens* as the horizon of this projection is already interpreted here in an ontologically secondary sense, as presence-at-hand, as the positivity of the object, as "now." The primal impression, the intentional content of the now-consciousness *marks* the object *with the stamp of existence.*

Thus, the difference of the "now"-consciousness, on the one hand, and its retentional and protentional modification, on the other hand, serves as the fundamental condition of the difference of being-what and being-how of the present-at-hand. And this difference depends in its turn on the difference of the temporal ecstases of primordial temporality and the horizons belonging to them.

2.6. *Absolute Temporal Flow and Primordial Temporality*

The presence "in" the "now"-phase of consciousness as a special form (*energeia*) of the manifestation ("manifestedness") of the manifested is absolute positivity, and therefore, according to Husserl, the source of all presence of the object for consciousness. Yet, according to Heidegger's reasoning, the positivity is only one of the

---

[38] A more detailed analysis of the phenomenological concept of *kinaesthesis* one can find in L. Landgrebe, "The phenomenology of corporeality," in *Phenomenology of E. Husserl. Six Essays*, ed. D. Welton (Cornell Univ. Press, 1981); U. Claesges, *E. Husserløs Theorie der Raumkonstitution*, Köln, 1963; A. G. Chernyakov, "Husserl's 'Genealogy of Logic,' Space-Constitution, and Noetic Geometry," *Recherches husserliennes*, 7 (1977), pp. 61–85. All these researches lean upon the outline of the theory of corporeality given by Husserl himself in the Second and the Third volumes of the *Ideas* (*Ideen II, III* = *Ideen zu einer reinen Phänomenologie und Phänomenologischen Philosophie*, 2. und 3. *Buch*. hrsg v. M. Biemel, Hua IV–V, Den Haag, 1952).

[39] Exactly in the sense we discussed in chap. 1 in connection with Aristotle's theory of time.

many directional senses (*Richtungssinne*) connected with the *ecstasis* of the present. In the horizon of positivity a certain mode of the presence, belonging to the *noemata*, is schematised as a definite *energeia of presence* irrespective of that which is present. In chapter 2 we have called this *energeia noetic*. We could say now that positivity is the definition of the internal form of noetic *energeia*, not of the form of the object (i.e. not of the *noema*). Positivity is schematised beforehand as the horizon of the presence-at-hand. And nonetheless it is only a particular case of schematisation.

Of course, speaking about retention we could also define it in a certain sense as the primordial *going-back-to*, a shift of consciousness towards... i. e., as Heidegger's *Entrückung*, ἔκ-στασις. And this "back-to" is determined in relation to the initial and primordial temporal position (*Zeitstelle*) of the pure *ego*. We have seen that, according to Husserl, this temporal position, which, strictly speaking, eludes the look of reflection but can still be guessed at, is the immobile self-identity of the *proto-ego*, στάσις, νῦν τὸ στάσιμον, *nunc stans,* out of which time issues forth, temporalizes itself as τὸ ἐκστατικόν. This holds true for protention: why not call protention a primordial *coming-towards-my-own-possibilities* (especially if we take into account the phenomenon of the kinaesthetic body) or a shift towards the forthcoming? In a word, could we not say that Husserl's primordial triad (now-consciousness, or in-tention in the strict sense, re-tention and pro-tention) is precisely the initial "outside-oneself" in relation to the *nunc stans* of the pure *ego*, i. e., exactly the ecstatic structure meant by Heidegger? Yet this would be an unacceptable narrowing of the Heideggerian concept of ecstatic temporality for two reasons.

First, one cannot presuppose a self-identical supra-temporal position of the pure *ego* behind the *ecstases*. Heidegger mentions a likely misunderstanding to be fended off. It consists in the idea, "that temporality is indeed a threefold being-carried-away, but one which operates in such a way, that these three *ecstases* flow together somehow in one substance" (MAL 267). In fact, "the whole of this three ways of being-carried-away does not centre in a kind of thing which would of itself lack any being-carried-away, something present on hand unecstatically and which would be the common centre for initiating and unfolding the *ecstases*. Moreover, the unity of the *ecstases* is itself ecstatic. It needs no support and pillars, as does the arch of a bridge" (MAL 267f.).

Second, retention is a holding back, and protention an anticipation of the *objective presence*. Now, finding ourselves in a new horizon of phenomena and possessing a new system of concepts, we could say that the projective meaning of retention and protention is schematised beforehand; its schema is positivity, or, to be more precise, positivity modified in the "primal memory" or, correspondingly, in the "primal anticipation." The schematisation of temporal horizons is not explicitly discussed in Husserl's phenomenology, but is outlined beforehand in Husserl's phenomenological project as something derived from the absolute positivity of the "now." Even before time consciousness becomes the most difficult and most important theme of investigation, this schematisation delimits the horizon of the phenomenology itself.

For Heidegger, the *ecstases* of temporality are equally primordial (more than that, the *ecstasis* of the future possesses a special advantage which will be discussed further on). Now we could say: the horizon towards which the retention projects itself (and the same is true concerning the protention) is only a particular case (and, according to Heidegger, a secondary, a derived one) of the primordial horizons of temporality.

The presence-in-retention (in protention) as a mode of givenness, as a special *energeia* of manifestation of what is manifested, is, according to Husserl, a modification of the primordial *energeia* of presence in the present, which means — in the "now." The problem of conditions and mechanisms of such a modification is the ultimate problem of Husserl's phenomenology. These conditions and mechanisms (although Husserl has a tendency to call them, by means of *oxymoron*, the "passive activity" of the pure *ego*) cannot, as we have seen in chapter 5, become an object of reflection, because the objectifying reflection is a prisoner of the temporal flow; it adheres to it and cannot pose as its own object the "where-from" of the flow and the "how" of its immobile streaming (*das stehende Strömen*).[40] One may speak of the proto-phenomenon of the absolute temporal flow issuing forth from its "non-temporal" or "supratemporal" source, of the initial passive synthesis; one may ask *what* or *who* performs this synthesis or (if the pure *ego* acting in concealment is already assumed as the performer) in what sense the *ego* temporalizes the time, yet within the limits of the phenomenology of *consciousness* it is impossible to get answers reliable by the criteria set by phenomenology itself. The answers can just testify to the paradoxical nature of the situation: "The stream is always ahead, but the *ego* is also ahead." The sought-for phenomena elude the glance of reflection.

Nevertheless, phenomenological reflection becomes a prisoner of the temporal flow insofar only as it is itself schematised beforehand and projected towards positivity: it pursues the positivity of the ob-ject. The existential analytic of *Dasein* (phenomenology of existence), by constructing quite a different space of meanings and "destroying" the history of ontology, allows us to come close to the same topos of paradox and to say something new:

The primordial ecstatic-horizontal temporality is the condition of the temporal flow of consciousness. Here the ecstasis of the making-present underlies positivity, the "now"-presence as a specific energeia (manifestation) of that which is manifested. The ecstases of the forthcoming and of the having-been-ness are the conditions of its protentional and retentional modifications.

Dasein as the primordial self-projection is capable, in a special mode of its being (a contemplative, "theoretical" relation to the object), of being directed towards the present *object* of consciousness (the content of the "now"-phase), to anticipate the future *objective* presence and to retain the past. Heidegger wants to say that the ecstatic horizontal temporality of *Dasein* serves as the source and the underlying structure of the temporal flow.

The Copernican turn in the ontological hierarchy of the modes of being, which Heidegger attempts to accomplish in *Being in Time*, is expressed in the fundamental

[40] One more *oxymoron*.

thesis concerning the so called *existential ontological modification*: being-present-at-hand, the extantness, is a *derivative* mode of being in relation to being-ready-to-hand, the result of a certain modification. Phenomenology as ontology asks about the most primordial ways of our encounter with beings in the world. Heidegger insists that proximally and for the most part *Dasein* encounters the entity, that is to say — discloses the entity in its being, due to the fact that it understands, and therefore has already understood, the totality of involvements.

When an entity within the world has already been proximally freed for its being, that being is its involvement. (SZ 84)

Through the practical "interpretation," the ready-to-hand comes explicitly into the sight of understanding, and this sight (*Sicht*) is nothing but circumspection (*Umsicht*). Heidegger describes the operation of interpretation, as we have seen, in the following way: "we take apart (*legen auseinander*) in its 'in-order-to' that which is circumspectively ready to hand" (SZ 189). That which has been circumspectively taken apart within the referential manifold of involvements has the structure of *something as something-in-order-to*. This τὶ ἕνεκά τινος comes to succeed within the scope of ontological principles to the classical Aristotelian formula of logical *apophansis*: τὶ κατά τινος. Interpretation is carried out primordially not in an *apophantic* statement but in an *action of circumspective concern*. Heidegger describes what is going on here in his famous narration about a hammer (SZ 157). The categorical statement: "The hammer is heavy," being deciphered as "This thing — a hammer — has the property of heaviness," is not the primordial and the most fundamental expression of being. Heidegger writes: "In concernful circumspection there are no such assertions 'at first'." The hammer reveals its presence-at-hand as a thing possessing the property of heaviness only when a certain obstacle takes place within the inconspicuous flow of dealing. Circumspection, facing this obstacle, has its specific ways of interpreting, and these may take some such form as: "The hammer is too heavy," or rather just "Too heavy!" — "Hand me another hammer!" The action of circumspective concern goes first, and a theoretical statement appears according to Heidegger as its derivative mode or substitution. And in this case a change-over in the mode of understanding happens.

Something ready-to-hand, with which we have to do or perform something, turns into that "about which" the assertion that points to it is made. Our fore-sight is aimed at something present-at-hand in what is ready to hand. (*ibid.*)

The ready-to-hand understood *as something*-in-order-to... becomes veiled and the present-at-hand becomes manifest *as something* objective (positive).

Whatever one might think about the cogency of this particular narration and its phenomenological authenticity, the remarkable interplay of two ontological principles — the classical principle of form or whatness, on the one hand, and the existential principle of care, of the "for-the-sake-of-which," on the other hand — comes to light here very distinctly.

The form or whatness is shown in the assertion *via* predication: something (the subject) is asserted *as* being something (predicate): for example, a hammer — as having the property of heaviness. But first, says Heidegger, the hammer must be understood as something-in-order-to... and its heaviness must disclose itself as an obstacle or an advantage in order to...

Thus assertion cannot disown its ontological origin from an interpretation which understands. The primordial "as" of an interpretation (ἑρμηνεία) which understands circumspectively we call the *existential-hermeneutical* "as" in distinction from the *apophantical* "as" of the assertion. (SZ 158)

After this long detour we come back to ecstatic horizontal temporality. Quite in keeping with Heidegger's terminology one could call the temporal flow in the sense of Husserl's phenomenology the "existential ontological modification" of primordial temporality, though now it is not the thing ready-to-hand that is singled out from the sphere of concerned dealing-with and posited subsequently as the object. Instead it is *Dasein* itself that now objectifies itself, posits itself as the transcendental subject or as "consciousness" understood as the totality of determinations, "transcendental properties" of the transcendental subject. The Change-Over occurs now in the primordial self-interpretation or self-projection: the care-full unindifference of my existence (which is an issue for *Dasein*) results in attentive objectifying phenomenological reflection.

Of course, Heidegger's existential analytic of *Dasein* also singles out this entity (sc. *Dasein* itself) from the living of life and posits it as the object of thought. But this happens in "reflection" of quite a different kind, in a system of essentially different conceptuality and methods of interpretation. The positing of the entity as present-at-hand, as an object, takes place simultaneously with another process, about which we spoke in chapter 6: *the human person assumes the role of the subject.* Besides, as I believe, the new system of concepts in question allows us at the present stage of our investigation to establish a more essential ontological relationship between the *existentiale of care and the transcendental subject, than that of negative analogy.*

The transcendental ego is to be considered the existential-ontological modification of the primordial self-projection, which is the meaning of care.

Aristotle asserted that the *energeia* of the intellect and the *energeia* of the intelligible form (*noema*) are one and the same. Husserl discovers this field of *energeiai* as being phenomenologically the first and the most fundamental, as what precedes the objectifying of the objective and the subjectifying of the subjective, as the no man's land of the manifest character of what is manifested. Yet he immediately reinterprets the field of phenomena discovered by him after a Cartesian model, by declaring the modes of givenness to be moments of the life of consciousness (*modi cogitandi*), and consciousness to be a property (or rather *the* property) of the pure *ego*. The transcendental subject, the possessor of consciousness, cannot be objectified in reflection, but *Dasein,* in the "absolute responsibility" of *transcendental philosophical* self-interpretation, in the action which the tradition turns into the ἦθος of philosophising, projects itself towards subjectivity understood in the Cartesian manner. *Dasein*, as philosophical "being-there," schematises in reflection its presence for itself (self-awareness) in a certain way, *viz.* in compliance with the "basic axiom of classical ontology," i. e., in imitation of absolute positivity. The absolute temporal flow is the form, the trace of this special self-projection, the ultimate form of self-objectifying *Dasein* is capable of. Yet it is the most fundamental condition of self-projecting as such, the primordial ecstatic horizontal temporality, that underlies the projection-towards-subjectivity.

2.7. *Possibility of All That Which Is Possible*

The "there" (*das Da)* belonging to *Dasein* is not, as we have seen, a *praesentia — praesentia, praesens* is just the schematised horizon of one of the three ecstases of temporality — but rather the *proximitas,* the attainable near. It is disclosed and gets its internal topography in a certain initial effort named by Heidegger ἔκστασις, the "*ecstasis* of temporality." It is true that the word *effort* is misleading since it has the classical connotations — δύναμις, *potentia,* i. e., refers to possibility, which has yet to become actuality. In a sense, the *ecstasis* already *is* an actuality, ἐνέργεια of the being of *Dasein*[41] and at the same time its impossibility of being objectified = actualised as an *essentia* understood traditionally. The tripartite unity of the temporal *ecstases* is not a *potentia* related to an *actualitas*, it is not a possibility, and yet it underlies every "I can" and the ability-to-be (*Seinkönnen*) as its root: indeed, the existential "I can be" underlies every "I can."

As a modal category of the presence-at-hand, possibility signifies what is *not yet* actual and what is *not at any time* necessary. It characterizes the *merely* possible. Ontologically it is on a lower level than actuality and necessity. On the other hand, possibility as an *existentiale* is the most primordial and the ultimate positive way in which *Dasein* is characterized ontologically. (SZ 143f.)

Thomas Aquinas defines being (*esse*) as actuality of all activities (*actualitas omnium actuum*).[42] The temporality of *Dasein* is the possibility of all that is possible or, perhaps, to give more expression to the phrase, the ability of all possibilities.

If we look back now, we shall discover that the answer to the question about the foundation and the primordial meaning of the difference of *being-what* and *being-how* for the entity-within-the-world is already found — more than that, it is already articulated. The distinction of being-what and being-how depends on singling out the *ecstasis* of the present and its horizontal schema (*praesens*) against the primal horizontal ecstatic *unity* of temporality. Being-how of the present-at-hand and of the ready-to-hand is outlined towards the *praesens* as the horizon of the *ecstasis* of the present.

The difference of the temporal ecstases and of the horizons belonging to them serves as the fundamental condition of the difference of being-what and being-how of the ready-to-hand. The difference of the "now"-consciousness and its retentional and protentional modifications serves as the fundamental condition of the difference of being-what and being-how of the present-at-hand. This difference depends, in its turn, on the difference of the temporal ecstases of the primordial temporality and of the horizons belonging to them.

---

[41] In the fundamental sense corresponding to the Aristotelian ζῇ καὶ ἔζηκεν [ἅμα] (*Metaph.* 1048b27). Cf. chap. 2, sect. 2.

[42] *De potentia* 7, 2.

Because of its ecstatic character, temporality makes possible the being of an entity which as a Self deals existently (*existierend*) with others and, as thus existent, deals with entities as ready-to-hand or as present-at-hand. Temporality makes possible *Dasein*'s comportment as a comportment towards entities, whether towards itself, towards others, or towards the ready-to-hand or the present-at-hand. Because of the unity of the horizontal schemata that belongs to its ecstatic unity, temporality makes possible the understanding of being, so that it is only in the light of this understanding of being that Dasein comports itself towards its own self, towards others as entities, and towards the present-at-hand as entity. (GP 452)

The distinction between being and entity *is there* (*ist da*), latent in *Dasein* and its existence, even if not in explicit awareness. The distinction is there; that is to say, it has the mode of being of *Dasein*: it belongs to existence. Existence means, as it were, "to be in the performance of this distinction." Only a soul that can make this distinction has the aptitude, going beyond the animal's soul, to become the soul of a human being. The distinction between being and entity is temporalized in the temporalizing of temporality. (GP 454)

2.8. *Differentia Differens*

Already Kant asserted that the temporal difference is a condition of all difference experienced.[43] In whatever way we may understand the fundamental meaning of our Self, be it according to Husserl, as the transcendental *ego*, or according to Heidegger, as existence, we must recognize that temporality is not something discovered by the subject (understood by *Dasein*) one way or another, but the condition itself of all discovering and understanding, the internal form of the life of consciousness (the fundamental structure of *Existenz*). Like the Schoolmen opposing the form forming to the form formed, we could say that the temporal primal distinction is *a distinguishing distinction, differentia differens.*

It remains to be understood what the distinction of being-what and being-how could mean for the being of *Dasein* itself. Yet the existential analytic of *Dasein* begins with an answer to this question (not yet asked): "The 'essence' of this entity lies in its 'to be.' Its being-what [-it-is] (*essentia*) must, so far as we can speak of it at all, be conceived in terms of its being (*existentia*)" (SZ 42). Thus, *Dasein* is, strictly speaking, an entity for which *being-what* and *being-how* coincide. After the existential analytic has grown to become fundamental ontology and thematizes primordial temporality, a new layer of meaning of this statement is disclosed to us. On the one hand, primordial temporality is the deepest foundation of existence, for temporality is the ontological meaning of the central *existentiale* of care. On the other hand, the primordial temporality, as temporalizing the temporal *ecstases*, is the *differentia differens*, the ultimate condition of possibility of any difference. It follows hence that it is impossible to posit the ontological difference (the difference between being-what-it-is and being-how-it-is) for *Dasein* itself (if it is understood in the primordial way), because it is itself in its being the "performance" (*Vollzug*) of this difference, the *actus differendi.*[44]

---

[43] *Critique of Pure Reason*, A 99.

[44] Because *Dasein* is itself the performance of the ontological difference, it is the only being for which we can state the ontological indifference, i. e., the identity of *being-what and being-how.* Imitating Hegel's language one might say that *Dasein* is the identity (indifference) of difference and indifference.

One reservation, however, is necessary here. *Dasein* itself asks the "existential question" about the "who" of existence (SZ §25) and thereby posits its own "whoness" as questioned. It is not without reason that Heidegger terms this question "existential," for it implies the "anticipatory resoluteness," i. e., by asking this question *Dasein* dares to bear witness to itself. The question about the "who" of existence is always thrown into the future. It is out of the open horizon of the future that the answer will (or will not) come. "The 'essence' of this entity lies in its 'to-be' (*in seinem Zu-sein*)..." On his copy of *Being and Time* Heidegger notes in the margin next to this "to be": ... *daß es zu seyn "hat"; Bestimmung! Dasein has to be* as the future, has to set forth as the forthcoming. That is why the *ecstasis* of the future is singled out from the original unity of the temporal *ecstases* while posing the existential question: "who?" *Dasein* in the form of the existential question poses the difference of being-who-it-is and being-how = existence as such. And this difference, questioned, implies a singling-out of the *ecstasis* of the future and, as a condition of it, a dividing, a differentiating the temporal *ecstases* and the horizons belonging to them.

The primordial difference of the temporal *ecstases*, temporalizing itself, is a distinction before that which is distinguished. The temporal form, the possibility of distinguishing between *before* and *after* is born of this proto-difference, of the distinguishing distinction.

Language, in a wonderful way, allows us to hear this proto-difference: the Latin verb *differre* means at once "to carry in different directions," "to distinguish" and "to postpone til another time" (*differre rem in aliud tempus*), i. e., *to temporize. Differentia differt.* The difference temporizes and temporalizes (itself) in the ecstatic temporality of *Dasein*. Heidegger says: *Der Unterschied von Sein und Seiendem ist in der Zeitigung der Zeitlichkeit gezeitigt.*

BIBLIOGRAPHY

Aubenque, P. *Le problème d'être chez Aristote*. Paris: PUF, 1960.

Barbotin, E. *La théorie Aristotélicienne de l'intellect d'après Théophraste*. Paris: PUF, 1954.

Bassenge, F. "Das τὸ ἑνὶ εἶναι, τὸ ἀγαθῷ εἶναι etc. etc. und das τὸ τί ἦν εἶναι bei Aristoteles." *Philologus*, **104** (1960), pp. 14–47, 201–222.

Boer, K. de, *Thinking in the Light of Time. Heidegger's Encounter with Hegel*. New York: SUNY Press, 2000.

Boer, Th. de, *The Rationality of Transcendence. Studies in Philosophy of Emmanuel Levinas*. Amsterdam: J. C. Gieben, 1997.

Boethius. "Liber de persona et duabus naturis." In *Traités théologiques* (édition bilingue). Paris: GF Flammarion, 2000.

Boethius. "Quomodo substantiae in eo quod sint bonae sint cum non sint substantialia bona." In *Traités théologiques* (édition bilingue). Paris: GF Flammarion, 2000.

Brague, R. *Aristote et la question du monde. Essais sur le contexte cosmologique et anthropologique de l'ontologie*. Paris: PUF, 1988.

Brague, R. *Du temps chez Platon et Aristote. Quatre études*. Paris : PUF, 1982.

Brentano, F. *Von der mannigfachen Bedeutung des Seienden nach Aristoteles*. Freiburg im Breisgau: Herder'sche Verlagshandlung, 1862.

Cavagnaro, E. *Aristotele e il tempo*. Leeuwarden, 1995.

Chernyakov, A. G. "Das Schicksal des Subjektbegriffs in *Sein und Zeit*." *Reihe der Österreichischen Gesellschaft für Phänomenologie*, Bd. 3 (1998): "Siebzig Jahre *Sein und Zeit*." Ed. H. Vetter. Frankfurt a.M., etc.: Peter Lang, pp. 175–188.

Chernyakov, A. G. "Husserl's 'Genealogy of Logic,' Space-Constitution, and Noetic Geometry." *Recherches husserliennes*, **7** (1997), pp. 61–85.

Claesges, U. E. *Husserl's Theorie der Raumkonstitution*. Köln, 1963.

Conen, P. F. *Die Zeittheorie des Aristoteles*. München: C. H. Beck'sche Verlagsbuchhandlung, 1964.

Coujolle-Zaslawsky, Fr. "Aristote: Sur quelques traductions récents du TO TI HN EINAI." *Revue de Théologie et de Philosophie*, **113** (1981), pp. 61–75.

Coxon, A. H. *The Fragments of Parmenides. A Critical Text with Introduction, Translation, the Ancient Testimonia and Commentary by A. H. Coxon*. (*Phronesis*: Supplementary Volumes, Vol. III.) Assen, etc.: Van Gorcum, 1986.

Craig, E. *The Mind of God and the Works of Man*. Oxford: Clarendon Press, 1987.

Derrida, J. "*Ousia* and *gramme*: Note on a Note from *Being and Time*." Trans. A. Bass. In *Margins of Philosophy*. New York, etc.: Harvester Wheatsheaf, 1982, pp.29–67.

Derrida, J. "White Mythology: Metaphor in the Text of Philosophy." Trans. A. Bass. *Ibid.*, pp. 207–271.

Fichte, J. G. *Wissenschaftslehre nova methodo*. Kollegnachschrift K. Chr. Fr. Krause. Hamburg: F. Meiner Verlag, 1982.

Figal, G. *M. Heidegger. Phänomenologie der Freiheit*. Athenäum, 1988.

Gadamer, H.-G. "Martin Heidegger und die Marburger Theologie." In *Heidegger. Perspektiven zur Deutung seiner Werks*. Ed. Otto Pöggeler. 3. Aufl. Weinheim, 1994.

Haardt, A. "Selbstbesinnung und Selbstbewußtsein in Edmund Husserls *Cartesianische Meditationen*." In *Cognitio humana – Dynamik des Wissens und der Werte*. Ed. Ch. Hubig. Berlin: Akademie Verlag, 1997, pp. 433–450.

Hegel, G. W. F. *Ästhetik*. Bd. I. Ed. Fr. Bassenge. Berlin: Aufbau-Verlag, 1976.

Hegel, G. W. F. *Die Vernunft in der Geschichte*. Sämtliche Werke. Ed. G. Lasson. Bd. 8. Leipzig: F. Meiner Verlag, 1923.

Hegel, G.W.F. *Phänomenologie des Geistes*. Ed. J. Hoffmeister. Berlin: Akademie Verlag, 1971.

Heidegger, M. *Aristoteles, Metaphysik Θ 1 –3. Von Wesen und Wirklichkeit der Kraft*. GA 35. 2. Aufl. Frankfurt a. M.: V. Klostermann, 1990.

– *Beiträge zur Philosophie*. GA 65. Frankfurt a. M.: V. Klostermann, 1994.

– "Das Ende der Philosophie und die Aufgabe des Denkens." In *Zur Sache des Denkens*. Tübingen: M. Niemeyer Verlag, 1969, pp. 61–80.

– *Denkerfahrungen 1910–1976*. Frankfurt a. M.: V. Klostermann, 1983.

– *Die Grundprobleme der Phänomenologie*. GA 24. Frankfurt a. M.: V. Klostermann, 1989.

– *Einführung in die Metaphysik*. GA 40. Frankfurt a. M.: V. Klostermann, 1983.

– *Erläuterung der "Einleitung" zu Hegels "Phänomenologie des Geistes."* GA 68. Frankfurt a. M.: V. Klostermann, 1993.

– *Kant und das Problem der Metaphysik*. Frankfurt a. M.: V. Klostermann, 1973.

– *Metaphysische Anfangsgründe der Logik im Ausgang von Leibniz*. GA 26. Frankfurt a. M.: V. Klostermann, 1978.

– *Parmenides*. GA 54. Frankfurt a. M.: V. Klostermann, 1982.

– *Platon: Sophistes*. GA 19. Frankfurt a. M.: V. Klostermann, 1992.

– *Prolegomena zur Geschichte des Zeitbegriffs*. GA 20. 2. Aufl. Frankfurt a. M.: V. Klostermann, 1988.

– *Sein und Zeit*. 16. Aufl. Tübingen: Max Niemeyer Verlag, 1986.

– "Zeit und Sein." In *Zur Sache des Denkens*. Tübingen: Max Niemeyer Verlag, 1969.

Held, K. "Husserls Rückgang auf das φαινόμενον und die geschichtliche Stellung der Phänomenologie." *Phänomenologische Forschungen*, **10** (1980), pp. 89–145.

– "Phänomenologie der Zeit nach Husserl." *Perspektiven der Philosophie. Neues Jahrbuch*, **7** (1981), pp. 185–221.

Hoche, H.-U. *Handlung, Bewußtsein und Leib. Vorstudien zu einer rein noematischen Phänomenologie*. Freiburg/München: Karl Alber, 1973.

Husserl, E. *Analysen zur passiven Synthesis. Aus Vorlesungs- und Forschungsmanuskripten, 1918–1926*. Ed. M. Fleischer. Hua XI. Den Haag, 1966.

– *Die Idee der Phänomenologie. Fünf Vorlesungen*. Ed. W. Biemel. Hua II. Den Haag, 1950.

– *Die Krisis der europäischen Wissenschaften und die transzendentale Phänomenologie. Eine Einleitung in die phänomenologische Philosophie*. Ed. W. Biemel. Hua VI. Den Haag 1954 (2. Aufl. 1962).

– *Erfahrung und Urteil*. Ed. L. Landgrebe. Hamburg: F. Meiner Verlag, 1985.

– *Erste Philosophie*. 2. Teil: *Theorie der phänomenologischen Reduktion*. Ed. R. Boehm. Hua VIII. Den Haag, 1959.

– *Formale und transzendentale Logik*. Halle, 1929.

– *Ideen zu einer reinen Phänomenologie und phänomenologischen Philosophie*. 5. Aufl. Tübingen: Max Niemeyer Verlag, 1993.

– *Ideen zu einer reinen Phänomenologie und phänomenologischen Philosophie*. Zweites Buch: *Phänomenologische Untersuchungen zur Konstitution*. Ed. M. Biemel. Hua IV. Den Haag, 1952.

– *Ideen zu einer reinen Phänomenologie und phänomenologischen Philosophie*. Drittes Buch: *Die Phänomenologie und die Fundamente der Wissenschaften*. Ed. M. Biemel. Hua V. Den Haag, 1952.

– *Logische Untersuchungen. Zweiter Band*, 2. Teil. Ed. U. Panzer. Hua XIX/2. Den Haag, etc., 1984.

– *Phänomenologische Psychologie*. Ed. W. Biemel. Hua IX. Den Haag, 1962.

– *Zur Phänomenologie des inneren Zeitbewußtsein* (1893–1917). Ed. R. Boehm. Hua X. Den Haag, 1959.

Kahn, Ch. "Questions and Categories." In *Questions*. Ed. H. Hiż. Dordrecht, 1978.

Kant, I. *Kritik der reinen Vernunft*. Ed. R. Schmidt. Hamburg: F. Meiner Verlag, 1990.

Kaulbach, F. *Der philosophische Begriff der Bewegung*. Köln-Graz: Böhlau Verlag, 1965.

Landgrebe, L. "The Phenomenology of Corporeality and the Problem of Matter." In *Phenomenology of E. Husserl. Six Essais*. Ed. D. Welton. Cornell Univ. Press, 1981, pp. 33–49.

Landgrebe, L. "Husserl's Departure from Cartesianism." *Ibid.*, pp. 66–121

Landgrebe, L. "The Problem of Passive Constitution." *Ibid.*, pp. 50–65.

Leibniz G. W. F. "Meditationes de cognitione, veritate et ideis." In *Die philosophische Schriften von G. W. F. Leibniz*. Ed. G. L. Gerhardt. Bd. 4. Hildesheim: Georg Olms, 1960, pp. 422–426.

Levinas, E. *Autrement qu'être ou au-delà de l'essence*. Phaenomenologica 54. Den Haag: M. Nijhoff, 1974.

Lossky, V. *The vision of God*. St.Vladimir Seminary Press, 1983.

Marion J.-L. "The Final Appeal of the Subject." In *Deconstructive Subjectivities*. Eds. S. Critchley and P. Dews. New York: SUNY Press, 1996.

Mertens, K. *Zwischen Letztbegründung und Skepsis. Kritische Untersuchungen zum Selbstverständnis der Phänomenologie Edmund Husserls*. Orbis Phaenomenologicus. Abt. VI, Bd. I. Freiburg/München: Karl Alber, 1996.

Miller, I. *Husserl, Perception, and Temporal Awareness*. Bradford Books. Cambridge, Mass.: MIT Press, 1984.

Oehler, K. *Die Lehre vom noetischen und dianoetischen Denken bei Platon und Aristoteles*. Hamburg: F. Meiner Verlag, 1985.

Patočka, J. "Der Subjektivismus der Husserlschen und die Möglichkeit einer 'asubjektiven' Phänomenologie." In *Philosophische Perspektiven*. Bd. 2. Frankfurt a. M., 1970.

Pöggeler, O. *Heidegger in seiner Zeit*. München: W. Fink Verlag, 1999.

Ricœur, P. *Oneself as Another*. Chicago&London: Univ. of Chicago Press, 1992.

Ricœur, P. *The Rule of Metaphor*. London: Routledge, 1994.

Ricœur, P. *Time and Narrative*. Vol. I–II. Chicago&London: Univ. of Chicago Press, 1984–1985.

Ruello, Fr. *Les "noms divins" et leur "raisons" selon St. Albert le Grand, commentateur du "De divinis nominibus."* Paris: Librairie philosophique J. Vrin, 1963.

Schmitz, H. *Die Ideenlehre des Aristoteles*. Bd. I, 1. Teil: *Kommentar zum 7. Buch der Metaphysik*. Bonn: Bouvier Verlag, 1985.

Schmitz, H. *Die Ideenlehre des Aristoteles*. Bd. I, 2. Teil: *Ontologie. Noologie. Theologie*. Bonn: Bouvier Verlag, 1985.

Snell, B. *Die Entdeckung des Geistes. Studien zur Entstehung des europäischen Denken bei den Griechen*. 3. Aufgabe. Hamburg, 1955.

Sokolowski, R. *Husserlian Meditations. How Words Present Things*. Evanston: North-western Univ. Press, 1974.

Suarez Fr. *Disputationes metaphysicae*. Opera omnia. Ed. C. Berton. T. XXV, XXVI. Paris, 1861.

Thomas Aquinas, *De ente et essentia*. Ed. 3. Torino: Marietti, 1957.

Torstrik, A. "ὅ ποτε ὄν: Ein Beitrag zur Kenntnis des aristotelischen Sprachgebrauch." *Rheinisches Museum für Philologie*, **12** (1857), pp. 161–173.

Tugendhat, E. *Der Wahrheitsbegriff bei Husserl und Heidegger*. 2. Aufl. Berlin, 1970

Tugendhat, E. *Selbstbewußtsein und Selbstbestimmung. Sprachanalytische Interpretationen*. Frankfurt a. M., 1979.

Tugendhat, E. TI KATA TINOΣ, *Eine Untersuchung zu Struktur und Ursprung aristotelischer Grundbegriffe*. Freiburg/München: Karl Alber, 1968.

Waterlow, B. S. "Aristotle's Now." *Philosophical Quarterly*, **34** (1982), # 135, pp. 104–128.

Wolff, Ch. *Philosophia prima sive ontologia*. Frankfurt, 1730.

INDEX OF NAMES

Aristotle, Husserl and Heidegger are not included in this index
as their names appear throughout the entire volume

SUBJECT INDEX

Phaenomenologica

1. E. Fink: *Sein, Wahrheit, Welt.* Vor-Fragen zum Problem des Phänomen-Begriffs. 1958 ISBN 90-247-0234-8
2. H.L. van Breda and J. Taminiaux (eds.): *Husserl et la pensée moderne / Husserl und das Denken der Neuzeit.* Actes du deuxième Colloque International de Phénoménologie / Akten des zweiten Internationalen Phänomenologischen Kolloquiums (Krefeld, 1.–3. Nov. 1956). 1959 ISBN 90-247-0235-8
3. J.-C. Piguet: *De l'esthétique à la métaphysique.* 1959 ISBN 90-247-0236-4
4. *E. Husserl: 1850–1959.* Recueil commémoratif publié à l'occasion du centenaire de la naissance du philosophe. 1959 ISBN 90-247-0237-2

5/6. H. Spiegelberg: *The Phenomenological Movement.* A Historical Introduction. 3rd revised ed. with the collaboration of Karl Schuhmann. 1982 ISBN Hb: 90-247-2577-1; Pb: 90-247-2535-6

7. A. Roth: *Edmund Husserls ethische Untersuchungen.* Dargestellt anhand seiner Vorlesungsmanuskripte. 1960 ISBN 90-247-0241-0
8. E. Levinas: *Totalité et infini.* Essai sur l'extériorité. 4th ed., 4th printing 1984 ISBN Hb: 90-247-5105-5; Pb: 90-247-2971-8
9. A. de Waelhens: *La philosophie et les expériences naturelles.* 1961 ISBN 90-247-0243-7
10. L. Eley: *Die Krise des Apriori in der transzendentalen Phänomenologie Edmund Husserls.* 1962 ISBN 90-247-0244-5
11. A. Schutz: *Collected Papers, I.* The Problem of Social Reality. Edited and introduced by M. Natanson. 1962; 5th printing: 1982 ISBN Hb: 90-247-5089-X; Pb: 90-247-3046-5
    Collected Papers, II *see* below under Volume 15
    Collected Papers, III *see* below under Volume 22
    Collected Papers, IV *see* below under Volume 136
12. J.M. Broekman: *Phänomenologie und Egologie.* Faktisches und transzendentales Ego bei Edmund Husserl. 1963 ISBN 90-247-0245-3
13. W.J. Richardson: *Heidegger. Through Phenomenology to Thought.* Preface by Martin Heidegger. 1963; 3rd printing: 1974 ISBN 90-247-02461-1
14. J.N. Mohanty: *Edmund Husserl's Theory of Meaning.* 1964; reprint: 1969 ISBN 90-247-0247-X
15. A. Schutz: *Collected Papers, II.* Studies in Social Theory. Edited and introduced by A. Brodersen. 1964; reprint: 1977 ISBN 90-247-0248-8
16. I. Kern: *Husserl und Kant.* Eine Untersuchung über Husserls Verhältnis zu Kant und zum Neukantianismus. 1964; reprint: 1984 ISBN 90-247-0249-6
17. R.M. Zaner: *The Problem of Embodiment.* Some Contributions to a Phenomenology of the Body. 1964; reprint: 1971 ISBN 90-247-5093-8
18. R. Sokolowski: *The Formation of Husserl's Concept of Constitution.* 1964; reprint: 1970 ISBN 90-247-5086-5
19. U. Claesges: *Edmund Husserls Theorie der Raumkonstitution.* 1964 ISBN 90-247-0251-8
20. M. Dufrenne: *Jalons.* 1966 ISBN 90-247-0252-6
21. E. Fink: *Studien zur Phänomenologie, 1930–1939.* 1966 ISBN 90-247-0253-4
22. A. Schutz: *Collected Papers, III.* Studies in Phenomenological Philosophy. Edited by I. Schutz. With an introduction by Aaron Gurwitsch. 1966; reprint: 1975 ISBN 90-247-5090-3
23. K. Held: *Lebendige Gegenwart.* Die Frage nach der Seinsweise des transzendentalen Ich bei Edmund Husserl, entwickelt am Leitfaden der Zeitproblematik. 1966 ISBN 90-247-0254-2
24. O. Laffoucrière: *Le destin de la pensée et 'La Mort de Dieu' selon Heidegger.* 1968 ISBN 90-247-0255-0
25. E. Husserl: *Briefe an Roman Ingarden.* Mit Erläuterungen und Erinnerungen an Husserl. Hrsg. von R. Ingarden. 1968 ISBN Hb: 90-247-0257-7; Pb: 90-247-0256-9
26. R. Boehm: *Vom Gesichtspunkt der Phänomenologie* (I). Husserl-Studien. 1968 ISBN Hb: 90-247-0259-3; Pb: 90-247-0258-5
    For *Band II* see below under Volume 83

Phaenomenologica

27. T. Conrad: *Zur Wesenslehre des psychischen Lebens und Erlebens.* Mit einem Geleitwort von H.L. van Breda. 1968 ISBN 90-247-0260-7
28. W. Biemel: *Philosophische Analysen zur Kunst der Gegenwart.* 1969 ISBN Hb: 90-247-0263-1; Pb: 90-247-0262-3
29. G. Thinès: *La problématique de la psychologie.* 1968 ISBN Hb: 90-247-0265-8; Pb: 90-247-0264-X
30. D. Sinha: *Studies in Phenomenology.* 1969 ISBN Hb: 90-247-0267-4; Pb: 90-247-0266-6
31. L. Eley: *Metakritik der formalen Logik.* Sinnliche Gewissheit als Horizont der Aussagenlogik und elementaren Prädikatenlogik. 1969 ISBN Hb: 90-247-0269-0; Pb: 90-247-0268-2
32. M.S. Frings: *Person und Dasein.* Zur Frage der Ontologie des Wertseins. 1969 ISBN Hb: 90-247-0271-2; Pb: 90-247-0270-4
33. A. Rosales: *Transzendenz und Differenz.* Ein Beitrag zum Problem der ontologischen Differenz beim frühen Heidegger. 1970 ISBN 90-247-0272-0
34. M.M. Saraiva: *L'imagination selon Husserl.* 1970 ISBN 90-247-0273-9
35. P. Janssen: *Geschichte und Lebenswelt.* Ein Beitrag zur Diskussion von Husserls Spätwerk. 1970 ISBN 90-247-0274-7
36. W. Marx: *Vernunft und Welt.* Zwischen Tradition und anderem Anfang. 1970 ISBN 90-247-5042-3
37. J.N. Mohanty: *Phenomenology and Ontology.* 1970 ISBN 90-247-5053-9
38. A. Aguirre: *Genetische Phänomenologie und Reduktion.* Zur Letztbegründung der Wissenschaft aus der radikalen Skepsis im Denken E. Husserls. 1970 ISBN 90-247-5025-3
39. T.F. Geraets: *Vers une nouvelle philosophie transcendentale.* La genèse de la philosophie de Maurice Merleau-Ponty jusqu'à la 'Phénoménologie de la perception.' Préface par E. Levinas. 1971 ISBN 90-247-5024-5
40. H. Declève: *Heidegger et Kant.* 1970 ISBN 90-247-5016-4
41. B. Waldenfels: *Das Zwischenreich des Dialogs.* Sozialphilosophische Untersuchungen in Anschluss an Edmund Husserl. 1971 ISBN 90-247-5072-5
42. K. Schuhmann: *Die Fundamentalbetrachtung der Phänomenologie.* Zum Weltproblem in der Philosophie Edmund Husserls. 1971 ISBN 90-247-5121-7
43. K. Goldstein: *Selected Papers/Ausgewählte Schriften.* Edited by A. Gurwitsch, E.M. Goldstein Haudek and W.E. Haudek. Introduction by A. Gurwitsch. 1971 ISBN 90-247-5047-4
44. E. Holenstein: *Phänomenologie der Assoziation.* Zu Struktur und Funktion eines Grundprinzips der passiven Genesis bei E. Husserl. 1972 ISBN 90-247-1175-4
45. F. Hammer: *Theonome Anthropologie?* Max Schelers Menschenbild und seine Grenzen. 1972 ISBN 90-247-1186-X
46. A. Pažanin: *Wissenschaft und Geschichte in der Phänomenologie Edmund Husserls.* 1972 ISBN 90-247-1194-0
47. G.A. de Almeida: *Sinn und Inhalt in der genetischen Phänomenologie E. Husserls.* 1972 ISBN 90-247-1318-8
48. J. Rolland de Renéville: *Aventure de l'absolu.* 1972 ISBN 90-247-1319-6
49. U. Claesges und K. Held (eds.): *Perspektiven transzendental-phänomenologischer Forschung.* Für Ludwig Landgrebe zum 70. Geburtstag von seiner Kölner Schülern. 1972 ISBN 90-247-1313-7
50. F. Kersten and R. Zaner (eds.): *Phenomenology: Continuation and Criticism.* Essays in Memory of Dorion Cairns. 1973 ISBN 90-247-1302-1
51. W. Biemel (ed.): *Phänomenologie Heute.* Festschrift für Ludwig Landgrebe. 1972 ISBN 90-247-1336-6
52. D. Souche-Dagues: *Le développement de l'intentionnalité dans la phénoménologie husserlienne.* 1972 ISBN 90-247-1354-4
53. B. Rang: *Kausalität und Motivation.* Untersuchungen zum Verhältnis von Perspektivität und Objektivität in der Phänomenologie Edmund Husserls. 1973 ISBN 90-247-1353-6
54. E. Levinas: *Autrement qu'être ou au-delà de l'essence.* 2nd. ed.: 1978 ISBN 90-247-2030-3
55. D. Cairns: *Guide for Translating Husserl.* 1973 ISBN Pb: 90-247-1452-4

Phaenomenologica

56. K. Schuhmann: *Die Dialektik der Phänomenologie, I.* Husserl über Pfänder. 1973 ISBN 90-247-1316-1
57. K. Schuhmann: *Die Dialektik der Phänomenologie, II.* Reine Phänomenologie und phänomenologische Philosophie. Historisch-analytische Monographie über Husserls 'Ideen I'. 1973 ISBN 90-247-1307-2
58. R. Williame: *Les fondements phénoménologiques de la sociologie compréhensive: Alfred Schutz et Max Weber.* 1973 ISBN 90-247-1531-8
59. E. Marbach: *Das Problem des Ich in der Phänomenologie Husserls.* 1974 ISBN 90-247-1587-3
60. R. Stevens: *James and Husserl: The Foundations of Meaning.* 1974 ISBN 90-247-1631-4
61. H.L. van Breda (ed.): *Vérité et Vérification / Wahrheit und Verifikation.* Actes du quatrième Colloque International de Phénoménologie / Akten des vierten Internationalen Kolloquiums für Phänomenologie (Schwabisch Hall, Baden-Württemberg, 8.–11. September 1969). 1974 ISBN 90-247-1702-7
62. Ph.J. Bossert (ed.): *Phenomenological Perspectives.* Historical and Systematic Essays in Honor of Herbert Spiegelberg. 1975. ISBN 90-247-1701-9
63. H. Spiegelberg: *Doing Phenomenology.* Essays on and in Phenomenology. 1975 ISBN 90-247-1725-6
64. R. Ingarden: *On the Motives which Led Husserl to Transcendental Idealism.* 1975 ISBN 90-247-1751-5
65. H. Kuhn, E. Avé-Lallemant and R. Gladiator (eds.): *Die Münchener Phänomenologie.* Vorträge des Internationalen Kongresses in München (13.–18. April 1971). 1975 ISBN 90-247-1740-X
66. D. Cairns: *Conversations with Husserl and Fink.* Edited by the Husserl-Archives in Louvain. With a foreword by R.M. Zaner. 1975 ISBN 90-247-1793-0
67. G. Hoyos Vásquez: *Intentionalität als Verantwortung.* Geschichtsteleologie und Teleologie der Intentionalität bei Husserl. 1976 ISBN 90-247-1794-9
68. J. Patočka: *Le monde naturel comme problème philosophique.* 1976 ISBN 90-247-1795-7
69. W.W. Fuchs: *Phenomenology and the Metaphysics of Presence.* An Essay in the Philosophy of Edmund Husserl. 1976 ISBN 90-247-1822-8
70. S. Cunningham: *Language and the Phenomenological Reductions of Edmund Husserl.* 1976 ISBN 90-247-1823-6
71. G.C. Moneta: *On Identity.* A Study in Genetic Phenomenology. 1976 ISBN 90-247-1860-0
72. W. Biemel und das Husserl-Archiv zu Löwen (eds.): *Die Welt des Menschen – Die Welt der Philosophie.* Festschrift für Jan Patočka. 1976 ISBN 90-247-1899-6
73. M. Richir: *Au-delà du renversement copernicien.* La question de la phénoménologie et son fondement. 1976 ISBN 90-247-1903-8
74. H. Mongis: *Heidegger et la critique de la notion de valeur.* La destruction de la fondation métaphysique. Lettre-préface de Martin Heidegger. 1976 ISBN 90-247-1904-6
75. J. Taminiaux: *Le regard et l'excédent.* 1977 ISBN 90-247-2028-1
76. Th. de Boer: *The Development of Husserl's Thought.* 1978 ISBN Hb: 90-247-2039-7; Pb: 90-247-2124-5
77. R.R. Cox: *Schutz's Theory of Relevance.* A Phenomenological Critique. 1978 ISBN 90-247-2041-9
78. S. Strasser: *Jenseits von Sein und Zeit.* Eine Einführung in Emmanuel Levinas' Philosophie. 1978 ISBN 90-247-2068-0
79. R.T. Murphy: *Hume and Husserl.* Towards Radical Subjectivism. 1980 ISBN 90-247-2172-5
80. H. Spiegelberg: *The Context of the Phenomenological Movement.* 1981 ISBN 90-247-2392-2
81. J.R. Mensch: *The Question of Being in Husserl's Logical Investigations.* 1981 ISBN 90-247-2413-9
82. J. Loscerbo: *Being and Technology.* A Study in the Philosophy of Martin Heidegger. 1981 ISBN 90-247-2411-2
83. R. Boehm: *Vom Gesichtspunkt der Phänomenologie II.* Studien zur Phänomenologie der Epoché. 1981 ISBN 90-247-2415-5

Phaenomenologica

84. H. Spiegelberg and E. Avé-Lallemant (eds.): *Pfänder-Studien*. 1982 ISBN 90-247-2490-2
85. S. Valdinoci: *Les fondements de la phénoménologie husserlienne*. 1982 ISBN 90-247-2504-6
86. I. Yamaguchi: *Passive Synthesis und Intersubjektivität bei Edmund Husserl*. 1982 ISBN 90-247-2505-4
87. J. Libertson: *Proximity*. Levinas, Blanchot, Bataille and Communication. 1982 ISBN 90-247-2506-2
88. D. Welton: *The Origins of Meaning*. A Critical Study of the Thresholds of Husserlian Phenomenology. 1983 ISBN 90-247-2618-2
89. W.R. McKenna: *Husserl's 'Introductions to Phenomenology'*. Interpretation and Critique. 1982 ISBN 90-247-2665-4
90. J.P. Miller: *Numbers in Presence and Absence*. A Study of Husserl's Philosophy of Mathematics. 1982 ISBN 90-247-2709-X
91. U. Melle: *Das Wahrnehmungsproblem und seine Verwandlung in phänomenologischer Einstellung*. Untersuchungen zu den phänomenologischen Wahrnehmungstheorien von Husserl, Gurwitsch und Merleau-Ponty. 1983 ISBN 90-247-2761-8
92. W.S. Hamrick (ed.): *Phenomenology in Practice and Theory*. Essays for Herbert Spiegelberg. 1984 ISBN 90-247-2926-2
93. H. Reiner: *Duty and Inclination*. The Fundamentals of Morality Discussed and Redefined with Special Regard to Kant and Schiller. 1983 ISBN 90-247-2818-6
94. M.J. Harney: *Intentionality, Sense and the Mind*. 1984 ISBN 90-247-2891-6
95. Kah Kyung Cho (ed.): *Philosophy and Science in Phenomenological Perspective*. 1984 ISBN 90-247-2922-X
96. A. Lingis: *Phenomenological Explanations*. 1986 ISBN Hb: 90-247-3332-4; Pb: 90-247-3333-2
97. N. Rotenstreich: *Reflection and Action*. 1985 ISBN Hb: 90-247-2969-6; Pb: 90-247-3128-3
98. J.N. Mohanty: *The Possibility of Transcendental Philosophy*. 1985 ISBN Hb: 90-247-2991-2; Pb: 90-247-3146-1
99. J.J. Kockelmans: *Heidegger on Art and Art Works*. 1985 ISBN 90-247-3102-X
100. E. Lévinas: *Collected Philosophical Papers*. 1987 ISBN Hb: 90-247-3272-7; Pb: 90-247-3395-2
101. R. Regvald: *Heidegger et le problème du néant*. 1986 ISBN 90-247-3388-X
102. J.A. Barash: *Martin Heidegger and the Problem of Historical Meaning*. 1987 ISBN 90-247-3493-2
103. J.J. Kockelmans (ed.): *Phenomenological Psychology*. The Dutch School. 1987 ISBN 90-247-3501-7
104. W.S. Hamrick: *An Existential Phenomenology of Law: Maurice Merleau-Ponty*. 1987 ISBN 90-247-3520-3
105. J.C. Sallis, G. Moneta and J. Taminiaux (eds.): *The Collegium Phaenomenologicum. The First Ten Years*. 1988 ISBN 90-247-3709-5
106. D. Carr: *Interpreting Husserl*. Critical and Comparative Studies. 1987. ISBN 90-247-3505-X
107. G. Heffernan: *Isagoge in die phänomenologische Apophantik*. Eine Einführung in die phänomenologische Urteilslogik durch die Auslegung des Textes der *Formalen und transzendenten Logik* von Edmund Husserl. 1989 ISBN 90-247-3710-9
108. F. Volpi, J.-F. Mattéi, Th. Sheenan, J.-F. Courtine, J. Taminiaux, J. Sallis, D. Janicaud, A.L. Kelkel, R. Bernet, R. Brisart, K. Held, M. Haar et S. IJsseling: *Heidegger et l'idée de la phénoménologie*. 1988 ISBN 90-247-3586-6
109. C. Singevin: *Dramaturgie de l'esprit*. 1988 ISBN 90-247-3557-2
110. J. Patočka: *Le monde naturel et le mouvement de l'existence humaine*. 1988 ISBN 90-247-3577-7
111. K.-H. Lembeck: *Gegenstand Geschichte*. Geschichtswissenschaft in Husserls Phänomenologie. 1988 ISBN 90-247-3635-8
112. J.K. Cooper-Wiele: *The Totalizing Act*. Key to Husserl's Early Philosophy. 1989 ISBN 0-7923-0077-7

Phaenomenologica

113. S. Valdinoci: *Le principe d'existence.* Un devenir psychiatrique de la phénoménologie. 1989 ISBN 0-7923-0125-0
114. D. Lohmar: *Phänomenologie der Mathematik.* 1989 ISBN 0-7923-0187-0
115. S. IJsseling (Hrsgb.): *Husserl-Ausgabe und Husserl-Forschung.* 1990 ISBN 0-7923-0372-5
116. R. Cobb-Stevens: *Husserl and Analytic Philosophy.* 1990 ISBN 0-7923-0467-5
117. R. Klockenbusch: *Husserl und Cohn.* Widerspruch, Reflexion und Telos in Phänomenologie und Dialektik. 1989 ISBN 0-7923-0515-9
118. S. Vaitkus: *How is Society Possible?* Intersubjectivity and the Fiduciary Attitude as Problems of the Social Group in Mead, Gurwitsch, and Schutz. 1991 ISBN 0-7923-0820-4
119. C. Macann: *Presence and Coincidence.* The Transformation of Transcendental into Ontological Phenomenology. 1991 ISBN 0-7923-0923-5
120. G. Shpet: *Appearance and Sense.* Phenomenology as the Fundamental Science and Its Problems. Translated from Russian by Th. Nemeth. 1991 ISBN 0-7923-1098-5
121. B. Stevens: *L'apprentissage des signes.* Lecture de Paul Ricœur. 1991 ISBN 0-7923-1244-9
122. G. Soffer: *Husserl and the Question of Relativism.* 1991 ISBN 0-7923-1291-0
123. G. Römpp: *Husserls Phänomenologie der Intersubjektivität.* Und Ihre Bedeutung für eine Theorie intersubjektiver Objektivität und die Konzeption einer phänomenologischen Philosophie. 1991 ISBN 0-7923-1361-5
124. S. Strasser: *Welt im Widerspruch.* Gedanken zu einer Phänomenologie als ethischer Fundamentalphilosophie. 1991 ISBN Hb: 0-7923-1404-2; Pb: 0-7923-1551-0
125. R.P. Buckley: *Husserl, Heidegger and the Crisis of Philosophical Responsibility.* 1992 ISBN 0-7923-1633-9
126. J.G. Hart: *The Person and the Common Life.* Studies in a Husserlian Social Ethics. 1992 ISBN 0-7923-1724-6
127. P. van Tongeren, P. Sars, C. Bremmers and K. Boey (eds.): *Eros and Eris.* Contributions to a Hermeneutical Phenomenology. Liber Amicorum for Adriaan Peperzak. 1992 ISBN 0-7923-1917-6
128. Nam-In Lee: *Edmund Husserls Phänomenologie der Instinkte.* 1993 ISBN 0-7923-2041-7
129. P. Burke and J. Van der Veken (eds.): *Merleau-Ponty in Contemporary Perspective.* 1993 ISBN 0-7923-2142-1
130. G. Haefliger: *Über Existenz: Die Ontologie Roman Ingardens.* 1994 ISBN 0-7923-2227-4
131. J. Lampert: *Synthesis and Backward Reference in Husserl's* Logical Investigations. 1995 ISBN 0-7923-3105-2
132. J.M. DuBois: *Judgment and Sachverhalt.* An Introduction to Adolf Reinach's Phenomenological Realism. 1995 ISBN 0-7923-3519-8
133. B.E. Babich (ed.): *From Phenomenology to Thought, Errancy, and Desire.* Essays in Honor of William J. Richardson, S.J. 1995 ISBN 0-7923-3567-8
134. M. Dupuis: *Pronoms et visages. Lecture d'Emmanuel Levinas.* 1996 ISBN Hb: 0-7923-3655-0; Pb 0-7923-3994-0
135. D. Zahavi: *Husserl und die transzendentale Intersubjektivität.* Eine Antwort auf die sprachpragmatische Kritik. 1996 ISBN 0-7923-3713-1
136. A. Schutz: *Collected Papers, IV.* Edited with preface and notes by H. Wagner and G. Psathas, in collaboration with F. Kersten. 1996 ISBN 0-7923-3760-3
137. P. Kontos: *D'une phénoménologie de la perception chez Heidegger.* 1996 ISBN 0-7923-3776-X
138. F. Kuster: *Wege der Verantwortung.* Husserls Phänomenologie als Gang durch die Faktizität. 1996 ISBN 0-7923-3916-9
139. C. Beyer: *Von Bolzano zu Husserl.* Eine Untersuchung über den Ursprung der phänomenologischen Bedeutungslehre. 1996 ISBN 0-7923-4050-7
140. J. Dodd: *Idealism and Corporeity.* An Essay on the Problem of the Body in Husserl's Phenomenology. 1997 ISBN 0-7923-4400-6
141. E. Kelly: *Structure and Diversity.* Studies in the Phenomenological Philosophy of Max Scheler. 1997 ISBN 0-7923-4492-8

Phaenomenologica

142. J. Cavallin: *Content and Object*. Husserl, Twardowski and Psychologism. 1997 ISBN 0-7923-4734-X
143. H.P. Steeves: *Founding Community*. A Phenomenological-Ethical Inquiry. 1997 ISBN 0-7923-4798-6
144. M. Sawicki: *Body, Text, and Science*. The Literacy of Investigative Practices and the Phenomenology of Edith Stein. 1997 ISBN 0-7923-4759-5; Pb: 1-4020-0262-9
145. O.K. Wiegand: *Interpretationen der Modallogik*. Ein Beitrag zur phänomenologischen Wissenschaftstheorie. 1998 ISBN 0-7923-4809-5
146. P. Marrati-Guénoun: *La genèse et la trace*. Derrida lecteur de Husserl et Heidegger. 1998 ISBN 0-7923-4969-5
147. D. Lohmar: *Erfahrung und kategoriales Denken*. 1998 ISBN 0-7923-5117-7
148. N. Depraz and D. Zahavi (eds.): *Alterity and Facticity*. New Perspectives on Husserl. 1998 ISBN 0-7923-5187-8
149. E. Øverenget: *Seeing the Self*. Heidegger on Subjectivity. 1998 ISBN Hb: 0-7923-5219-X; Pb: 1-4020-0259-9
150. R.D. Rollinger: *Husserls Position in the School of Brentano*. 1999 ISBN 0-7923-5684-5
151. A. Chrudzimski: *Die Erkenntnistheorie von Roman Ingarden*. 1999 ISBN 0-7923-5688-8
152. B. Bergo: *Levinas Between Ethics and Politics*. For the Beauty that Adorns the Earth. 1999 ISBN 0-7923-5694-2
153. L. Ni: *Seinsglaube in der Phänomenologie Edmund Husserls*. 1999 ISBN 0-7923-5779-5
154. E. Feron: *Phénoménologie de la mort*. Sur les traces de Levinas. 1999 ISBN 0-7923-5935-6
155. R. Visker: *Truth and Singularity*. Taking Foucault into Phenomenology. 1999 ISBN Hb: 0-7923-5985-2; Pb: 0-7923-6397-3
156. E.E. Kleist: *Judging Appearances*. A Phenomenological Study of the Kantian *sensus communis*. 2000 ISBN Hb: 0-7923-6310-8; Pb: 1-4020-0258-0
157. D. Pradelle: *L'archéologie du monde*. Constitution de l'espace, idéalisme et intuitionnisme chez Husserl. 2000 ISBN 0-7923-6313-2
158. H.B. Schmid: *Subjekt, System, Diskurs*. Edmund Husserls Begriff transzendentaler Subjektivität in sozialtheoretischen Bezügen. 2000 ISBN 0-7923-6424-4
159. A. Chrudzimski: *Intentionalitätstheorie beim frühen Brentano*. 2001 ISBN 0-7923-6860-6
160. N. Depraz: *Lucidité du corps*. De l'empirisme transcendantal en phénoménologie. 2001 ISBN 0-7923-6977-7
161. T. Kortooms: *Phenomenology of Time*. Edmund Husserl's Analysis of Time-Consciousness. 2001 ISBN 1-4020-0121-5
162. R. Boehm: *Topik*. 2002 ISBN 1-4020-0629-2
163. A. Chernyakov: *The Ontology of Time*. Being and Time in the Philosophies of Aristotle, Husserl and Heidegger. 2002 ISBN 1-4020-0682-9
164. D. Zahavi and F. Stjernfelt (eds.): *One Hundred Years of Phenomenology*. Husserl' Logical Investigations Revisited. 2002 ISBN 1-4020-0700-0
165. B. Ferreira: *Stimmung bei Heidegger*. Das Phänomen der Stimmung im Kontext von Heideggers Existenzialanalyse des Daseins. 2002 ISBN 1-4020-0701-9

Previous volumes are still available

Further information about *Phenomenology* publications are available on request

Kluwer Academic Publishers – Dordrecht / Boston / London

MIX
Papier aus verantwortungsvollen Quellen
Paper from responsible sources
FSC® C105338

If you have any concerns about our products,
you can contact us on
ProductSafety@springernature.com

In case Publisher is established outside the EU,
the EU authorized representative is:
Springer Nature Customer Service Center GmbH
Europaplatz 3, 69115 Heidelberg, Germany

Printed by Libri Plureos GmbH
in Hamburg, Germany